SELKIRK

ADVANCES IN
CHROMATOGRAPHY
Volume 16

ADVANCES IN
CHROMATOGRAPHY
Volume 16

Edited by

J. CALVIN GIDDINGS
EXECUTIVE EDITOR

DEPARTMENT OF CHEMISTRY
UNIVERSITY OF UTAH
SALT LAKE CITY, UTAH

ELI GRUSHKA
GAS CHROMATOGRAPHY AND LIQUID CHROMATOGRAPHY

DEPARTMENT OF CHEMISTRY
STATE UNIVERSITY OF NEW YORK AT BUFFALO
BUFFALO, NEW YORK

JACK CAZES
MACROMOLECULAR CHROMATOGRAPHY

WATERS ASSOCIATES, INC.
MILFORD, MASSACHUSETTS

PHYLLIS R. BROWN
BIOCHEMICAL CHROMATOGRAPHY

DEPARTMENT OF CHEMISTRY
UNIVERSITY OF RHODE ISLAND
KINGSTON, RHODE ISLAND

MARCEL DEKKER, Inc., New York and Basel

Library of Congress Cataloging in Publication Data

Advances in chromatography. v. 1-
1965-
New York, M. Dekker.

v. illus. 24 cm.
Editors: v. 1-   J. C. Giddings and R. A. Keller.

1. Chromatographic analysis--Addresses, essays, lectures. I. Giddings, John Calvin, 1930- ed. II. Keller, Roy A., 1928- ed.
QD271.A23   544.92   65—27435

COPYRIGHT © 1978 by MARCEL DEKKER, INC.

ALL RIGHTS RESERVED

Neither this book nor any part may be reproduced or transmitted in any form or by any means, electronic or mechanical, including photocopying, microfilming, and recording, or by any information storage and retrieval system, without permission in writing from the publisher.

MARCEL DEKKER, INC.
270 Madison Avenue, New York, New York 10016

LIBRARY OF CONGRESS CATALOG CARD NUMBER 65-27435

ISBN 0-8247-6659-8

Current printing (last digit):
10 9 8 7 6 5 4 3 2 1

PRINTED IN THE UNITED STATES OF AMERICA

CONTRIBUTORS TO VOLUME 16

MAHMOUD M. ABDEL-MONEM, Department of Medicinal Chemistry, College of Pharmacy, University of Minnesota, Minneapolis, Minnesota

HENRI L. O. De CLERCQ,[*] Farmaceutisch Instituut, Vrije Universiteit Brussel, Jette, Belgium

F. A. FITZPATRICK, Department of Physical and Analytical Chemistry, Drug Metabolism Research Section, The Upjohn Company, Kalamazoo, Michigan

IRA S. KRULL,[†] Environmental Biology Department, Boyce Thompson Institute, Yonkers, New York

DESIRE L. MASSART, Farmaceutisch Instituut, Vrije Universiteit Brussel, Jette, Belgium

JON F. PARCHER, Chemistry Department, University of Mississippi, University, Mississippi

WILLIAM PLUNKETT, Department of Developmental Therapeutics, M. D. Anderson Hospital and Tumor Institute, The University of Texas System Cancer Center, Houston, Texas

OLOF SAMUELSON, Department of Engineering Chemistry, Chalmers University of Technology, Goteborg, Sweden

JAMES K. SELKIRK, Cancer and Toxicology Section, Biology Division, Oak Ridge National Laboratory, Oak Ridge, Tennessee

---

[*]Current affiliation: Academisch Ziekenhuis, Vrije Universiteit Brussel, Brussels, Belgium

[†]Current affiliation: Cancer Research Center, Thermo Electron Corporation, Waltham, Massachusetts

CONTENTS

Contributors to Volume 16   iii

Contents of Other Volumes   xi

Chapter 1
ANALYSIS OF BENZO(a)PYRENE METABOLISM BY
HIGH-PRESSURE LIQUID CHROMATOGRAPHY 1
James K. Selkirk

    I. Introduction 2
   II. Metabolism of Benzo(a)pyrene 4
  III. Methods of Analysis for Polycyclic Hydrocarbons 6
  IV. Effect of Enzyme Inhibitors on Benzo(a)pyrene Metabolism ... 10
   V. Epoxides as Intermediates in Benzo(a)pyrene Metabolism 12
  VI. Metabolism of Benzo(a)pyrene in Human Tissue 17
 VII. Benzo(a)pyrene Metabolism in Malignantly
       Transformable Cells 21
VIII. High-Resolution HPLC (Liquid Chromatography
       by Recycling) 27
  IX. Identification of Four Benzo(a)pyrene Phenols
       as Metabolites 30
       Acknowledgments 33
       References 33

Chapter 2

HIGH-PERFORMANCE LIQUID CHROMATOGRAPHY
OF THE STEROID HORMONES 37

F. A. Fitzpatrick

I. Introduction 38
II. Steroids—A Brief Review 39
III. Scope of High-Performance Liquid
  Chromatography of Steroids 45
IV. Initial Developments 46
V. Steroids in Biological Fluids 51
VI. High-Performance Liquid Chromatography
  of Steroids in Pharmaceuticals 61
VII. Chemistry Before, During, and After
  HPLC of Steroids 65
VIII. The Future of High-Performance Liquid
  Chromatography of Steroid Hormones 68
IX. Conclusions 70
  Acknowledgments 70
  Appendix 71
  References 71

Chapter 3

NUMERICAL TAXONOMY IN CHROMATOGRAPHY 75

Desire L. Massart and Henri L. O. De Clercq

I. Introduction 76
II. Numerical Taxonomic Techniques 77
III. Applications 87
  Acknowledgments 110
  References 110

Chapter 4
CHROMATOGRAPHY OF OLIGOSACCHARIDES AND RELATED COMPOUNDS ON ION-EXCHANGE RESINS 113

Olof Samuelson

I.	Introduction	114
II.	Partition Chromatography of Oligomeric Sugars and Alditols	114
III.	Separations of Oligosaccharides as Borate Complexes	130
IV.	Permeation Chromatography of Oligosaccharides in Aqueous Solution	131
V.	Anion-Exchange Chromatography of Oligomers with a Terminal Aldonic Acid Group	134
VI.	Anion-Exchange Chromatography of Oligomers with One Uronic Acid Moiety	143
VII.	Anion-Exchange Chromatography of Dicarboxylic Oligomers	144
VIII.	Anion-Exchange Chromatography of Oligogalacturonic Acids	146
	References	147

Chapter 5
APPLICATIONS AND THEORY OF FINITE CONCENTRATION FRONTAL CHROMATOGRAPHY 151

Jon F. Parcher

I.	Introduction	152
II.	Frontal Chromatography	153
III.	Theory of Chromatography at Finite Concentrations	155
IV.	Previous Chromatographic Investigations Involving Simplifying Assumptions	160
V.	Discussion	164
VI.	Practical Applications of Frontal Chromatography	165

VII.	Summary	170
	Acknowledgments	171
	References	171

Chapter 6

THE LIQUID-CHROMATOGRAPHIC RESOLUTION OF ENANTIOMERS ... 175

Ira S. Krull

I.	Introduction and Background	176
II.	Liquid-Chromatographic Separation of Diastereomers	177
III.	A Potential Optically Active Resolving Column for Liquid Chromatography	183
IV.	Direct Resolutions by Liquid Chromatography Using Optically Active Eluents	184
V.	Direct Resolutions by Liquid Chromatography Using an Optically Active Substrate	186
VI.	Direct Resolutions Using Ion-Pair Partition Chromatography	202
VII.	Conclusions	203
	Acknowledgments	204
	References	204

Chapter 7

THE USE OF HIGH-PRESSURE LIQUID CHROMATOGRAPHY IN RESEARCH ON PURINE NUCLEOSIDE ANALOGS ... 211

William Plunkett

I.	Introduction	211
II.	Methods	212
III.	Results and Discussion	215
IV.	Conclusions	245
	Acknowledgements	246
	References	246

Chapter 8

THE DETERMINATION OF DI- AND POLYAMINES BY
HIGH-PRESSURE LIQUID AND GAS CHROMATOGRAPHY 249

    Mahmoud M. Abdel-Monem

    I. Introduction 249
    II. High-Pressure Liquid Chromatography 251
    III. Gas Chromatography 261
    IV. Conclusions 264
        Acknowledgments 265
        References 265

Author Index 269

Subject Index 283

CONTENTS OF OTHER VOLUMES

Volume 1

Ion-Exchange Chromatography —F. Helfferich
Chromatography and Electrophoresis on Paper and Thin Layers: A Teacher's Guide —Ivor Smith
The Stationary Phase in Paper Chromatography —George H. Stewart
The Techniques of Laminar Chromatography —E. V. Truter
Qualitative and Quantitative Aspects of the Separation of Steroids —E. C. Horning and W. J. A. VandenHeuvel
Capillary Columns: Trials, Tribulations, and Triumphs —D. H. Desty
Gas Chromatographic Characterization of Organic Substances in the Retention Index System —E. sz. Kováts
Inorganic Gas Chromatography —Richard S. Juvet, Jr., and Franjo Zado
Lightly Loaded Columns —Barry L. Karger and W. D. Cooke
Interactions of the Solute with the Liquid Phase —Daniel E. Martire and Luigi Z. Pollara

Volume 2

Ion Exchange Chromatography of Amino Acids: Recent Advances in Analytical Determinations —Paul B. Hamilton
Ion Mobilities in Electrochromatography —John T. Edward
Partition Paper Chromatography and Chemical Structure —J. Green and D. McHale
Gradient Techniques in Thin-Layer Chromatography —A. Niederwieser and C. C. Honegger
Geology — An Inviting Field to Chromatographers —Arthur S. Ritchie
Extracolumn Contributions to Chromatographic Band Broadening —James C. Sternberg
Gas Chromatography of Carbohydrates — James W. Berry
Ionization Detectors for Gas Chromatography —Arthur Karmen
Advances in Programmed Temperature Gas Chromatography —Louis Mikkelsen

Volume 3

The Occurrence and Significance of Isotope Fractionation during Analytical Separations of Large Molecules —Peter D. Klein
Adsorption Chromatography —Charles H. Giles and I. A. Easton
The History of Thin-Layer Chromatography —N. Pelick, H. R. Bolliger, and H. K. Mangold
Chromatography as a Natural Process in Geology —Arthur S. Ritchie
The Chromatographic Support —D. M. Ottenstein
Electrolytic Conductivity Detection in Gas Chromatography —Dale M. Coulson
Preparative-Scale Gas Chromatography —G. W. A. Rijnders

Volume 4

R_F Values in Thin-Layer Chromatography on Alumina and Silica —Lloyd R. Snyder
Steroid Separation and Analysis: The Technique Appropriate to the Goal —R. Neher
Some Fundamentals of Ion-Exchange-Cellulose Design and Usage in Biochemistry —C. S. Knight
Adsorbents in Gas Chromatography —A. V. Kiselev
Packed Capillary Columns in Gas Chromatography —István Halász and Erwin Heine
Mass-Spectrometric Analysis of Gas-Chromatographic Eluents —William McFadden
The Polarity of Stationary Liquid Phases in Gas Chromatography —L. Rohrschneider

Volume 5

Prediction and Control of Zone Migration Rates in Ideal Liquid-Liquid Partition Chromatography —Edward Soczewiński
Chromatographic Advances in Toxicology —Paul L. Kirk
Inorganic Chromatography on Natural and Substituted Celluloses —R. A. A. Muzzarelli
The Quantitative Interpretation of Gas Chromatographic Data —H. Wilson Johnson, Jr.
Atmospheric Analysis by Gas Chromatography —A. P. Altshuller
Non-Ionization Detectors and Their Use in Gas Chromatography —J. D. Winefordner and T. H. Glenn

Volume 6

The Systematic Use of Chromatography in Structure Elucidation of Organic Compounds by Chemical Methods —Jiři Gasparič
Polar Solvents, Supports, and Separation —John A. Thoma
Liquid Chromatography on Lipophilic Sephadex: Column and Detection Techniques —Jan Sjövall, Ernst Nyström, and Eero Haahti
Statistical Moments Theory of Gas-Solid Chromatography: Diffusion-Controlled Kinetics —Otto Grubner
Identification by Retention and Response Values —Gerhard Schomburg (translated by Roy A. Keller)
The Use of Liquid Crystals in Gas Chromatography —H. Kelker and E. von Schivizhoffen
Support Effects on Retention Volumes in Gas-Liquid Chromatography —Paul Urone and Jon F. Parcher

Volume 7

Theory and Mechanics of Gel Permeation Chromatography —K. H. Altgelt
Thin-Layer Chromatography of Nucleic Acid Bases, Nucleosides, Nucleotides, and Related Compounds — György Pataki
Review of Current and Future Trends in Paper Chromatography —V. C. Weaver
Chromatography of Inorganic Ions — G. Nickless
Process Control by Gas Chromatography —I. G. McWilliam
Pyrolysis Gas Chromatography of Involatile Substances —S. G. Perry
Labeling by Exchange on Chromatographic Columns —Horst Elias

Volume 8

Principles of Gel Chromatography —Helmut Determann
Thermodynamics of Liquid-Liquid Partition Chromatography —David C. Locke
Determination of Optimum Solvent Systems for Countercurrent Distribution and Column Partition Chromatography from Paper Chromatographic Data —Edward Soczewiński
Some Procedures for the Chromatography of the Fat-Soluble Chloroplast Pigments —Harold H. Strain and Walter A. Svec
Comparison of the Performances of the Various Column Types Used in Gas Chromatography —Georges Guiochon
Pressure (Flow) Programming in Gas Chromatography —Leslie S. Ettre, László Mázor, and József Takács
Gas Chromatographic Analysis of Vehicular Exhaust Emissions —Basil Dimitriades, C. F. Ellis, and D. E. Seizinger
The Study of Reaction Kinetics by the Distortion of Chromatographic Elution Peaks —Maarten van Swaay

Volume 9

Reversed-Phase Extraction Chromatography in Inorganic Chemistry —E. Cerrai and G. Ghersini
Determination of the Optimum Conditions to Effect a Separation by Gas Chromatography —R. P. W. Scott
Advances in the Technology of Lightly Loaded Glass Bead Columns —Charles Hishta, Joseph Bomstein, and W. D. Cooke
Radiochemical Separations and Analyses by Gas Chromatography —Stuart P. Cram
Analysis of Volatile Flavor Components of Foods —Phillip Issenberg and Irwin Hornstein

Volume 10

Porous-Layer Open Tubular Columns— Theory, Practice, and Applications —Leslie S. Ettre and John E. Purcell
Resolution of Optical Isomers by Gas Chromatography of Diastereomers —Emanuel Gil-Av and David Nurok
Gas-Liquid Chromatography of Terpenes —E. von Rudloff

Volume 11

Quantitative Analysis by Gas Chromatography —Josef Novák
Polyamide Layer Chromatography —Kung-Tsung Wang, Yau-Tang Lin, and Iris S. Y. Wang
Specifically Adsorbing Silica Gels —H. Bartels and B. Prijs
Nondestructive Detection Methods in Paper and Thin Layer Chromatography —G. C. Barrett

Volume 12

The Use of High-Pressure Liquid Chromatography in Pharmacology and Toxicology - Phyllis R. Brown
Chromatographic Separation and Molecular-Weight Distributions of Cellulose and Its Derivatives - Leon Segal
Practical Methods of High-Speed Liquid Chromatography - Gary J. Fallick
Measurement of Diffusion Coefficients by Gas-Chromatography Broadening Techniques: A Review - Virgil R. Maynard and Eli Grushka
Gas-Chromatography Analysis of Polychlorinated Diphenyls and Other Nonpesticide Organic Pollutants - Joseph Sherma
High-Performance Electrometer Systems for Gas Chromatography - Douglas H. Smith
Steam Carrier Gas-Solid Chromatography - Akira Nonaka

Volume 13

Practical Aspects in Supercritical Fluid Chromatography — T. H. Gouw and Ralph E. Jentoft
Gel Permeation Chromatography: A Review of Axial Dispersion Phenomena, their Detection, and Correction — Nils Friis and Archie Hamielec
Chromatography of Heavy Petroleum Fractions — Klaus H. Altgelt and T. H. Gouw
Determination of the Adsorption Energy, Entropy, and Free Energy of Vapors on Homogeneous Surfaces by Statistical Thermodynamics — Claire Vidal-Madjar, Marie-France Gonnord, and Georges Guiochon
Transport and Kinetic Parameters by Gas Chromatographic Techniques — Motoyuki Suzuki and J. M. Smith
Qualitative Analysis by Gas Chromatography — David A. Leathard

Volume 14

Nutrition: An Inviting Field to High-Pressure Liquid Chromatography — Andrew J. Clifford
Polyelectrolyte Effects in Gel Chromatography — Bengt Stenlund
Chemically Bonded Phases in Chromatography — Imrich Sebestian and István Halász
Physicochemical Measurements Using Chromatography — David C. Locke
Gas-Liquid Chromatography in Drug Analysis — W. J. A. VandenHeuvel and A. G. Zacchei
The Investigation of Complex Association by Gas Chromatography and Related Chromatographic and Electrophoretic Methods — C. L. de Ligny
Gas-Liquid-Solid Chromatography — Antonio De Corcia and Arnaldo Liberti
Retention Indices in Gas Chromatography — J. K. Haken

Volume 15

Detection of Bacterial Metabolites in Spent Culture Media and Body Fluids by Electron Capture Gas-Liquid Chromatography — John B. Brooks
Signal and Resolution Enhancement Techniques in Chromatography — Raymond Annino
The Analysis of Organic Water Pollutants by Gas Chromatography and Gas Chromatography-Mass Spectrometry — Ronald A. Hites
Hydrodynamic Chromatography and Flow-Induced Separators — Hamish Small
The Determination of Anticonvulsants in Biological Samples by Use of High-Pressure Liquid Chromatography — Reginald F. Adams

The Use of Microparticulate Reversed-Phase Packing in High-Pressure
  Liquid Chromatography of Compounds of Biological Interest — John A.
  Montgomery, Thomas P. Johnston, H. Jeanette Thomas, James R.
  Piper, and Carroll Temple, Jr.
Gas-Chromatographic Analysis of the Soil Atmosphere — K. A. Smith
Kinematics of Gel Permeation Chromatography — A. C. Ouano
Some Clinical and Pharmacological Applications of High-Speed Liquid
  Chromatography — J. Arly Nelson

Volumes 1 and 3 are out of print

ADVANCES IN
CHROMATOGRAPHY
Volume 16

ROUTING AND TRANSMITTAL SLIP

Date

TO: (Name, office symbol, room number, building, Agency/Post)	Initials	Date
1.		
2.		
3.		
4.		
5.		

Action	File	Note and Return
Approval	For Clearance	Per Conversation
As Requested	For Correction	Prepare Reply
Circulate	For Your Information	See Me
Comment	Investigate	Signature
Coordination	Justify	

REMARKS

DO NOT use this form as a RECORD of approvals, concurrences, disposals, clearances, and similar actions

FROM: (Name, org. symbol, Agency/Post)	Room No.—Bldg.
	Phone No.

5041-102

☆ GPO : 1984 O -421-529 (408)

OPTIONAL FORM 41 (Rev. 7-76)
Prescribed by GSA
FPMR (41 CFR) 101-11.206

Chapter 1

ANALYSIS OF BENZO(a)PYRENE METABOLISM BY
HIGH-PRESSURE LIQUID CHROMATOGRAPHY

James K. Selkirk

Cancer and Toxicology Section
Biology Division
Oak Ridge National Laboratory[*]
Oak Ridge, Tennessee

I.	INTRODUCTION	2
II.	METABOLISM OF BENZO(a)PYRENE	4
III.	METHODS OF ANALYSIS FOR POLYCYCLIC HYDROCARBONS	6
IV.	EFFECT OF ENZYME INHIBITORS ON BENZO(a)PYRENE METABOLISM	10
V.	EPOXIDES AS INTERMEDIATES IN BENZO(a)PYRENE METABOLISM	12
VI.	METABOLISM OF BENZO(a)PYRENE IN HUMAN TISSUE	17
VII.	BENZO(a)PYRENE METABOLISM IN MALIGNANTLY TRANSFORMABLE CELLS	21
VIII.	HIGH-RESOLUTION HPLC (LIQUID CHROMATOGRAPHY BY RECYCLING)	27

*Operated for the Energy Research and Development Administration under contract with Union Carbide Corporation.

This work was supported in part by NCI-ERDA interagency agreement YOL-CP-50200 and by EPA contract EPA-IAG-D5-E681-AO.

IX. IDENTIFICATION OF FOUR BENZO(a)PYRENE
PHENOLS AS METABOLITES........................ 30

ACKNOWLEDGMENTS............................. 33

REFERENCES 33

I. INTRODUCTION

The last decade of carcinogenesis research has seen a new burst of energy and knowledge of the biological and chemical systems involved with the study of malignant transformation. This is especially important today since the human population is being exposed to polycyclic aromatic hydrocarbons, hydrocarbon carcinogens, at an increasing rate [1]. This is a direct consequence of industrial growth and relates back to the original observation of Percival Pott (1775) that skin cancer, found in chimney sweeps, was probably caused by some agent in soot. However, it was not until 1918 that tumors were experimentally produced in rabbit skin by painting with coal tar extracts [2]. This marked the beginning of the search for the active carcinogenic substance in soot which culminated in the isolation and identification of benzo(a)pyrene by British investigators two decades later [3]. During the next quarter century the major emphasis of polyaromatic research was the synthesis of new and interesting compounds, which were tested in vivo for tumorigenicity [4]. It was hoped that through this empirical approach there would become apparent structure-activity relationships that would explain the role of these molecules in carcinogenesis. Theory centered around the study of the relative physical size of the molecule, e.g., three-ring versus five-ring structure, as well as the sizes of ring substitution products, e.g., hydroxyl, acid, or alkyl. From this mode of research grew the K-region hypothesis of Pullman and Pullman [5]. This theory, based on molecular orbital calculations in conjunction with phenomenological observation of the compounds tested for carcinogenicity, was centered around the concept that those areas of the polyaromatic molecule possessing the most electrophilic nature would be the most attractive region for a substrate to bind to a molecule containing numerous nucleophilic sites. This would certainly be the case for any test tube reaction in which a strong nucleophilic agent was mixed with a polyaromatic hydrocarbon. It is reasonable to expect the biochemical processes at the molecular level to follow the same physical laws and it was postulated that this indeed was occurring in cells, that these regions of the polyaromatic molecule were responsible for the aberrant cellular behavior which would eventually become what we classify as malignancy. In actuality the known metabolites of polyaromatic hydrocarbons were substitution products exactly in those areas of the molecule. However, a major drawback to such a theoretical approach was the fact that synthetic techniques were limited, and only a few derivatives could be synthesized at that time; furthermore, the analytical methods for determination of the polyaromatic metabolites were insufficient for separation of positional isomers. Also, other types of chemical carcinogens having no structural similarities and possessing no demonstrable chemical relationship with polyaromatic hydrocarbons, e.g., azo dyes and nitrosamines, were tumorigenic and could not be categorized by the K-region hypothesis. We know now that there are many more metabolites of polyaromatic hydrocarbons at other sites on the molecule and far more synthetic derivatives available to act as test molecules for studying the mechanism of action of these carcinogenic chemicals. Elucidation of the metabolic mechanism by which these compounds transform normal cells into malignant cells necessarily rests in a complete description of their physicochemical similarities and the molecular conformation formed between the carcinogen and the target site itself. At present the ultimate species of these chemical carcinogens and the target site or sites at which transformation to malignancy occurs remain obscure. It is hoped that complete assemblance of the metabolic pathway will illuminate the mechanism of the activated carcinogenic species.

Salient to understanding steps in the metabolic pathway are the enzyme systems involved in activation and detoxification of these molecules. Polycyclic hydrocarbons are metabolized by the microsomal mixed function oxidase, aryl hydrocarbon hydroxylase (AHH), which is predominantly found in liver. This enzyme activity has been studied extensively in rodents and is the subject of several excellent reviews [6, 7]. Although it is known that this enzyme complex is involved with detoxification of xenobiotics in conjunction with cytochrome P_{450}, it is not yet apparent which part of this enzyme complex is involved with the activation process. A second microsomal enzyme, epoxide hydrase, converts epoxide into dihydrodiols [33]. Since epoxides are more active carcinogens than the parent hydrocarbon [8] this enzyme may play a critical role in carcinogenesis. Although epoxide hydrase has been primarily investigated using noncarcinogens [9], its activity has also been demonstrated in the formation of the three known dihydrodiols of benzo(a)pyrene [10].

It seems clear that subtle mechanisms and fast reactions are important steps in carcinogen activation, detoxification, and the physicochemical binding to the transformation receptor(s) in the cell. It is therefore critical to develop equally sensitive methodology to efficiently trap, isolate, and identify these elusive intermediates in order to determine their proper place in tumorigenesis.

II. METABOLISM OF BENZO(a)PYRENE

Polyaromatic hydrocarbons are hydrophobic molecules which are transported into cells through lipid solubilization, and metabolized to several types of oxygenated derivatives. The majority of metabolites are still significantly nonpolar and are readily extractable from tissue incubation mixture with organic solvents. Therefore, polyaromatic metabolites are arbitrarily divided into two groups: organic (solvent) soluble and water soluble. The organic-soluble group consists of ring hydroxylated products such as phenols and dihydrodiols [11-13], and hydroxymethyl derivatives for those polyaromatics with aliphatic side chains such as dimethylbenz(a)anthracene [14] and methylcholanthrene [15]. In addition to the hydroxylated metabolites are quinones which are produced enzymatically by microsomes and nonenzymatically from air oxidation of phenols. Labile metabolic intermediates such as epoxides can also be found in this fraction using special isolation conditions [16].

In the second group are the water-soluble products remaining after extraction with an organic solvent. Although it is believed that many of these derivatives are formed by conjugation of the hydroxylated products to glutathione [17-19] or other moieties that would render the compound more hydrophilic and presumably less toxic, this group of derivatives has not been rigorously studied.

The metabolite profile of benzo(a)pyrene under standard chromatographic conditions can be seen in Fig. 1 which consists of three groups of positional isomers. There are three dihydrodiols, three quinones, and two phenols. The major benzo(a)pyrene metabolite found is 3-hydroxybenzo(a)pyrene, with 9-hydroxypyrene present in lesser amounts. The 4,5-BP-epoxide has been isolated and identified as a precursor of the 4,5-BP-diol [16]. Other studies suggest that epoxides are the precursors of the 7,8-diol and 9,10-diol [10], and diol-epoxides may be further metabolites of diol intermediates [20, 21]. There have been no intermediates isolated as phenol precursors; however, evidence exists for epoxide involvement in the formation of 9-hydroxybenzo(a)pyrene [16]. There is little further knowledge of the order and kinetics of metabolite formation. The effect of various metabolites on microsomal enzyme activity has not been thoroughly studied. However, it has been demonstrated that there are at least two different types of hepatic oxygenase induction [22-24] which are dependent on the chemical nature of the inducer. This may have significant influence upon an active intermediate species of carcinogen formed, and its probability of reaching a target site. In addition, there appears to be a number of sulfur conjugation enzymes which possess specificities for diols, phenols, and epoxides [25]. Removal of reactive intermediates and toxic metabolites is certain to affect the probability of cell survival, both normal and transformed.

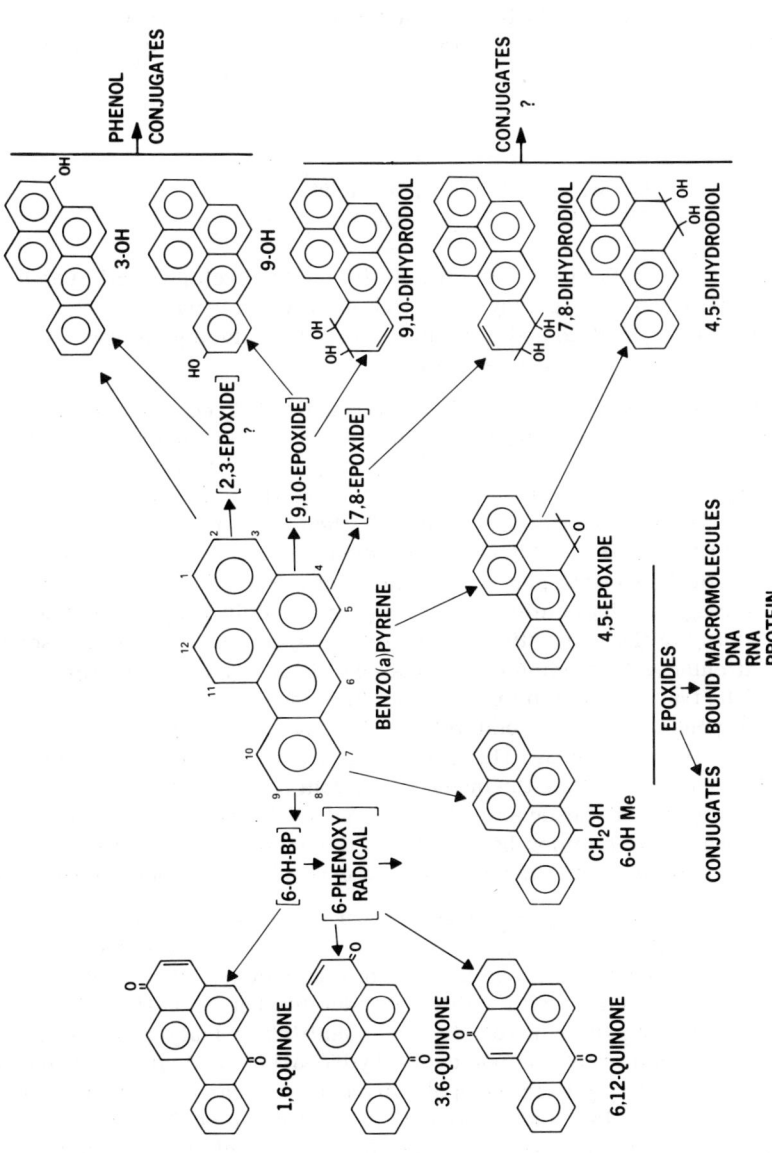

FIG. 1. Benzo(a)pyrene metabolism. Drawn structures are metabolites that have been isolated and characterized, with the exception of 1-OH-BP and 7-OH-BP [57]. Bracketed compounds are projected intermediates which have not yet been isolated or need further characterization. Molecular species bound to macromolecules and conjugated derivatives have not been characterized.

Significant progress has been made in the isolation and characterization of the aryl hydrocarbon hydroxylase enzyme complex [26, 27], and similar work is in progress for epoxide hydrase [28]. We can therefore expect considerable advances in the next several years in assembling the dynamic scheme of metabolite formation for carcinogenic polycyclic hydrocarbons.

III. METHODS OF ANALYSIS FOR POLYCYCLIC HYDROCARBONS

Chromatography with silicic acid was in its early developmental stages during the period when most polycyclic metabolites were reported. This has resulted in an incomplete and often inaccurate literature with either partial or tentative structural characterization.

Thin-layer chromatography was the major analytical technique and required no special instrumentation. A glass plate was uniformly covered with a layer of silicic acid or alumina (0.25 mm), which when dry became fixed to the plate. The compounds were spotted and the plate placed in a solvent tank. Separation was accomplished when the solvent ascended the plate by capillary action. Although this form of chromatography has a limited number of theoretical plates, indicative of a low resolving capacity, it has the advantages of the stationary phase (e.g., silicic acid) and the solvent phase, aiding in the separation of the solutes by a combination of absorption and partitioning. This combination has enabled fairly good separation for groups of derivatives containing different functional groups (e.g., dihydrodiols and phenols), but is relatively ineffective for resolving isomers (e.g., 9-OH-BP from 3-OH-BP) [11], and is destructive to labile metabolites [29]. In addition, neither the quantitative data nor the migration data possess a high degree of reproducibility. Recent modifications using packed columns and less acidic alumina have reduced the destruction of some labile polycyclic intermediates, but have not improved resolution or quantitation [30].

The major analytical technique used for drugs, pesticides, and other nonpolar organic molecules during the last decade has been gas chromatography. This method contains a high number of theoretical plates (~2000) yielding good peak resolution with highly reproducible retention time that qualifies it as both a superior qualitative and quantitative analytical system. However, gas chromatography in polycyclic hydrocarbon analysis has been confined almost exclusively to the nonpolar parent hydrocarbon and has been the method of choice for the study of benzo(a)pyrene as an environmental pollutant [1]. Since the major requirements for successful gas-chromatographic analysis are a reasonable degree of volatility and thermal stability, most oxygenated polycyclics are eliminated from analysis without prior protection of functional groups. Formation of silyloxy derivatives and subsequent gas chromatography for one polycyclic epoxide have been reported [31].

Since reactive intermediates are usually quite labile and short-lived, a rapid and nondestructive assay is necessary which can yield as good or better resolution for positional isomers as gas chromatography. It is also necessary that such a system be far more sensitive and rapid than conventional forms of chromatography and be a single-step process since it is quite probable that more labile polycyclic metabolites will, for the most part, be lost during separation or subsequent workup for structural analysis.

High-pressure liquid chromatography (HPLC) has many distinct advantages over other forms of chromatography and is ideally suited for the study of reactive intermediates in metabolism and carcinogenesis. Resolution of isomers is equivalent to gas chromatography with a large number of theoretical plates. Retention times are highly reproducible because of tightly packed small-bore columns with microparticle stationary phases (5-30 μm), which reduce meniscus effects and ensure the best chance of resolving closely eluting compounds. Unlike gas chromatography, there is no requirement for volatility or thermal stability. It is only necessary that the compounds be soluble in some solvent, either aqueous or organic. The HPLC system is amenable to biological systems since the column packing can be neutral, anionic, or cationic. In aqueous systems, salt concentration, salt anion, and pH can be readily manipulated. Two important features in mechanistic studies are the rapidity of the analysis and gradient elution. In most cases, HPLC analysis takes less than 1 hr with the sample completely recoverable for further reaction or structural analysis. HPLCs contain gradient devices which allow several gradient modes that facilitate separation of mixtures of very similar compounds.

One HPLC system currently used is schematically represented in Fig. 2. This system is the basis of the DuPont Model 830, and can be operated either in an isocratic mode, using a constant mixture of solvents A and B, or with a gradient mode where solvent B is added to solvent A either linearly, slowly (concave), or rapidly (convex) by means of preset proportioning valves. The design of this system also uses solvent A as a pressure head to drive solvent B from the holding coil to form the gradient mixture. A mixing chamber placed immediately in front of the column ensures uniform solvent composition, with the injection port located immediately after the mixing chamber to allow injection close to the front of the column and directly into the solvent flow. Separation of benzo(a)pyrene metabolites accomplished by HPLC using gradient elution and a solvent mixture of water-methanol is seen in Fig. 3.

A small unidentified peak is eluted first, then three clearly separated vicinal glycols: 9,10-diol, 4,5-diol, and 7,8-diol, and three quinones, two

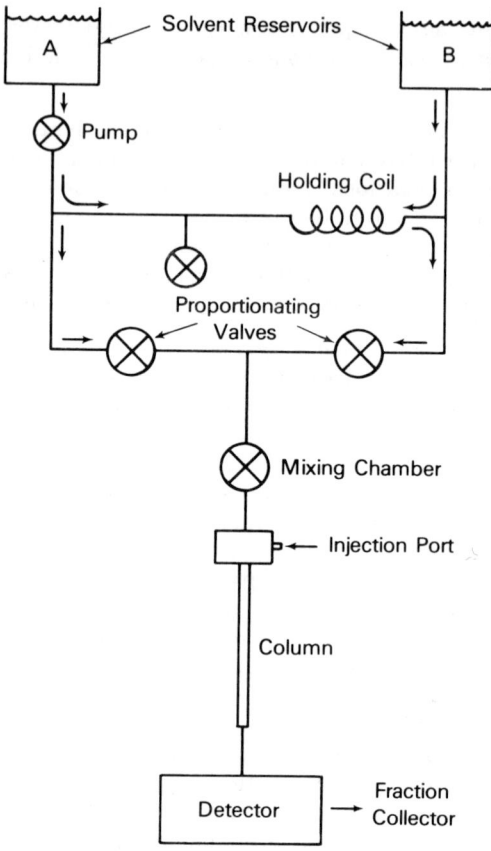

FIG. 2. Schematic representation of HPLC system. This flow diagram utilizes a single pump to drive both solvents as in the DuPont Model 830. Proportionating valves regulate solvent mixtures which are passed through a mixing chamber near the head of the column. Detectors routinely used with HPLC are uv, fluorescence, and refractometry.

of which are not completely separated, with these solvent proportions: 1,6-quinone, 3,6-quinone, and 6,12-quinone. These are followed by two phenols, 9-OH-BP and 3-OH-BP. The unused BP elutes considerably behind the metabolites, so that tailing effects common to thin-layer chromatography are minimized. Table 1 shows the retention times and molecular weights of the metabolites, which were determined from the peaks collected after passage through the flow cell.

ANALYSIS OF BENZO(a)PYRENE METABOLISM BY HPLC

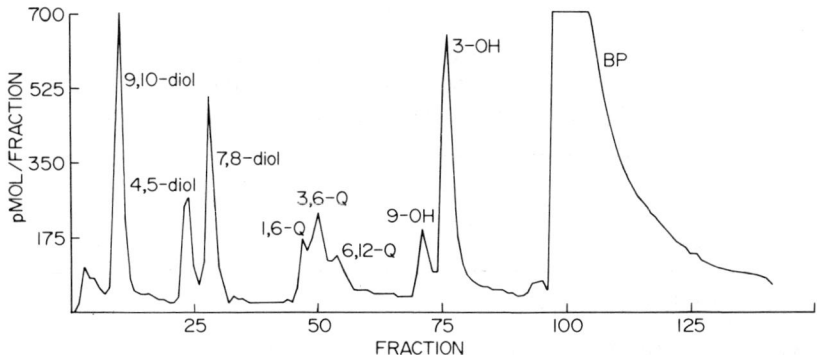

FIG. 3. Separation of metabolites of benzo(a)pyrene. The microsome incubation technique used to form BP metabolites throughout this report (except as indicated) is as follows: Male Sprague-Dawley rats (160-180 g) were injected intraperitoneally with 5 mg of 3-methylcholanthrene in 0.5 ml of corn oil and killed 40 hr later. Liver microsomes were prepared as described [12]. The metabolites were formed by incubating rat liver microsomes with [^{3}H]benzo(a)pyrene (specific activity 200 mCi/mmol) or [^{14}C]benzo(a)pyrene (specific activity 21 mCi/mmol) (Amersham/Searle, Arlington Heights, Illinois) as follows. Each flask contained a total volume of 1.0 ml: 100 µg of microsomal protein, 0.36 µmol of reduced nicotinamide adenine dinucleotide phosphate, 3 µmol of $MgCl_2$, 50 µmol of tris-HCl buffer, pH 7.5, and 100 µmol of benzo(a)pyrene dissolved in 0.040 ml of methanol. The flasks were incubated for 10 min at 37°C under red light illumination, and the reaction was stopped by the addition of 1.0 ml of acetone. The mixture was then extracted with 2.0 ml of ethyl acetate; five extracts were pooled and dried over 1.0 g of anhydrous magnesium sulfate, the solvent was evaporated under vacuum, and the residue (metabolites) was dissolved in 0.1 ml of methanol. The metabolite separation was performed on a high-pressure liquid chromatograph (DuPont Model 830) fitted with a Zipax Permaphase column (1-m ODS). The metabolites were eluted with a reverse-phase gradient system, with methanol and water (30:70 initially, and 70:30 at final composition). The gradient rate of change was 3%/min, the column temperature 50°C, the pressure 350 psi, and the flow rate 0.6 ml/min. The eluate was monitored by uv absorption at 254 nm. Fractions were collected at 20-sec intervals, and the radioactivity was determined in a Beckman 350 scintillation counter with Aquasol (New England Nuclear) as the counting medium [10].

Comparison of BP metabolite profiles between thin-layer chromatography and HPLC is seen in Fig. 4, in which A shows the [^{3}H]BP metabolite pattern with the uv absorbance (254 nm) superimposed. There is no discernible lag between the appearance of a peak and collection of the compound,

TABLE 1

The Retention Time and Molecular Weight of
Benzo(a)pyrene Metabolites Separated by
High-Pressure Liquid Chromatography

	Metabolite	Retention time (min)	m/e[a]
1.	9,10-Dihydrodihydroxy-BP	8.5	286
2.	4,5-Dihydrodihydroxy-BP	15.5	286
3.	7,8-Dihydrodihydroxy-BP	18.0	286
4.	1,6-Quinone-BP	25.5	282
5.	3,6-Quinone-BP	26.0	282
6.	6,12-Quinone-BP[b] (tentative)	28.0	—
7.	9-Hydroxy-BP	35.0	268
8.	3-Hydroxy-BP	37.0	268
9.	BP	48.0	252

[a]Molecular weight determinations performed on a JeoL JMS-01SG-2 at 70 eV with a solid probe. Temperature ranged from 90 to 150°C.
[b]Insufficient material for complete analysis.

because of the small volume size of the flow cell (8 μl) and the bore size of tubing carrying the eluate (0.019 in.). Individual metabolites in B through G were from a [^{14}C]BP microsomal incubation that was chromatographed on silica gel thin-layer plates. The spots (benzene and methanol) were eluted and further purified by HPLC. The two peaks in C show 9-OH-BP and 3-OH-BP which were not resolved by thin-layer chromatography. This was the first unequivocal demonstration that 9-OH-BP is a metabolic product [13].

IV. EFFECT OF ENZYME INHIBITORS ON BENZO(a)PYRENE METABOLISM

7,8-Benzoflavone (7,8-BF) is a strong inhibitor of AHH [32], as determined by fluorescence studies measuring formation of phenolic metabolites. However, monitoring the entire metabolic profile with HPLC can accurately

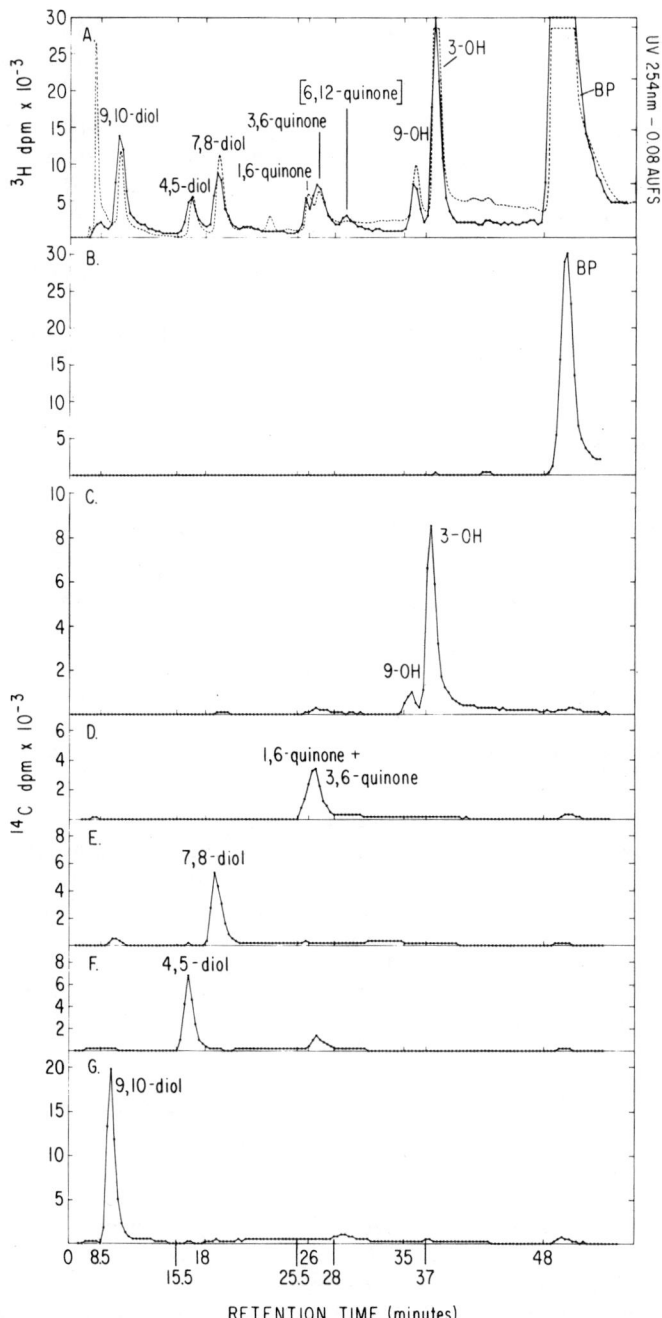

FIG. 4. HPLC identification of total and individual BP metabolites [13]. A shows uv scan superimposed on radioactivity of collected eluate fractions. B-G were isolated by thin-layer chromatography and reanalyzed by high-pressure liquid chromatography.

measure the specificity of 7,8-BF. Table 2 shows the effect of 7,8-BF on BP metabolite formation, and confirms a large inhibitory effect on phenol formation as measured by HPLC. However, the broad analytical range of HPLC also shows inhibition of diol and quinone formation. Furthermore, there was a decrease in the amount of BP metabolized (5.9%) as compared to the control (18.3%). It is clear from this type of analysis that 7,8-BF apparently interacts with the enzyme complex to inhibit oxygenation at all sites of the benzo(a)pyrene molecule. However, diol formation was inhibited by 7,8-BF to a relatively greater extent than the phenol (76-79%) and quinone formations (55-68%). Since diols represent the products of hydrase action on epoxide intermediates, 7,8-BF would seem to have somewhat greater effect on epoxide formation than on phenol formation.

TCPO (1,2-epoxy-3,3-trichloropropane) is a potent inhibitor of the epoxide hydrase enzyme [33]. In contrast to 7,8-BF, TCPO had a more selective effect on the metabolism of BP. As shown in Table 2, and in Fig. 5, TCPO completely inhibited the formation of all three diols and reduced total BP metabolism by almost 50%, suggesting that TCPO may interfere with BP oxygenation as well as with epoxide hydration. However, the formation of 3-OH-BP was not affected by 2 mmol TCPO as determined by total radioactivity of each metabolite formed. This result has been confirmed by fluorescence studies [34], suggesting that the enzymatic mechanism for phenol formation at position 3 follows more complicated kinetics or may utilize a different enzymatic mechanism.

In the presence of TCPO, the ratio of 9-OH-BP relative to 3-OH-BP was increased to greater than twice the control value. It is probable that the increase of 9-OH-BP was the result of nonenzymatic rearrangement of the 9,10-BP-epoxide. Using selective enzyme inhibition and HPLC analysis it should be possible to isolate portions of the metabolic scheme and determine the number of steps involved in the activation and detoxification of carcinogenic polyaromatics including precursor metabolites for later enzymatic steps or rearrangements.

V. EPOXIDES AS INTERMEDIATES IN BENZO(a)PYRENE METABOLISM

One of the keystones in the reactive intermediate theory of benzo(a)pyrene carcinogenesis is proof that epoxides are, indeed, metabolic intermediates. In the preceding discussion it was quite apparent that epoxides were present in microsomal incubation mixtures. Although epoxides were not isolated from incubations containing an epoxide hydrase inhibitor, the relative increase of 9-OH-BP suggested that the 9-10-BP-epoxide was being produced but was unstable and rearranged to 9-OH-BP. Also, a BP metabolite characteristic of an epoxide had been partially characterized [35, 36].

TABLE 2

The Effect of 7,8–BF and TCPO on BP Metabolite Formation[a,b]

	9,10-Diol	4,5-Diol	7,8-Diol	Quinones	9-OH-BP	3-OH-BP	Total metabolites	BP	Percent of metabolism
Control	215.0	81.4	146.4	172.2	51.2	248.8	915.0	4035.0	18.3
7,8-BF	48.5 (22.5)[c]	19.6 (24.1)	31.5 (21.5)	77.7 (45.1)	15.7 (30.7)	95.8 (38.5)	288.8	4650.8	5.9
TCPO	0.0 (0.0)	0.0 (0.0)	0.0 (0.0)	95.8 (55.6)	123.1 (240)	250.8 (100.8)	469.4	4360.4	9.4

[a] Data expressed as pmol × 10^{-2}. Each experiment consisted of five pooled incubations containing a total of 5000 × 10^{-2} pmol BP as substrate.
[b] Ethyl acetate–extractable material. The water-soluble radioactivity was <1% in all three incubations.
[c] Numbers in parentheses represent percentage of control value.

Utilizing the water-methanol solvent gradient with alkaline buffering to pH 9.0 to help stabilize epoxides and inhibit epoxide hydrase, it was possible to trap an intermediate epoxide. Figure 5 shows the BP metabolite spectrum in the presence of 2 mmol TCPO. As was previously seen, there was complete inhibition of dihydrodiol formation and an increase in 9-OH-BP. In addition, the adjusted solvent system revealed a new peak (retention time 29.2 min) between the quinone and phenol regions. This peak was isolated and characterized as the BP-4,5-epoxide.

Table 3 summarizes the analytical data. The metabolite formed in the presence of TCPO migrated on HPLC with a retention time identical to authentic BP-4,5-epoxide. The uv spectra of the metabolite and standard BP-4,5-epoxide were almost identical (Fig. 6), as was the molecular weight and mass spectral fracture pattern. The epoxide metabolite was then incubated with liver microsomes and the products analyzed on HPLC, together with a mixture of known [^{14}C]BP metabolites.

Figure 7 shows that the metabolite was converted to a [^{3}H]diol product which cochromatographed with [^{14}C]BP-4,5-dihydrodiol. No ^{3}H peak was found that corresponded to the 7,8- or the 9,10-dihydrodiol.

TCPO inhibition of dihydrodiol formation and the appearance of the 4,5-epoxide suggested that the other two diols passed through epoxide intermediates but were too unstable to be isolated under the conditions used.

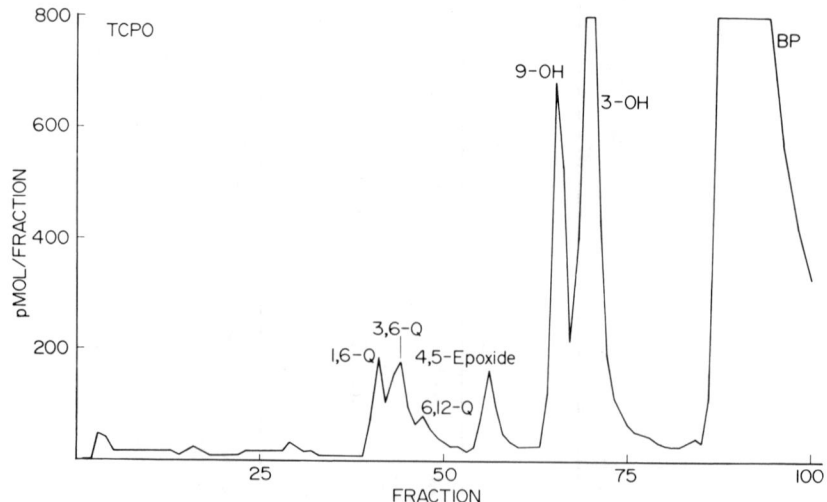

FIG. 5. HPLC pattern of BP metabolites formed in the presence of 2 mmol TCPO [16].

TABLE 3

Characterization of a Metabolite of Benzo(a)pyrene as Benzo(a)pyrene-4,5-epoxide

	Metabolite	Synthetic BP-4,5-epoxide
Mass spectrum m/e	268	268
Major fragments[a]	252 (6.0), 239 (46)	252 (4.9), 239 (47)
	134 (25), 119.4 (42)	134 (21), 119.5 (45)
Uv absorbance, λ_{max}[b]	328, 315, 303, 275, 265	328 (3.8), 315 (4.0)
		303 (4.0), 275 (5.0), 265 (4.8)
HPLC retention time	29.2 min	29.2 min
HPLC retention time of dihydrodiol product	13.2 min	13.2 min

[a]Numbers in parentheses indicate percentage of base peak.
[b]Numbers in parentheses indicate molar extinction coefficient.

This hypothesis was reinforced by the fact that synthetic BP-7,8-epoxide was unstable to those analytical conditions (unpublished results).

The isolation and characterization of an epoxide of BP lends additional support to the role of epoxides as significant intermediates in polyaromatic hydrocarbon metabolism. There is an increasing amount of data [20, 21] supporting the possible existence of a diol-epoxide intermediate [7,8-dihydro-7,8-dihydroxybenzo(a)pyrene 9,10-oxide] where presumably the intermediate diol is a substrate for enzymatic formation of a second epoxide group on the same molecule.

The renewed interest in carcinogenic polyaromatic hydrocarbons as a major environmental contaminant has produced a host of new polyaromatic derivatives synthesized by new and innovative chemistry [37, 38], with structural analysis based on more reliable analytical procedures than used even a decade ago. It is therefore probable that in the near future carcinogenesis experiments and tumorigenesis screening will be performed on these new chemical species, resulting in a clearer picture of which components of the metabolite spectrum are the reactive forms critical to malignant transformation.

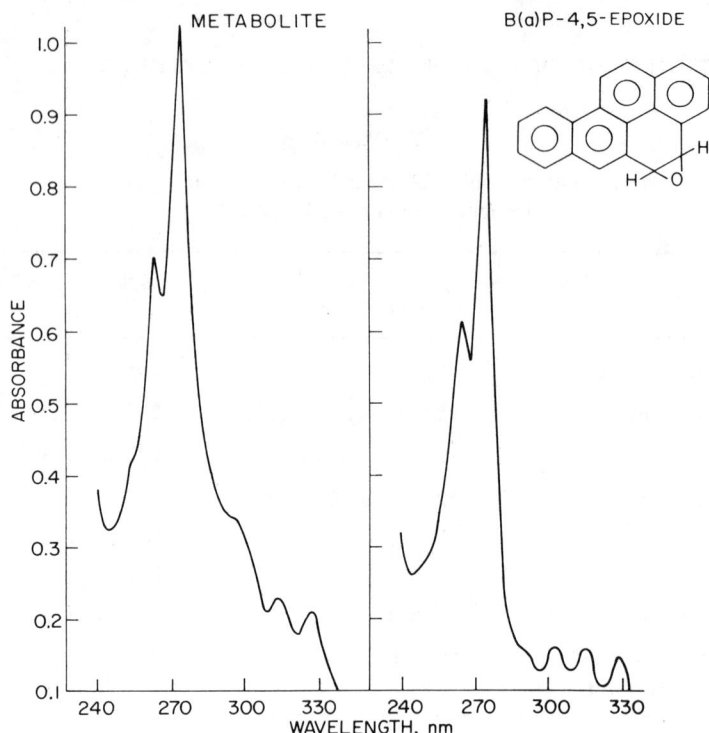

FIG. 6. Ultraviolet spectra of metabolite and synthetic BP-4,5-epoxide [16].

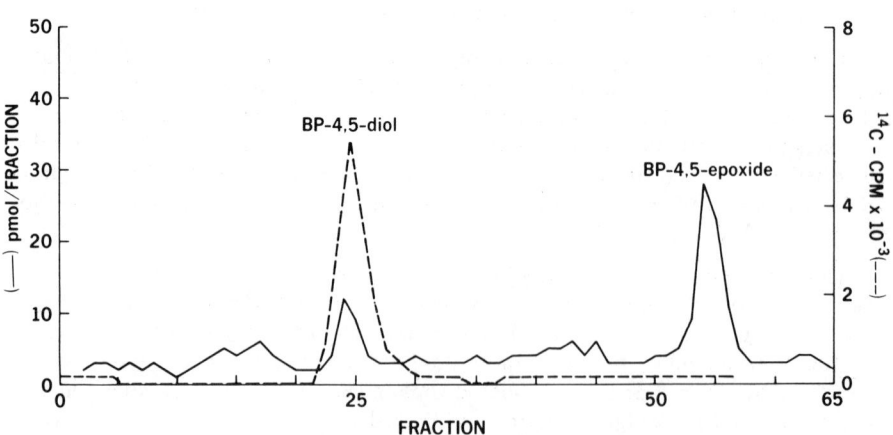

FIG. 7. Conversion of metabolite to BP-4,5-dihydrodiol by liver microsomes. Dashed line represents [^{14}C]BP-4,5-diol added to the organic extract [16].

VI. METABOLISM OF BENZO(a)PYRENE IN HUMAN TISSUE

Environmental contamination by polyaromatic hydrocarbons is a consequence of continued industrial and urban expansion, since they are formed by incomplete pyrolysis of fossil fuels used for heat, power generation and transportation sources. Some of these chemicals are cancer-causing agents in experimental animals, and are becoming increasingly suspect as human carcinogens.

Accordingly, studies dealing with the known chemical and biochemical components associated with either detoxification or activation of chemical carcinogens are timely and extremely important. Identification of carcinogen patterns in humans at various risk levels (e.g., rural versus industrial) may assist in identifying the active metabolites and differentiate variability in human susceptibility.

A comparison between the metabolism of BP by microsomes from human liver and those from rat liver is shown in Fig. 8 [39]. The human liver profile presented additional metabolites not observed in the rodent metabolite pattern. In the diol region, a large peak (I) appeared just after 9,10-diol (fractions 13-15), and a small peak (II) appeared after 7,8-diol (fractions 31-34). The quinone region presented a major peak (III) at the region of the 6,12-quinone. We found synthetic 6-hydroxymethyl-BP cochromatographed with 6,12-quinone in the system and peak III may have contained that derivative [40, 41]. A small peak (IV) followed immediately after (fractions 49-51) with a peak (V) in the epoxide region (fractions 53-56) where the 4,5-BP-epoxide was isolated. However, peak V probably represented another type of derivative since the known epoxides were unstable under such conditions. The increasing background (fractions 31-80) results from a small quantity of BP leaching from the column, which occurred when using reversed-phase elution when low metabolism created a large unmetabolized BP residue. The background was removed by elimination of most of the BP before analysis (Fig. 9).

Lymphocyte metabolism over a 30-min period, seen in Fig. 9, also showed some marked differences from both human and rat liver. All three dihydrodiols were absent, and with the exception of small peaks II and IV, which were also absent, the metabolites (I, III, V, VI) were present as with human liver. However, the relative amount of peak III in the quinone region was reduced, the 3-OH-BP/9-OH-BP ratio was altered, and there was a relative increase in peak VI.

The metabolite profile shown in Fig. 10 for a 24-hr incubation of [^{3}H]BP and human lymphocytes in culture, showed the presence of all three dihydrodiols with the 7,8-dihydrodiol as the major peak. This result generally agreed with previously reported results [42], although the authors did not report the presence of quinones, nor resolve phenol peaks with their mode

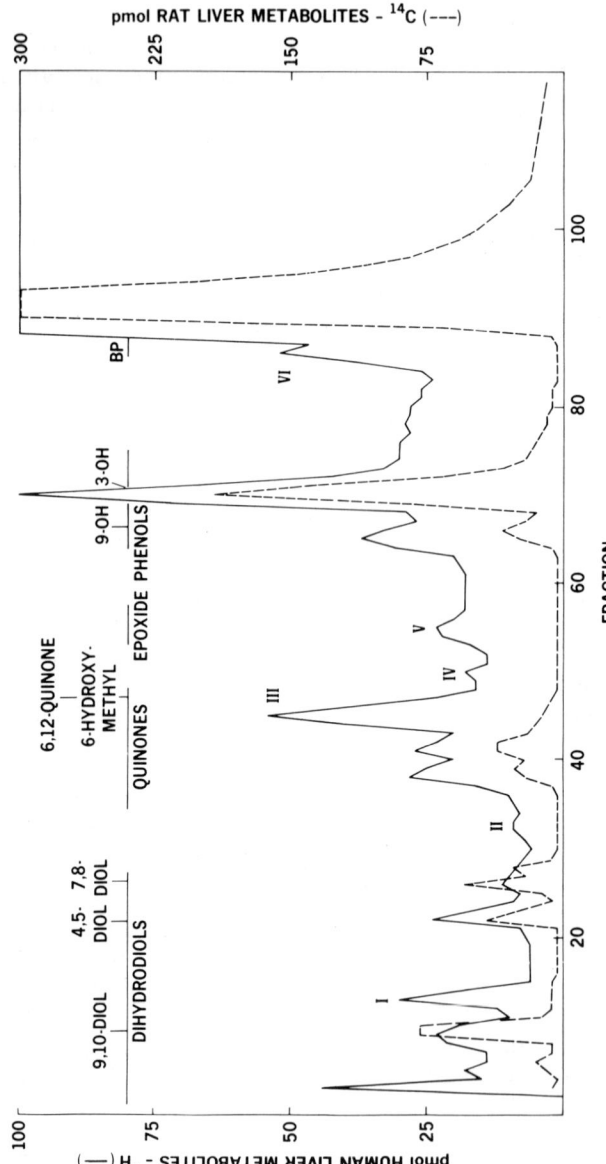

FIG. 8. Pattern of BP metabolites formed by incubation with human liver microsomes. Solid line represents picomoles of hydrocarbon formed by human microsomes with [³H]BP. Dashed line represents picomoles of hydrocarbon formed by rat liver microsomes with [¹⁴C]BP. Roman numerals indicate metabolites produced only by human microsomes [39].

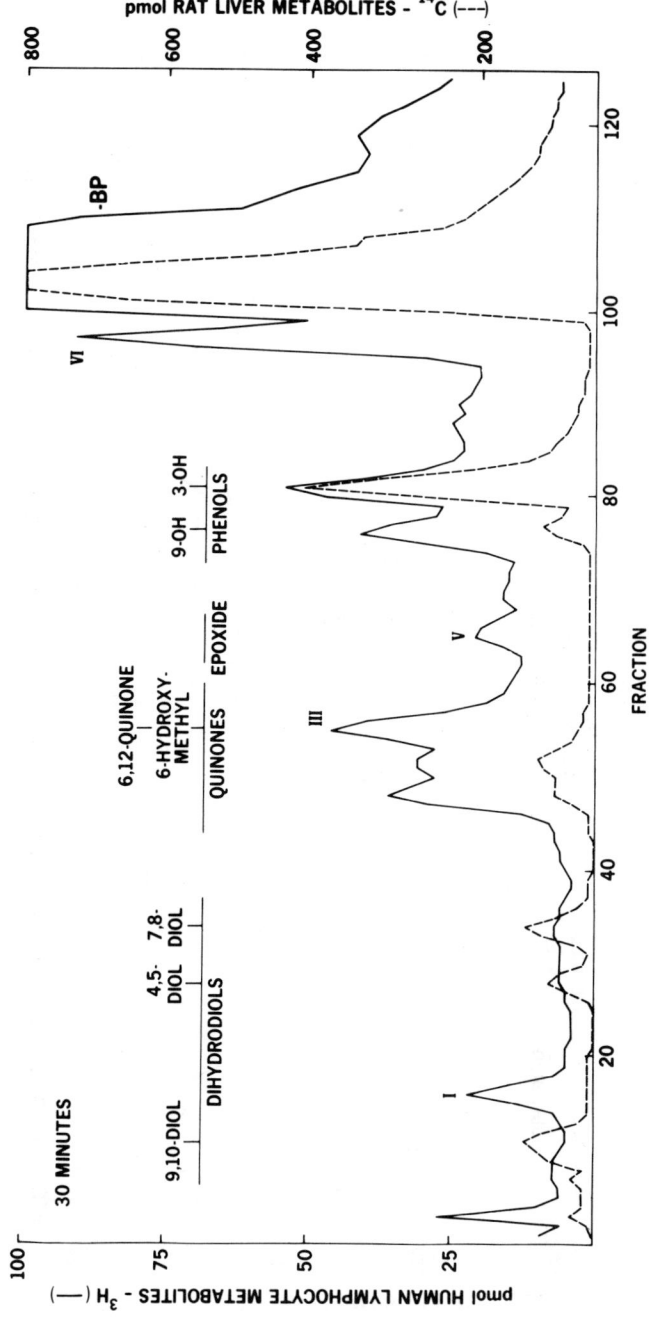

FIG. 9. Pattern of BP metabolites formed by incubation with human lymphocytes for 30 min. Lymphocytes were isolated by the shock-lysing procedure. For complete incubation conditions see Selkirk et al. [39].

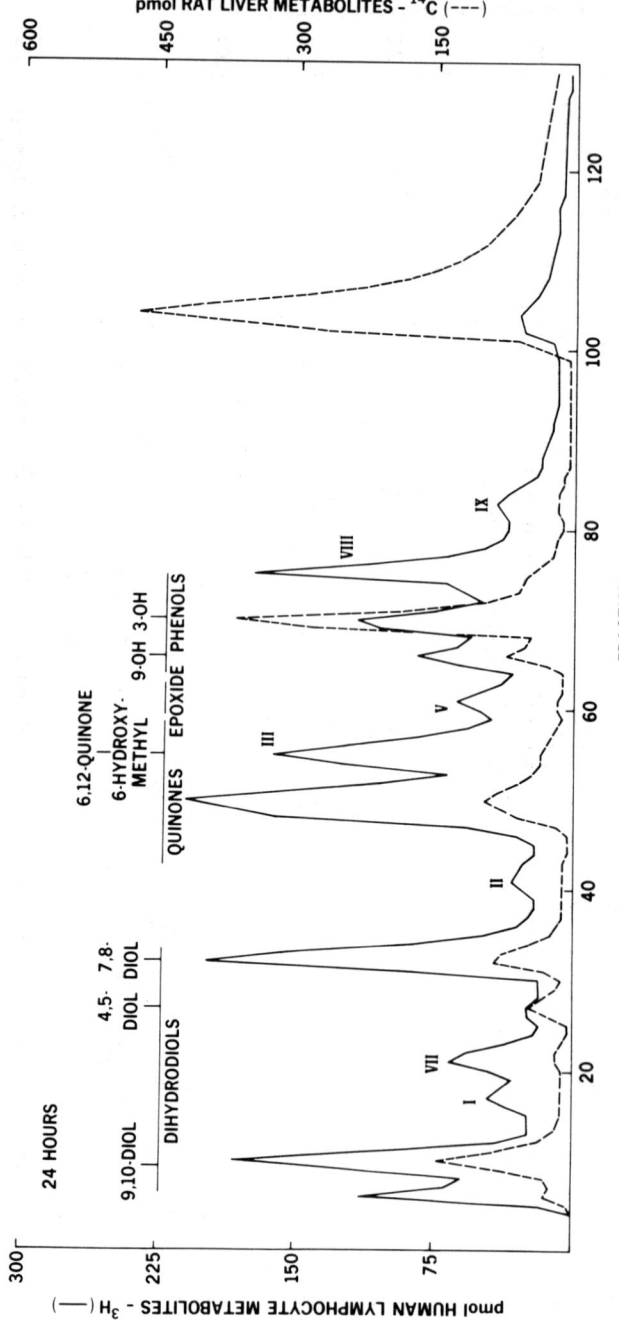

FIG. 10. Pattern of BP metabolites formed by incubation with human lymphocytes for 24 hr [39].

of separation. Peaks I to III and V represented the new metabolites seen in human liver and not in rat liver. Peak IV possibly is hidden between peaks III and V. Peak VI and most of the BP were removed before HPLC analysis.

In addition to peaks I to VI, several more new peaks were observed during the longer incubation period: one peak in the diol region (VII) and two more peaks in the phenol region (VIII, IX). Peak VII was large enough to allow partial analytical characterization and had a molecular weight (m/e 268) corresponding to a phenol, with its uv spectrum similar to, but distinct from, the other known BP phenols.

Human tissue studies of polycyclic aromatic hydrocarbons have been largely confined to determining the amount, specificity, and inducibility of the drug-metabolizing enzymes [43-50]. The fluorimetric analysis which measures conversion of BP to 3-OH-BP is sensitive and adequate to discriminate slight variations in enzyme content. However, it cannot determine the flux of the total metabolite profile with time, nor can it differentiate pattern changes of metabolites between various tissues and species. Although the human studies reported herein have not completely characterized the new metabolites, they nevertheless show salient differences between human and rat tissue metabolism of BP. This finding may be especially important since most carcinogenesis data have been derived from rodent studies.

Lymphocyte metabolism of BP, which produced relatively the same metabolites as human liver, also showed the epoxide hydrase activity in the cells not to be as rapid as the rat epoxide hydrase during short-term incubation. Longer exposure to the hydrocarbon was required to reach an active level of hydrase activity.

It is anticipated, based on the known lability of reactive epoxides, that reduced epoxide hydrase activity could allow epoxides in lymphocytes to possess a greater potential to alkylate target sites. This has already been shown for binding to DNA and for enhancement of tumor formation in mouse skin [51].

VII. BENZO(a)PYRENE METABOLISM IN MALIGNANTLY TRANSFORMABLE CELLS

Benzo(a)pyrene has been studied in many different biological systems. Since metabolite analyses were done with systems less refined than HPLC, incomplete and inaccurate profiles were observed and there has been a tendency to assume that polycyclic aromatic hydrocarbons (PAH) metabolism is generally equivalent in all systems. However, it was apparent that human metabolic patterns differed from rodent metabolic patterns and showed variance between tissues; this type of tissue heterogeneity can be seen in some of the rodent systems routinely used for carcinogenesis studies.

Since the mechanism of action of polycyclic aromatic hydrocarbons is not clear with respect to the actual target site for transformation, determination of the metabolite pattern in susceptible and refractory tissues may help to clarify which steps are necessary for the biosynthesis of sufficient quantities of the reactive intermediate(s) to induce malignancy. Also salient to the mechanism of action will be the development of precursor relationships and reaction kinetics of metabolite formation, since the critical transformation step may necessitate the accumulation of a reactive intermediate(s). Confirmation of epoxides as metabolic intermediates and as precursors to dihydrodiols, and subsequent characterization of the enzyme involved, epoxide hydrase, is an example of one such product-precursor relationship. The suggested 7,8-diol, 9,10-BP-epoxide, as a further reactive intermediate [20, 21] presumably of the 7,8-diol, would be indicative of an ability of the enzyme system to recycle metabolites to what may be even a more carcinogenic species than the previous reactive intermediate, the 7,8-epoxide. In addition, recent discovery of an epoxide reductase [52] appears to further the possibility of a cyclic metabolic pattern whereby polycyclic aromatic molecules may have several opportunities to be metabolically reactivated and attack a target site.

Comparison of the rodent liver metabolite profiles as seen in Figs. 11 and 12 and the total amounts of each metabolite as seen in Table 4 showed rat liver to have the highest levels of 9,10-diol, followed by mouse and

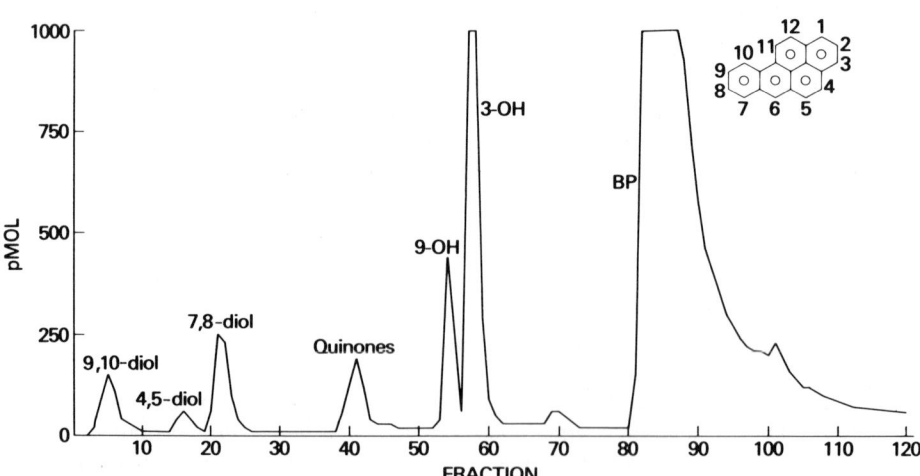

FIG. 11. BP metabolism by mouse liver microsomes. Mouse liver metabolites are identical to rat microsomal pattern (Fig. 3). However, there are percentage differences between individual peaks (Table 4) and an inverted 9,10-diol/9-OH ratio suggestive of a product-precursor relationship.

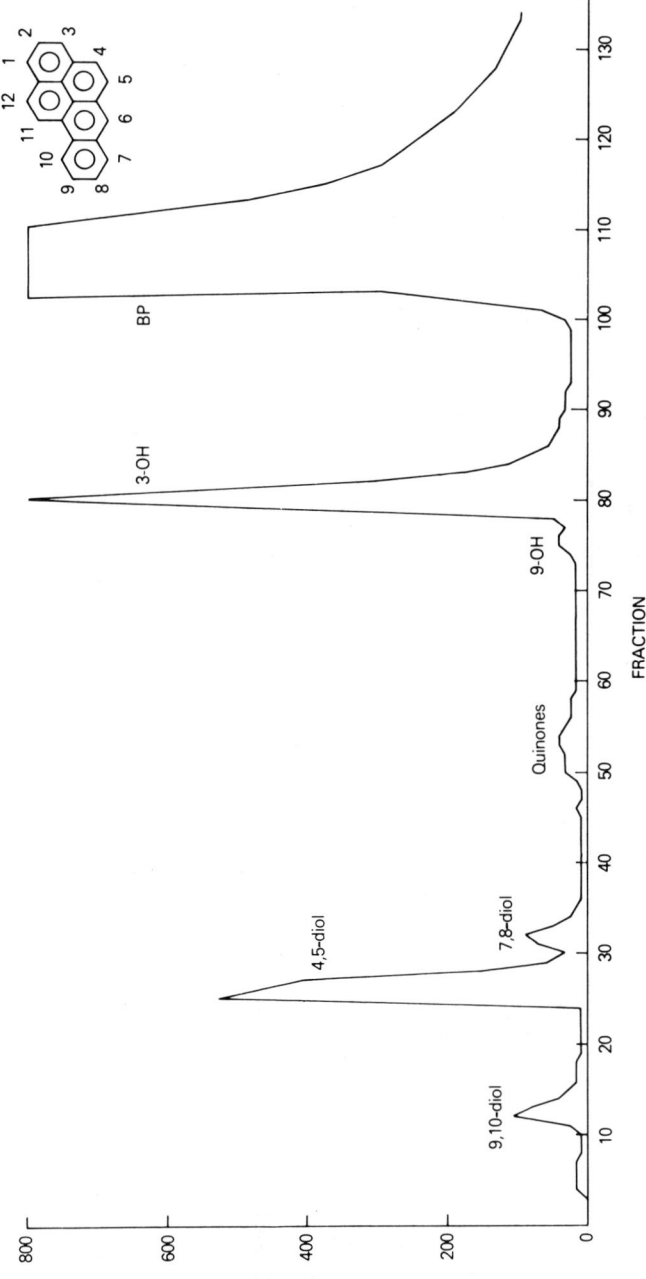

FIG. 12. BP metabolism by hamster liver microsomes. Hamster microsomes show the same pattern as both rat (Fig. 3) and mouse (Fig. 11) microsomes with a markedly different peak ratio (Table 4). The 4,5-diol is a major peak, with the other two diols and 9-OH significantly reduced [53].

TABLE 4
Percentage of Total Metabolites[a]

	Mouse embryo cells	Hamster embryo cells	Mouse liver	Hamster liver	Rat liver
9,10-Diol	7.5 ± 0.3[b]	5.7 ± 1.0	9.0 ± 0.8	5.4 ± 0.4	15.3 ± 0.9
4,5-Diol	0	0	1.2 ± 0.2	28.9 ± 1.9	6.2 ± 0.7
7,8-Diol	10.8 ± 0.5	4.6 ± 0.2	10.4 ± 0.9	4.6 ± 0.7	11.3 ± 0.2
Quinones	17.3 ± 4.0	19.1 ± 6.2	9.3 ± 0.4	3.7 ± 0.1	11.7 ± 0.3
9-OH-BP	39.5 ± 1.7	44.8 ± 0.5	15.0 ± 0.7	1.7 ± 0.2	8.0 ± 1.4
3-OH-BP	21.0 ± 2.2	18.0 ± 2.5	60.4 ± 3.0	55.6 ± 1.7	47.3 ± 0.6

[a]Mouse and hamster embryo cell values represent two experiments using cells derived from different embryos, each experiment consisting of 10 100-mm dishes containing approximately 10^8 cells/dish. Liver microsome values are based on three separate experiments using microsomes from different animals for each run.
[b]Mean ± standard deviation.

hamster liver, respectively [53]. However, the 4,5-diol (K-region) was significantly higher in hamster than in rat or mouse, comprising almost one-third of the total hamster liver metabolites which agreed with an earlier report showing the 4,5-diol as a major hamster liver metabolite using thin-layer chromatography [29]. Whereas the 7,8-diol, which was approximately equivalent in rat and mouse, was least for hamster, the proportion of diols to total metabolites was greatest for hamster (38.9%), followed by rat (32.8%), with the mouse producing only about two-thirds of the diols (20.6%) as the other species.

Hamster liver produced the lowest quinone volume, i.e., about one-third of the other species, which may be related to the fact that in this species the reaction kinetics was directed toward the K-region and therefore fewer molecules were available for activation at the 6-carbon of BP.

In the case of phenols, the 9-hydroxy is significantly greater in the mouse than in rat or hamster, with the hamster having less than 2% of its overall metabolism in the 9-hydroxy, also most likely due to a major K-region attack. In contrast to the marked variance in 9-hydroxy production, the 3-hydroxy shows a smaller percentage spread between the three species with the values clustering around 50% of total organic-soluble metabolites. The overall metabolism of BP between the three species varied between 15 and 20% in all three experiments.

Metabolism of BP by cells in culture as seen in Figs. 13 and 14 and Table 4 was markedly different than that by liver. Although the 9,10-diol and 7,8-diol were comparable to liver microsomal metabolism, there was no detectable formation of the 4,5-diol for either mouse or hamster embryo cells. In comparison to liver there was also an increased quinone yield which was probably due to the longer exposure to air during the 24-hr incubation period. There was also a significant alteration in the phenol production. Whereas 9-hydroxy was a minor component in liver metabolism, it was the major component of embryo cell metabolism with a simultaneous reduction in 3-OH formation. The overall BP metabolism in these embryo cell cultures under these conditions averaged about 5%.

These results clearly show species differences in terms of ratio of metabolites formed and even greater variation when cell types routinely used to study in vitro transformation are compared to the liver microsomal pattern. The metabolite profiles shown here suggest several metabolic approaches to the BP molecule with the enzyme preferentially making a "right side" attack (carbons 3, 4, 5) to form large amounts of K-region derivatives as with hamster liver, or having more substitution at the "left side" (carbons 7, 8, 9, 10) of the molecule as in rat and mouse. This is seen by greater formation of the 7,8- and 9,10-diol and 9-OH in rat and mouse as compared to hamster. Shifting to "left side" metabolism is more readily seen in both cell cultures studied. In both cases the level of the

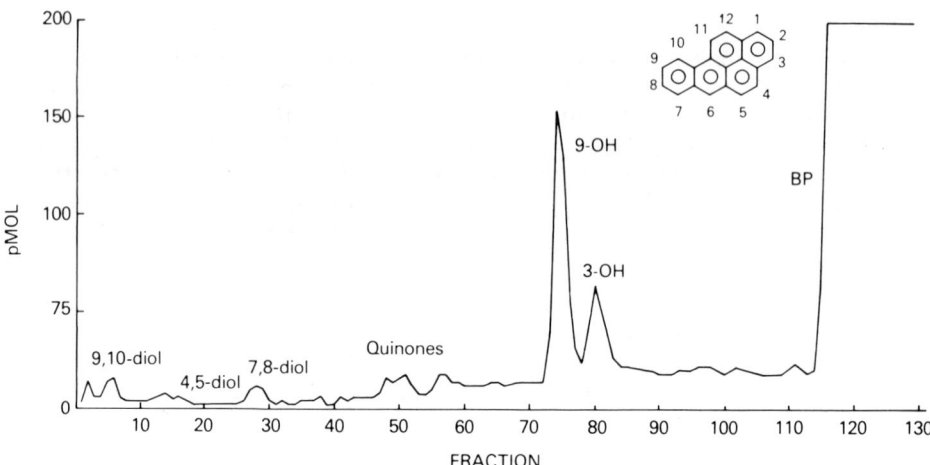

FIG. 13

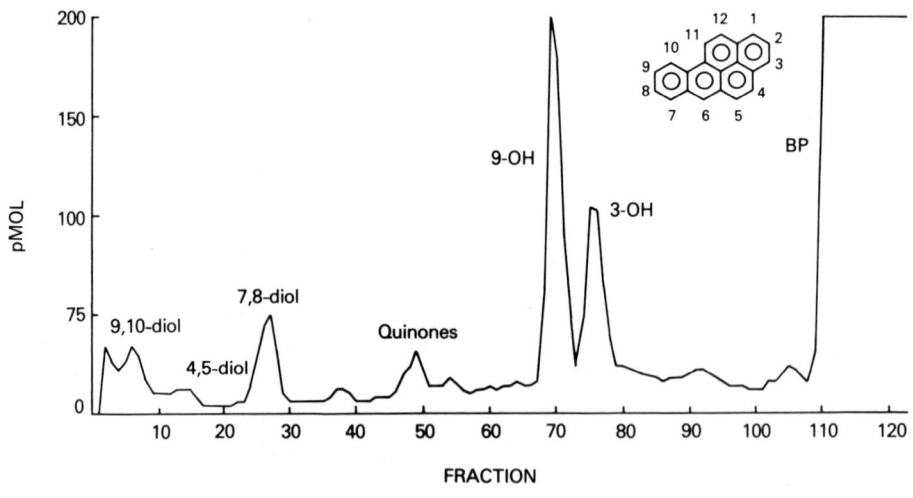

FIG. 14

FIGS. 13-14. BP metabolism by mouse and hamster embryo cells. Both patterns show a complete reorienting of the enzymatic attack on the molecule. The major product is the 1-OH rather than 3-OH as seen in liver microsomes. There are also significant reductions in yield of metabolites oxygenated at other regions of the molecule [53].

K-region 4,5-diol was too low to measure, and complete inversion of the 3-OH/9-OH ratio with 9-OH forming almost half of the total metabolites is indicative of an enzyme activity strongly oriented toward the 7 to 10 region of the BP molecule.

Several reports have appeared showing different DNA-bound products for BP and BP-4,5-epoxide when the compounds are incubated with either hamster or mouse embryo cells [54, 55]. These reports assume that a more involved biochemistry was taking place which was not utilizing the K-region 4,5-epoxide and suggested a new intermediate when in fact the K-region epoxide as evidenced by the absence of its product diol was either not being formed or was rapidly remetabolized. The salient question as suggested by the aforementioned DNA binding studies and the data presented here focuses on the fact that rodent livers are not common sites for tumor formation by BP whereas embryo cell systems are readily transformed by treatment with polycyclic hydrocarbons. It remains to be determined whether region-specific attack on BP as seen in the case of cell culture systems is necessary to in vitro transformation.

To date, the burgeoning carcinogenesis literature has not taken into account the inherent variability of biochemical attack on these chemicals. It will be incumbent upon future work that results and conclusions derived from one biological source should not be generalized as a universal scheme for the activation and detoxification of polycyclic hydrocarbons. It is hoped that this biochemical diversity can be utilized to chart the critical metabolic scheme for these carcinogens much the same as bacterial mutants have aided in gene mapping.

VIII. HIGH-RESOLUTION HPLC (LIQUID CHROMATOGRAPHY BY RECYCLING)

The methanol-water solvent gradient which was used to separate the BP metabolites is only one type of solvent regimen from an almost infinite number of permutations. We have utilized a different approach to the question of the possibility of small amounts of unknown metabolites, especially phenols cochromatographing with and therefore masked by the major peaks. This possibility was made apparent when it was shown that all 12 possible BP phenols chromatographed in two peaks in the methanol-water system [56].

We therefore decided to exploit the small steric or polarity differences between these 12 positional isomers. To accomplish this it would be necessary to perform more extensive chromatography without loss of partially resolved peaks by avoiding collection of large solvent volumes for reinjection into the column. Since water tends to negate the slight polarity variance between the phenols, it could not be used as a solvent. Accordingly, we modified the instrument to allow multiple rechromatography by means of two linked columns and a low dead volume, six-port valve. This system enabled continuous transferral between columns without the peaks passing out of the chromatography system [57].

Figure 15 shows a schematic representation of the recycle system. The system is a modification of the one developed by Henry et al. [58] and incorporates the same alternate pumping design with a six-port, two-position valve (Model CV-G-HPAX, Valco, Inc., Houston, Texas).

When the valve is set to position 1, compounds injected into the system pass first through ports 1 and 2 and into column 1. After eluting from column 1, they pass through the uv cell and back through ports 5 and 3 and enter column 2. At this point the valve is changed to position 2, so the eluting compounds from column 2 pass through ports 4 and 2 and reenter column 1. As the compounds elute from column 1 and pass through the uv flow cell, they may either be collected at port 6 after passing through three columns or be recycled again through another two columns by changing the valve back to position 1 and allowing the compounds to enter column 2 via ports 5 and 3. Once the compounds are transferred onto column 2, which

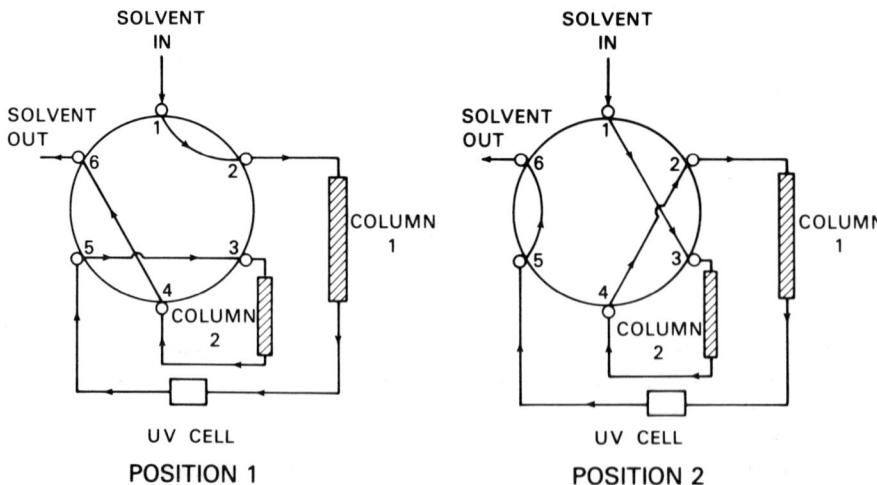

FIG. 15. Valve assembly for HPLC recycling. This diagram is a schematic representation of the six-port valve in both flow positions. After the peak enters column 2 in position 1 the valve is switched to position 2 for recycling back through column 1. When the peak appears in the uv cell, it can be collected at port 6 or the valve switched back to position 1 which recycles the peak onto column 2 [57].

is complete after the final peak has passed through the uv cell, the valve is changed to position 2. The compounds can be collected after five columns at port 6, or the valve can be changed once again to position 1 to transfer the compounds once again onto column 2, repeating the recycle procedure.

Compounds which separate adequately without recycling can be injected into the system with the valve in position 2 and collected after passing through two columns.

Figure 16 shows the separation of 10 isomeric benzo(a)pyrene phenols. The compounds were injected into the system with the valve in position 1 and are first detected after elution through column 1. 5-OH-BP is the first compound to be eluted followed by a peak containing both the 4 and 6 isomers. These peaks are not able to be recycled with the remaining compounds because they are eluted from column 2 before the last peak (8-OH, 9-OH) has

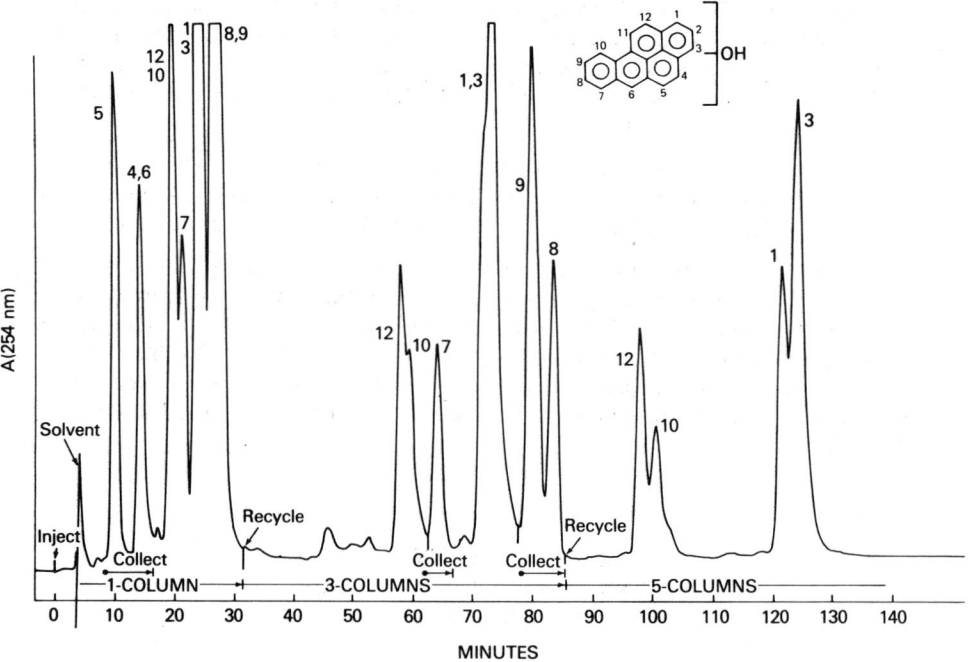

FIG. 16. HPLC of BP phenols. This chromatogram shows 10 synthetic BP phenols which were isocratically separated and collected during two recycle steps. Eluting solvent was hexane-dioxane 9:1 with 0.04 ml formic acid/200 ml solvent [57].

been eluted from column 1 and are passed out of the system before the valve is changed. The first peak which is recycled contains 12-OH and 10-OH-BP. The 7-OH appears as a shoulder on this peak. It is followed by two peaks containing the 3,1- and 8,9-phenol isomers, respectively. After elution of the 8- and 9-OH peak from column 1, the compounds are transferred onto column 2 and the valve is changed to position 2. The 1, 3, 7, 10, 8, 9, and 12 phenols pass through column 2 and back through column 1 before passing through the uv flow cell. The first recycled peak to appear after passing through three columns contained 12-OH with 10-OH as a shoulder. These two phenols were recycled again by changing the valve to position 1 when the peak first appeared, shunting it back onto column 2. As the peak containing 7-OH was detected, the valve is changed back to position 2 so that this peak, being fully resolved, was collected. The following peak containing 3- and 1-OH-BP was recycled in the same manner as the 10-OH and 12-OH peak. The valve was changed to position 1 as the peak was detected, transferring 1- and 3-OH onto column 2, and then changed back to position 2 after they were eluted from column 2. The next two peaks containing 9-OH and 8-OH-BP, respectively, were collected with the valve left in position 2 after passing through three columns. Leaving the valve in position 2, the peaks which were recycled a second time and passed through five columns were then collected. The 12- and 10-OH peaks elute first followed by the 1- and 3-OH peaks. Further resolution may be obtained by recycling them again following the method just described.

The addition of an acidic modifier, formic acid, stabilized the 1, 3, and 12 isomers. However, the most labile, 6-OH, still produced two small peaks (5%) identified as the 1,6- and 3,6-benzo(a)pyrene quinones.

A gradual deactivation of the column, presumably through water adsorption, takes place during prolonged use with this solvent system. The activity and separation efficiency may be restored by washing with methanol-isopropyl ether [57]. The method reported here combined with the reversed-phase method, which separates the 4- and 6-OH isomers, allows complete resolution and identification of the 10 phenol isomers of BP available for this study.

IX. IDENTIFICATION OF FOUR BENZO(a)PYRENE PHENOLS AS METABOLITES

The recycling chromatography system was utilized to reanalyze the 3-OH and 9-OH phenol peaks first isolated from the liver microsome incubation mixture by HPLC using the ODS/Permaphase (DuPont)-methanol/water system. Figure 17 shows the further resolution of the phenol region

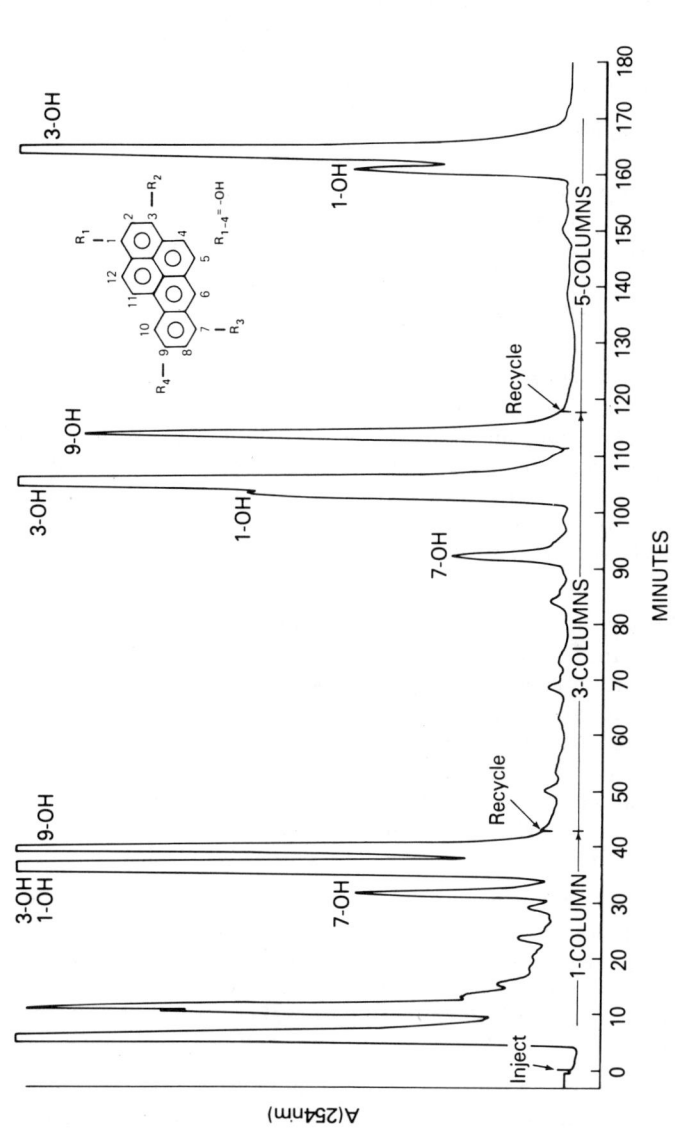

FIG. 17. HPLC of four BP phenol metabolites. The phenol peaks from the metabolite mixture isolated by HPLC with methanol-water were rechromatographed by the recycling procedure. 7-OH and 9-OH were collected after three columns and the 1-OH and 3-OH show peak resolution after five columns.

with recycle chromatography. The first two unidentified peaks are not metabolites but rather 254 absorbing materials which when collected show no uv spectrum related to benzo(a)pyrene. These peaks are derived from microsomal extracts. Three metabolite peaks are observed after passage through a single column. After passage through the entire five-column recycling system, four distinct peaks are resolved.

Each of the metabolites was isolated and characterized by cochromatography and uv spectra and compared to the 10 synthetic phenols. They were found to correspond to the 3-phenol and 9-phenol previously identified as metabolites and two newly identified metabolites, the 1-phenol (max, 413, 399, 387, 379, 298, 287, 267, 259 nm) and the 7-phenol (max, 398, 375, 356, 304, 290, 266, 257, 250 nm). Analysis of the 9-OH and 3-OH peaks separately indicated the two new metabolites were from the 3-OH peak.

Berenblum and Schoental [59] isolated and partially characterized a compound from rabbit, which they identified as equivalent to 1-OH-BP (IUPAC, 1957). However, an authentic 1-OH-BP standard was not available at that time and the data cannot be directly compared to the present work. The identification of 7-OH-BP as a metabolite reinforces the argument of a labile 7,8-epoxide intermediate. In addition, isolating 1-OH-BP as a metabolite introduces a new region of the BP molecule as a potential carcinogenic site.

This report has been designed to exemplify the ease with which this technique can obtain reproducible data concerning the critical biochemical structural information concerning BP metabolism. Although the experiments reported have dealt only with one polycyclic species, this same technique will work with equal facility for other chemical carcinogens. The efficiency of this new technique with its superior utility for the study of labile reactive molecules has made the task of understanding chemical carcinogenesis more approachable than was envisioned using previous methods with less analytical capacity.

The widespread occurrence of these experimental carcinogens is a result of industrialization and increased transportation reliance on engines which produce polyaromatic hydrocarbons by incomplete pyrolysis of fossil fuels. In addition, internal pollution by the presence of these compounds in cigarette smoke makes these chemicals increasingly suspect as human carcinogens.

Therefore the importance of understanding polycyclic aromatic carcinogenesis in experimental animals and cell systems cannot be overstated since the ultimate goal of these studies is to extrapolate them to understand and defeat malignancy in humans.

ACKNOWLEDGMENTS

The author would like to thank Dr. H. V. Gelboin, Chief of the Chemistry Branch of the National Cancer Institute, Bethesda, Maryland. Thanks is also extended to Mr. Robert G. Croy for assistance and collaboration during the course of this work, and to Sandra Vaughan for her very able assistance in the preparation of this manuscript.

REFERENCES

1. Committee on Biological Effects of Atmospheric Pollutants, Particulate Polycyclic Organic Matter, National Academy of Sciences, Washington, D.C., 1972.
2. K. Yamagiwa and K. Ichikawa, J. Cancer Res., 3, 1 (1918).
3. I. Heiger, J. Chem. Soc., 1933, 395.
4. Survey of Compounds Which Have Been Tested for Carcinogenic Activity, Public Health Service Publication No. 149.
5. A. Pullman and B. Pullman, Adv. Cancer Res., 3, 117 (1955).
6. H. V. Gelboin, Adv. Cancer Res., 10, 1 (1967).
7. A. H. Conney, Pharmacol. Rev., 19, 317 (1967).
8. E. Huberman, T. Kuroki, H. Marquardt, J. K. Selkirk, C. Heidelberger, P. L. Grover, and P. Sims, Cancer Res., 32, 1391 (1972).
9. F. Oesch, Xenobiotica, 3, 305 (1973).
10. J. K. Selkirk, R. G. Croy, and H. V. Gelboin, Cancer Res., 34, 3474 (1974).
11. P. Sims, Biochem. Pharmacol., 19, 795 (1970).
12. N. Kinoshita, B. Shears, and H. V. Gelboin, Cancer Res., 33, 1937 (1973).
13. J. K. Selkirk, R. G. Croy, and H. V. Gelboin, Science, 184, 169 (1974).
14. E. Boyland and P. Sims, Biochem. J., 95, 780 (1965).
15. P. Sims, Biochem. J., 98, 215 (1966).
16. J. K. Selkirk, R. G. Croy, and H. V. Gelboin, Arch. Biochem. Biophys., 168, 322 (1975).

17. J. Booth, E. Boyland, and P. Sims, Biochem. J., 79, 516 (1961).
18. N. Nemoto and H. V. Gelboin, Assay and properties of glutathione-s-benzo(a)pyrene-4,5-oxide transferase. Arch. Biochem. Biophys., 170, 739-742 (1975).
19. N. Nemoto, H. V. Gelboin, W. H. Habig, J. N. Ketley, and W. B. Jakoby, Nature (London), 255, 512 (1975).
20. P. Sims, P. L. Grover, A. Swaisland, K. Pal, and A. Hewer, Nature (London), 252, 326 (1974).
21. P. Daudel, M. Duquesesne, P. Vigny, P. L. Grover, and P. Sims, FEBS Lett., 57, 250 (1975).
22. A. P. Alvares, G. R. Schilling, W. Levin, and R. Kuntzman, Biochem. Biophys. Res. Commun., 29, 521 (1967).
23. A. P. Alvares, G. R. Schilling, and R. Kuntzman, Biochem. Biophys. Res. Commun., 30, 588 (1968).
24. F. J. Wiebel, J. C. Leutz, L. Diamond, and H. V. Gelboin, Arch. Biochem., 144, 78 (1971).
25. E. Boyland and L. F. Chasseaud, Adv. Enzymol., 32, 173 (1969).
26. A. Y. H. Lu and M. J. Coon, J. Biol. Chem., 243, 1331 (1968).
27. A. Y. H. Lu, R. Kuntzman, S. West, M. Jacobson, and A. H. Conney, J. Biol. Chem., 247, 1727 (1972).
28. P. Dansette, H. Yagi, D. M. Jerina, J. W. Daly, W. Levin, A. Y. H. Lu, R. Kuntzman, and A. H. Conney, Arch. Biochem. Biophys., 164, 511 (1974).
29. A. Borgen, H. Darvey, N. Castagnoli, T. T. Crocker, R. E. Rasmussen, and I. Y. Wang, J. Med. Chem., 16, 502 (1973).
30. P. L. Grover, A. Hewer, and P. Sims, Biochem. Pharmacol., 21, 2713 (1972).
31. T. Stoming, D. Knapp, and E. Bresnick, Life Sci., 12, 425 (1973).
32. F. J. Wiebel, H. V. Gelboin, N. P. Buu-Hoi, M. G. Stout, and W. S. Burnham, Chemical Carcinogenesis (P. O. P. Ts'o and J. A. DiPaolo, eds.), Marcel Dekker, New York, 1972, Part A, pp. 249-270.
33. F. Oesch, N. Kaubisch, D. M. Jerina, and J. W. Daly, Biochemistry, 10, 4858 (1971).
34. C. S. Yang and F. S. Strickhart, Biochem. Pharmacol., 24, 646 (1975).
35. P. L. Grover, A. Hewer, and P. Sims, Biochem. Pharmacol., 21, 2713 (1972).

36. I. V. Wang, J. F. Rasmussen, and T. T. Crocker, Biochem. Biophys. Res. Commun., 49, 1142 (1972).
37. S. H. Goh and R. G. Harvey, J. Am. Chem. Soc., 95, 242 (1973).
38. P. Dansette and D. M. Jerina, J. Am. Chem. Soc., 96, 1224 (1974).
39. J. K. Selkirk, R. G. Croy, J. P. Whitlock, Jr., and H. V. Gelboin, Cancer Res., 35, 3651 (1975).
40. J. W. Flesher and K. L. Sydnor, Int. J. Cancer, 11, 433 (1973).
41. J. W. Flesher and K. L. Sydnor, XI International Cancer Congress Florence, Italy, Panel, 2, 52 (1974).
42. J. Booth, G. R. Keysall, P. L. Kalyani, and P. Sims, FEBS Lett., 43, 341 (1974).
43. R. Kuntzman, L. C. Mark, L. Brand, M. Jacobson, W. Levin, and A. H. Conney, J. Pharmacol. Exp. Therp., 152, 151 (1966).
44. W. Levin, A. H. Conney, and A. P. Alvares, Science, 176, 419 (1972).
45. D. W. Nebert, J. Winkler, and H. V. Gelboin, Cancer Res., 29, 1763 (1969).
46. R. M. Welch, Y. E. Harrison, B. W. Gommi, P. T. Poppers, M. Ernster, and A. H. Conney, Clin. Pharmacol. Ther., 10, 100 (1959).
47. D. L. Busbee, C. R. Shaw, and E. T. Cantrell, Science, 178, 315 (1972).
48. J. P. Whitlock, Jr., H. L. Cooper, and H. V. Gelboin, Science, 177, 618 (1972).
49. R. C. Bast, Jr., J. P. Whitlock, Jr., H. Miller, H. J. Rapp, and H. V. Gelboin, Nature (London), 250, 664 (1974).
50. E. T. Cantrell, G. A. Warr, D. L. Busbee, and R. R. Martin, J. Clin. Invest., 52, 1881 (1973).
51. N. Kinoshita and H. V. Gelboin, Proc. Natl. Acad. Sci. U.S.A, 69, 824 (1972).
52. J. Booth, A. Hewer, G. R. Keysell, and P. Sims, Xenobiotica, 5, 197 (1975).
53. J. K. Selkirk, R. G. Croy, F. J. Wiebel, and H. V. Gelboin, Cancer Res., 36, 4476 (1976).
54. W. M. Baird, R. G. Harvey, and P. Brookes, Cancer Res., 35, 54 (1975).
55. H. S. King, M. H. Thompson, and P. Brookes, Cancer Res., 34, 1263 (1975).

56. G. Holder, H. Yagi, P. Dansette, D. M. Jerina, W. Levin, A. Y. H. Lu, and A. H. Conney, Proc. Natl. Acad. Sci. U.S.A., 71, 4356 (1974).

57. R. G. Croy, J. K. Selkirk, R. G. Harvey, J. F. Engel, and H. V. Gelboin, Biochemical Pharma., 25, 227 (1976).

58. R. A. Henry, S. H. Byre, and D. R. Hudson, DuPont Liquid Chromatography Technical Bulletin No. 73-2.

59. I. Berenblum and R. Schoental, Cancer Res., 6, 699 (1946).

Chapter 2

HIGH-PERFORMANCE LIQUID CHROMATOGRAPHY OF THE STEROID HORMONES

F. A. Fitzpatrick

Department of Physical and Analytical Chemistry
Drug Metabolism Research Section
The Upjohn Company
Kalamazoo, Michigan

I.	INTRODUCTION	38
II.	STEROIDS—A BRIEF REVIEW	39
	A. Corticosteroids	41
	B. Androgens	42
	C. Estrogens	43
	D. Progestins	43
III.	SCOPE OF HIGH-PERFORMANCE LIQUID CHROMATOGRAPHY OF STEROIDS	45
IV.	INITIAL DEVELOPMENTS	46
V.	STEROIDS IN BIOLOGICAL FLUIDS	51
	A. Estrogen Assays	51
	B. Corticosteroid Assays	56
	C. Androgen Assays	57
	D. Summary	59
VI.	HIGH-PERFORMANCE LIQUID CHROMATOGRAPHY OF STEROIDS IN PHARMACEUTICALS	61
	A. Development of Separations	61
	B. Corticosteroid Pharmaceuticals	63
	C. Estrogen Pharmaceuticals	64
	D. Summary	65

VII. CHEMISTRY BEFORE, DURING, AND
 AFTER HPLC OF STEROIDS 65

   A. Derivatization to Enhance Detectability 65
   B. Chemistry during Chromatography 67
   C. Chemistry after Chromatography 68

VIII. THE FUTURE OF HIGH-PERFORMANCE
  LIQUID CHROMATOGRAPHY OF
  STEROID HORMONES 68

   A. Instrumental Developments 68
   B. Column Technology 69
   C. Other Equipment 70

IX. CONCLUSIONS 70

  ACKNOWLEDGMENTS 70

  APPENDIX....................................... 71

  REFERENCES 71

I. INTRODUCTION

Chromatography has long been a useful tool in the field of steroid research. Early workers attempting to isolate and identify naturally occurring steroid hormones from biological fluids used adsorption chromatography to effect difficult separations of structurally similar compounds. Later, column partition chromatography, paper chromatography, and thin-layer chromatography were applied to the microscale separation of many steroids for both identification and assay. Conversely, steroids have played a substantial role in the evolution of improved chromatographic methods. Efforts to develop theoretical correlations between chromatographic behavior and chemical structure employed steroids as model compounds because of their planar ring formations and the variety of functional group combinations they contain. Steroids were among the first low-vapor-pressure compounds separated intact by gas chromatography, resulting in the reevaluation of the belief that poorly volatile substances could not be analyzed by this approach. Because of its resolution and speed, the latest improved chromatographic method, high-performance liquid chromatography (HPLC), has been used to separate many biochemical and pharmaceutical compounds. Rapid growth has made it pertinent to review its role in steroid separations since many of the papers published dealing with HPLC have focused on these compounds.

Recent advances at the interface of two broad areas, high-performance liquid chromatography and steroids, comprise the theme of this chapter. Complexity prevents a detailed treatment of either topic within the confines of this series, so useful background material is cited for the interested reader. The principles of modern high-performance liquid chromatography are covered in comprehensive works by Brown [1] and Kirkland [2], as well as in other chapters of this series and recently published reviews [3-5]. Key information on the chemistry of the steroids can be found in books by Fieser and Fieser [6], Shoppee [7], Klyne [8], and Heftman [9]. Steroid analysis and steroid chromatography exclusively are treated by Carstensen [10], Bush [11], and Neher [12, 13].

For convenience, this chapter has been divided into sections that include first, a review of the initial developments of steroid HPLC; second, HPLC of steroids in physiological fluids; and finally, the application of HPLC to the separation and analysis of steroids in pharmaceutical preparations. It will become evident that there are significantly different problems in each instance. First, it is important to summarize briefly some features of the structure, biochemistry, and nature of the steroid hormones.

II. STEROIDS—A BRIEF REVIEW

Steroids are a group of compounds having in common the perhydrocyclopentanophenanthrene fused ring system. The three cyclohexane rings are designated A, B, C and the cyclopentane ring is the D ring. The basic structure and the accepted numbering system are shown in Fig. 1. Mammalian endocrine glands have been identified as the source of several hundred naturally occurring steroids, many with biochemical, clinical, or therapeutic significance. In addition to those steroids and their metabolites already isolated and identified, a virtual flood of analogs are being constantly synthesized by enterprising organic chemists who seem to have two main goals.

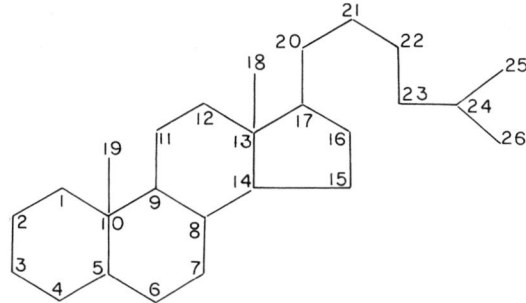

FIG. 1. Basic structure and accepted numbering system of steroid hormones.

The first is to delight the medical profession by producing compounds with such potency that lower and lower doses are sufficient to achieve a desired biological response. The second goal appears to be to perplex the analytical chemist faced with the task of analyzing the continually decreasing quantities present in pharmaceuticals, and the even more miniscule amounts found in the general circulation after dosing. All steroids are subdivided into four major categories based on their chemical structures and biological effects. These are the corticosteroids, androgens, estrogens, and progestins. Typical members of each group are shown in Figs. 2 to 4.

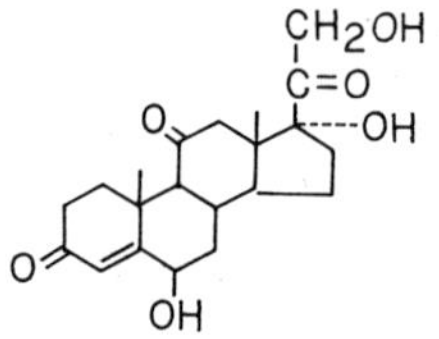

I

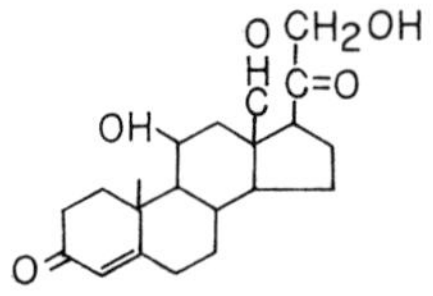

II

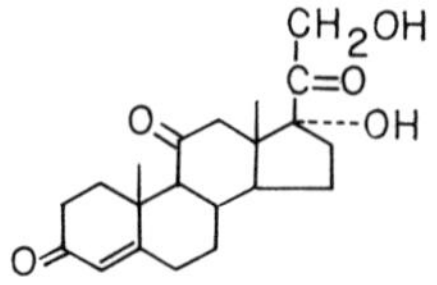

III

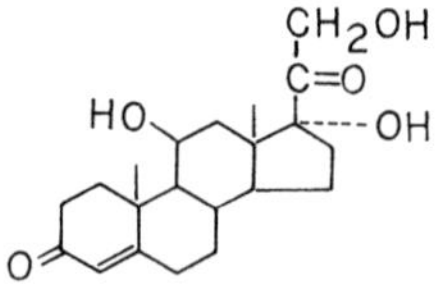

IV

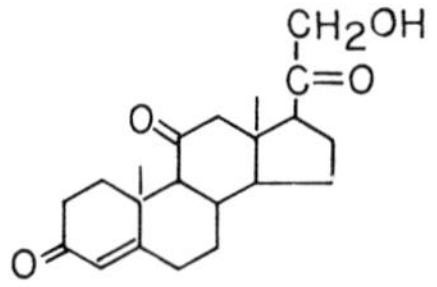

V

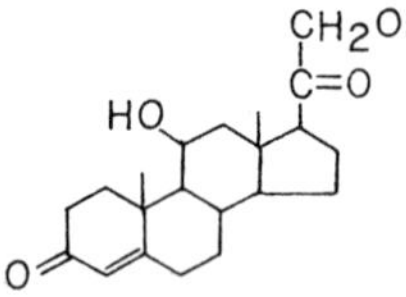

VI

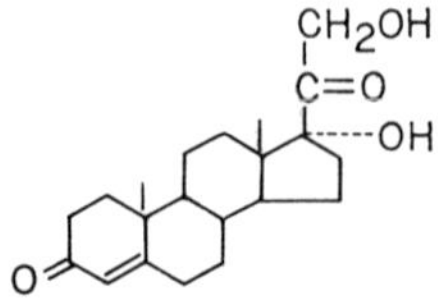

VII

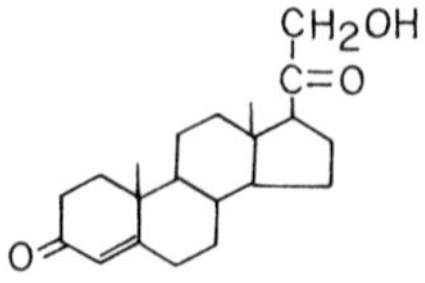

VIII

FIG. 2. Typical corticosteroids. I = 6-β-OH cortisol; II = aldosterone; III = cortisone; IV = cortisol; V = 11-dehydrocorticosterone; VI = corticosterone; VII = 11-deoxycortisol; VIII = 11-deoxycorticosterone.

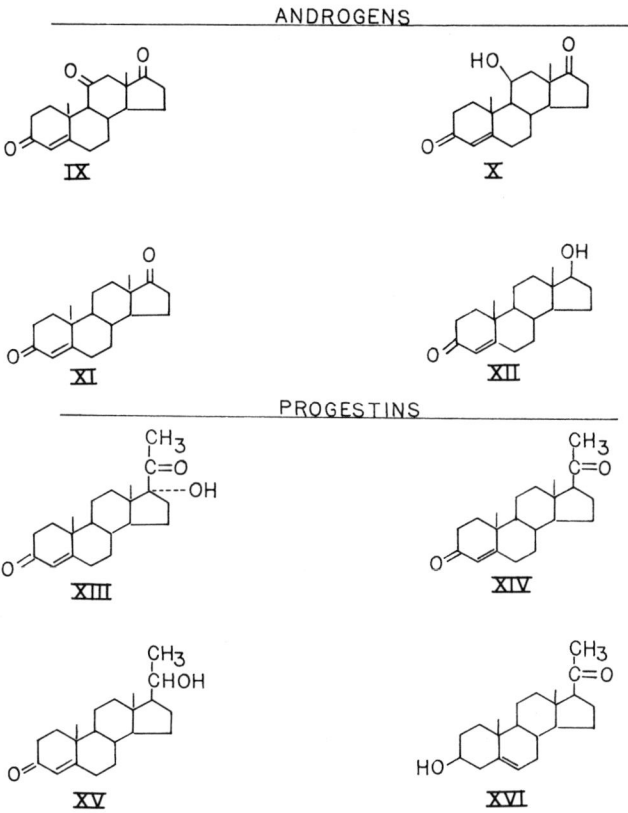

FIG. 3. Some typical androgens and progestins. Androgens: IX = 4-androstene-3,11,17-trione; X = 4-androstene-11β-ol-3,17-dione; XI = 4-androstene-11,17-dione; XII = testosterone. Progestins: XIII = 17α-hydroxyprogesterone; XIV = progesterone; XV = 4-pregnene-20β-ol-3-one; XVI = pregnenolone.

A. Corticosteroids

Corticosteroids are C-21 compounds with at least three oxygen atoms incorporated into the steroid nucleus. Hydroxyl groups at C_{21} and C_{17} are common, as well as a hydroxyl or carbonyl group at C_3. The corticosteroids are derived from the adrenal cortex and are often further classified as mineralocorticoids or glucocorticoids. The glucocorticoids, whose principal members are cortisone and cortisol, influence carbohydrate and protein metabolism; the mineralocorticoids, whose most potent members are aldosterone and 11-deoxycortisone, affect the electrolyte balance in man.

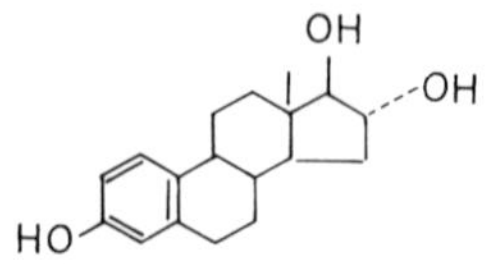

FIG. 4. Some typical estrogens: XVII = estriol; XVIII = 17-epiestriol; XIX = estradiol; XX = estrone.

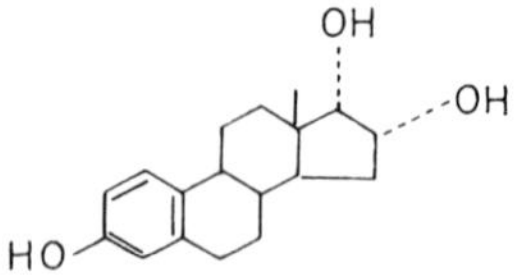

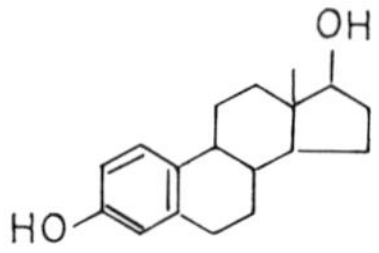

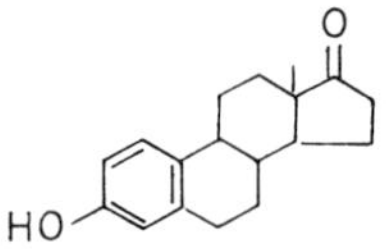

Urinary metabolites of corticosteroids which are clinically significant are tetrahydrocortisol, tetrahydrocortisone, pregnanetriol, and 17-hydroxypregnanolone. Table 1 lists some medical disorders in which normal corticosteroid levels are altered. Virtually all of the urinary steroids and their metabolites are excreted as salts of glycopyranosiduronic acid or sulfuric acid.

B. Androgens

Androgens are male sexual hormones which have in common the absence of a side chain at the C_{17} position. The most potent naturally occurring androgen, testosterone, is produced in the testes. The 17-ketosteroids, which are derived in part from the metabolism of the corticosteroids and in part from the metabolism of testosterone, are the principal androgens excreted in the urine. Biologically, androgens comprise any steroid promoting comb growth in capons or seminal vesicle and prostate growth in castrated rats. Most naturally occurring steroids eliciting this response have a carbonyl function at C_{17} but only androsterone and dehydroepiandrosterone are significantly potent in this respect.

TABLE 1
Disorders Associated with Altered Levels of Corticosteroids

Addison's disease	Lowered 17-hydroxycorticosteroids
Anterior pituitary hypofunction	Lowered 17-hydroxycorticosteroids
Severe hypertension	Increased 17-hydroxycorticosteroids
Virilism	Increased 17-hydroxycorticosteroids
Stress (surgery, infectious disease)	Increased 17-hydroxycorticosteroids
Third trimester of pregnancy	Increased 17-hydroxycorticosteroids
Cushing's syndrome	Substantially increased

C. Estrogens

The estrogens are female sexual hormones produced by the ovaries under normal conditions and by the placenta during pregnancy. The distinguishing feature shared by all estrogens is the benzenoid character of the A ring. As a result of this, the hydroxyl group on the C_3 carbon is phenolic and there is no angular methyl group attached at C_{10}. The primary physiological role of the estrogens is to stimulate the growth and maintenance of the uterus and to prepare it for the appropriate response to progesterone in order to initiate pregnancy. Estrogen levels in blood and urine rise continuously during pregnancy and sudden decreases especially during the third trimester are indicative of fetal distress. Estrogens of clinical significance include estrone, estriol, and estradiol. An intriguing observation has been made about the steroids responsible for the male and female secondary sexual characteristics, testosterone and estradiol, respectively. The chemical formula of testosterone is $C_{19}H_{28}O_2$ and the formula for estradiol is $C_{18}H_{24}O_2$. One immediately perceives that a factor contributing to the difference between men and women is a single methane molecule. This is undoubtedly responsible for fueling the fires of debate between male chauvinists and liberated women.

D. Progestins

Progestins are C21 steroids secreted by the adrenal cortex of both sexes and also by the corpus luteum during the menstrual cycle. Progesterone is partly responsible for the cyclical changes in cellular morphology during menstruation and it is necessary for the preparation and maintenance of the uterus in pregnancy. The relationship between increased progresterone levels after conception and inhibition of further ovulation has been recognized and exploited with profound impact by the introduction of birth control pills. Table 2 lists biological indicators whose normal status has been associated

TABLE 2

Biologically Indicative Test Results Altered by Oral Contraceptives[a]

Substance measured[b]	Change in level
Albumin	Decreased
Glucose	Elevated glucose level at 1 hr
Triglycerides (S)	Elevated
Thyroid-binding globulin (S)	Elevated
Thyroxine (S)	Elevated (free thyroxine normal)
Triiodothyronine resin uptake	Decreased (free thyroxine index normal)
Cortisol (hydrocortisone)-binding globulin (S)	Elevated
Hydrocortisone (P)	Elevated
17-OH-Corticosteroids (U)	Decreased
17-Ketogenic steroids (U)	Decreased
Metyrapone (Metopirone) test	Impaired responsiveness
Sulfobromophthalein (bromsulphalein) test	Impaired excretion
Platelets (B)	Elevated, mild
Blood procoagulants (B)	Elevated (not evident on routine testing)
Iron (S)	Elevated
Iron-binding capacity (S)	Elevated

Source: From J. H. Wendling, JAMA, 229, 1762 (1974).
[a]Listed are the altered laboratory measurements due to oral contraceptives with the most clinical significance.
[b]S = serum; P = plasma; U = urine; B = blood.

with "pill"-induced changes. Ordinarily, progesterone is metabolized so rapidly that circulating levels can only be determined by radioimmunoassay or electron capture gas chromatography. A urinary metabolite, pregnanetriol, occurs at levels accessible by other chromatographic techniques.

Their remarkable diversity and therapeutic potential ensure that active research on the steroid hormones will be sustained for many years. Some contributions to this research that can be made by high-performance liquid chromatography are outlined in the following section.

III. SCOPE OF HIGH-PERFORMANCE LIQUID CHROMATOGRAPHY OF STEROIDS

Rapid advances in high-performance liquid chromatography have made it difficult to define its boundaries even for a limited group of compounds such as the steroids. However, we may provide a crude outline by listing some problems which have been attacked with HPLC. The list is certainly not exhaustive.

1. During the synthesis of steroids, reaction kinetics can be easily monitored by HPLC. Often, precise quantitative results are unnecessary and the ratio of products to starting materials is adequate to optimize the synthetic yield. Once the conditions have been established, the desired product can be isolated and purified on a preparative scale for structure and identity confirmation. By extension, HPLC can be used to advantage for improving product yield and for product isolation after microbial conversions or other biosynthetic steroid reactions [14]. Because of its speed and improved resolution HPLC promises to supplant techniques such as TLC which have been commonly used for these purposes.

2. If a steroid appears promising in biological tests, information about its metabolism is generally desired. At this stage, HPLC is useful in separating and isolating metabolites from biological fluids, or for tissue residue analyses to determine the distribution and clearance rate of the administered drug. More comprehensive adsorption, distribution, metabolism, and excretion studies can also be designed around HPLC methodology. It must be emphasized that such applications are usually confined to radiolabeled steroids. The generic problem of trace level (ng/ml or g) detection of steroids which do not have appreciable uv extinction at 254 nm, the operating wavelength of common HPLC detectors, has not yet been solved.

3. After selection of a steroid for development into a drug, problems generally fall under the direct jurisdiction of the analytical chemist. Stability-indicating assays are needed to devise suitable formulations and to monitor

the active ingredient content and quality of manufactured products. HPLC is again ideal for solving problems arising at this stage. It seems probable that in the future, bioavailability studies, required by governmental agencies to determine the levels of ingredients in circulation after dosing with equivalent formulations, will rely on HPLC if other techniques are unsatisfactory.

4. After a steroidal drug has been in use, occasions will arise in which clinical significance might be attached to its plasma or blood levels, or those of a metabolite. HPLC assays used during the prior stages of research and development can be adapted to this problem.

The scope of high-performance liquid chromatography extends from the identification of promising therapeutic leads from natural starting materials, through the synthesis and isolation of the desired steroid, identification of metabolic pathways and residue levels, and finally through the development of methodology to ensure the quality and efficacy of steroidal drugs. At every stage, different problems will be encountered requiring different approaches for their solution. The analyst processing and analyzing hundreds of samples for a bioavailability study will have substantially different criteria than the analyst identifying and quantitating three degradation products at concentrations under 0.5% when the separation system is being devised. Similarly, the worker interested in preparative scale isolation faces different problems than the researcher evaluating interactions between steroids and chromatographic columns to assess influences of partition coefficients on the in vivo absorption rate of a compound. It will be instructive to survey the brief history of high-performance liquid chromatography of the steroid hormones to see what has been accomplished, what approaches have been fruitful for solving specific problems, and more significantly, why progress has been less substantial than one might have suspected given the demonstrated potential of this technique.

IV. INITIAL DEVELOPMENTS

The first published work describing the use of modern high-performance liquid chromatography for the separation of steroid hormones appeared in 1970 [15]. It is important to note that commercially developed instrumentation for this technique was not generally available so that the construction of instrumentation suitable for such studies was an achievement in its own right. Experience has since shown that laboratories building their own instruments are often as successful as those employing sophisticated chromatographic units.

The authors of the first paper sought to capitalize on apparent advantages inherent in HPLC suggesting its superiority over other steroid separation techniques. Conventional column chromatography suffers from many problems. Lack of speed, incomplete recovery of material applied to the column in most instances, and discontinuous detection make it virtually useless as a quantitative technique. Thin-layer chromatography, although faster than conventional gravity flow column chromatography, is still slow compared to HPLC. Even under ideal conditions only about eight compounds can be separated on a 20-cm TLC plate developed in one dimension for 1 to 2 hr. Few TLC systems approaching ideality have been reported. The similarity of many steroids prevents mixtures of fair complexity from being separated without multidimensional development in several solvent systems. Gas chromatography had become popular for steroid analysis because of its speed, sensitivity, and high resolution, but HPLC appeared to have advantages over it also. Primary among these is the fact that liquid chromatography exploits natural interactions between the column and the compounds to be separated under mild conditions in solution, whereas gas chromatography requires operation at high temperatures (greater than 200°C in most cases) and formation of thermally stable derivatives. Although derivatization solves one problem, it creates others, especially control of reaction conditions to obtain quantitative and reproducible conversion. Often derivatives are formed yielding multiple peaks from single steroids in the reaction mixture, for example, syn and anti isomers of ketosteroids. For detection of trace components in biological fluids, the removal of excess derivatizing agent without altering the derivative is a serious problem. Since these difficulties are not as trivial in practice as they appear to be on paper, HPLC seemed ideally suited for steroid chromatography.

With these potential advantages in mind, Siggia and Dishman [15] evaluated the effects of different solid supports and stationary partitioning phases on the resolution of mixtures of related steroids. In comparing totally porous diatomaceous earth (140-325 mesh), Zipax, superficially porous, spherical silica beads (37-44 µm), and Plaskon, CTFE tetrafluoroethylene polymer (140-325 mesh), the latter proved to be superior as a support. This superiority was ascribed mainly to its hydrophobic nature, which minimized tailing and also added an undetermined but favorable interaction of its own, since some steroids could be separated in the reversed-phase mode on uncoated CTFE support alone. Of several hydrophobic stationary phases evaluated, Amberlite LA-1, n-dodecenal trialkyl methylamine, proved most effective. This unusual coating was selected initially because of a suspected Schiff base type of interaction between the carbonyl groups on the steroids and the amino functionality of the stationary phase. In practice, however, it proved generally effective for all classes of steroids, indicating that the separations were a composite effect of support material interactions, ordinary liquid-liquid partitioning, and conceivably Schiff base-type interactions. Impressive separations of individual corticosteroids, androgens, estrogens, and progestins were reported, as shown in Figs. 5 to 7. Detection in all cases was accomplished by using a converted uv spectrophotometer (Beckman DU) adapted to the particular constraints of modern HPLC, namely low dead

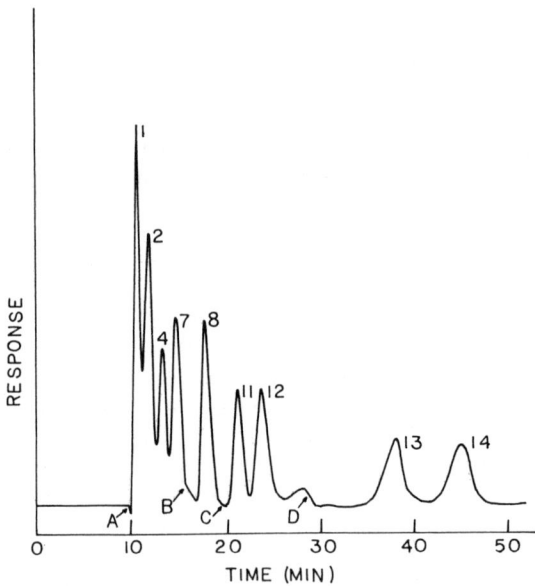

FIG. 5. 1 = 6β-hydroxycortisone; 2 = aldosterone; 4 = cortisone; 7 = 11-dehydrocorticosterone; 8 = corticosterone; 11 = 11-deoxycortisol; 12 = cortisone 21-acetate; 13 = 4-pregnene-20β,21-diol-3-one; 14 = deoxycorticosterone. Column 48.5 cm × 0.2 cm I.D., 23% LA-1 on CTFE; H_2O eluent; A = 0.13 ml/min; B = 0.21 ml/min; C = 0.26 ml/min; C = 0.44 ml/min.

volume between the column outlet and the end of the detection system. Quantitative analysis was attempted but was only partially successful, since the detector output was neither linearly nor logarithmically related to the absorption signal. This resulted mainly in poor precision, but results indicated that quantitation analogous to gas chromatography was possible.

Although this identical system has not been widely adopted, it laid the foundations for consideration of HPLC in the reversed-phase, partition mode as a useful approach for separating lipophilic compounds. The development of hydrolytically stable, permanently bonded, reversed-phase columns based on the formation of an octadecylsilyl polymer coating on a silica support was the fundamental reason behind its quick replacement. Permanently bonded columns have now supplanted most HPLC partition systems which had to deal with the unwieldy problem of maintaining a mobile phase saturated with stationary phase to replace material mechanically stripped from the support. In active laboratories it was common to repack partition columns on a weekly basis in the infancy of HPLC. Although amply demonstrating the potential of HPLC for steroid separations, the results of this work may

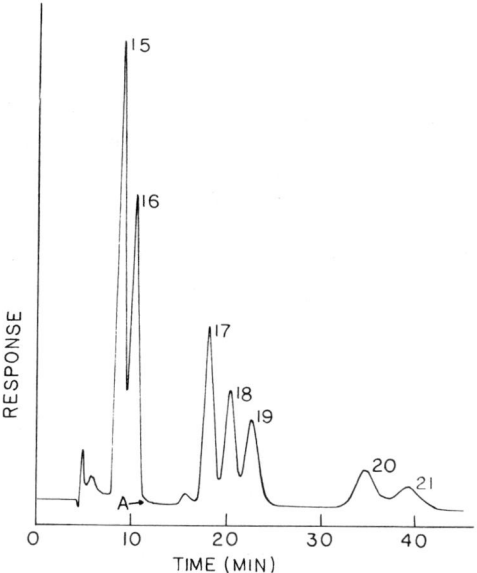

FIG. 6. 15 = 4-androstene-3,11,17-trione; 16 = 4-androstene-11β-ol-3,17-dione; 17 = $\Delta^{1,4}$-androstadiene-17β-ol-3-one; 18 = 19-nor-4-androstene-3,17-dione; 19 = 19-nor-testosterone; 20 = 4-androstene-3,17-dione; 21 = testosterone. Column 48.5 cm × 0.2 cm I.D., 23% LA-1 on CTFE; initial flow = 0.17 ml/min; A (flow) = 0.49 ml/min.

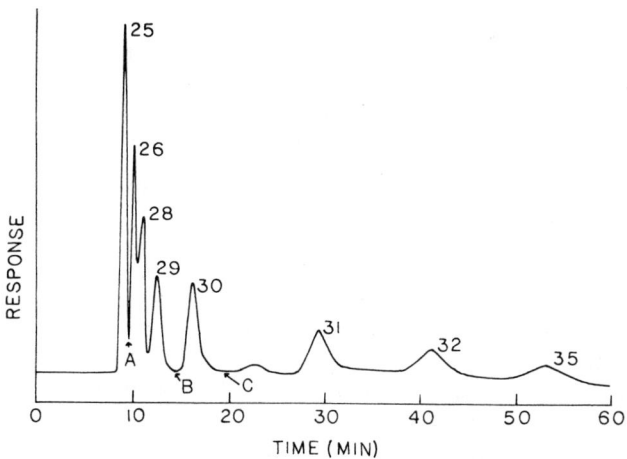

FIG. 7. 25 = estradiol-17α-glucosiduronic acid; 26 = estriol; 28 = 16-ketoestrone; 29 = 16-ketoestradiol; 30 = 16-epiestriol; 31 = equilinen; 32 = estradiol; 35 = estrone. Column 48.5 cm × 0.2 cm I.D., 28% LA-1 on CTFE; eluent = H2O (pH 11.5, NaOH); flow: A = 0.145 ml/min, B = 0.19 ml/min, C = 0.49 ml/min.

have been too elegant to have appeared at such an early stage in the evolution of a new technique. The successes of the initial report stimulated interest but equivalent successes did not follow rapidly. Nearly a year elapsed before another paper appeared on steroid HPLC [16]. In 1971, Henry and coworkers at DuPont described the resolution of steroids of all four classes by normal phase partition chromatography on β,β-oxydipropionitrile-coated Zipax, superficially porous silica columns, and reversed-phase partition chromatography on superficially porous, permanently bonded, octadecyl silane columns. In addition, steroid glucuronides were separated by high-performance ion-exchange chromatography for the first time. The paper by Henry was notable for the following accomplishments:

1. It emphasized the diversity of HPLC for performing separations of steroids ranging from extremely polar compounds such as the steroid conjugates, to very nonpolar steroids such as progesterone. In so doing, it illustrated that a particular problem can be attacked and solved by one mode of chromatography in nearly every case. Some general guidelines concerning the selection of a suitable chromatographic system could be extracted from the paper. These are that strongly polar compounds with salt groups, e.g., steroid conjugates, are best separated by ion-exchange chromatography at the first attempt. Intermediate polarity compounds, steroids containing two or more oxygen functionalities, e.g., testosterone, nortesterone, or corticosteroids and estradiol which are soluble in hydrocarbon/alcohol solutions, can generally be separated by either normal liquid-liquid partition chromatography or reversed-phase partition chromatography. Compounds such as progesterone, which are highly soluble in hydrocarbons, are best chromatographed by reversed-phase partition systems, since the eluent in normal phase systems elutes these steroids too near the chromatographic void volume for practical use.

2. It introduced the concept of derivative formation to enhance the absorption of uv radiation by otherwise nonabsorbing compounds. Since most commercially available HPLC detectors are single-wavelength, 254-nm spectrophotometric monitors, the limits of detection could be substantially improved by this route. While particular examples were given of the formation of 2,4-dinitrophenylhydrazones of carbonyl-containing steroids, the generic concept was extended by others to the formation of different steroid derivatives, for example, benzoates and nitrobenzoates, and subsequently to other compounds.

3. Advantages of superficially porous, often called pellicular or solid core supports, were presented. The ease of packing reproducible columns of high efficiency was stressed and proved to be valid in practice in many laboratories. The availability of these supports from commercial suppliers helped to minimize some of the problems that usually develop when trying to reproduce chromatographic separations reported in the literature.

4. A practical demonstration of the application of HPLC to the determination of steroids in a topical cream was shown, exemplifying the utility of HPLC for quality control and stability-indicating assays. Excipients and other stabilizing ingredients were shown not to interfere when chromatographic conditions were chosen judiciously.

V. STEROIDS IN BIOLOGICAL FLUIDS

With the potential of high-performance liquid chromatography firmly established, attention was immediately directed toward practical problems which had resisted solution by other chromatographic approaches or to problems for which HPLC offered an obvious improvement. Paramount among these were the separation and analysis of steroids in whole blood, plasma, or urine for both diagnostic and research purposes. Before solutions were realized, several obstacles had to be overcome. These are discussed in the following sections.

A. Estrogen Assays

Urinary estrogen levels are recognized as a valuable indicator of health and organ status in females. During pregnancy, sudden depression of these levels signals fetal distress or placental malfunction. Other estrogen therapy regimens (e.g., fertility drug treatments) require that the circulating levels be monitored directly or indirectly by urinary analysis to ensure the health and safety of the patient. Interest in specific, accurate, and rapid urinary assays has been abundant. Traditionally, nonspecific fluorometric or colorimetric methods based on the Kober reaction [17] have been used to measure the total urinary estrogen content. By incorporating preliminary separation steps such as thin-layer chromatography, individual estrogens can be determined but assay times are appreciably longer. The clinical significance of estrogen assays coupled with improvements in assay time and specificity prompted Huber to report the development of an HPLC method for their measurement [18].

Several important operations not directly related to HPLC had to precede the final separation and measurement step. Estrogens, like most steroids, are excreted into the urine conjugated as sulfate or glucuronide salts. A hydrolysis step is necessary to release the free steroid, which can then be extracted from the polar constituents of urine into an organic solvent. Conditions for the hydrolysis have been examined by many investigators [19, 20], and two alternatives are available, enzymatic or acid hydrolysis. In the former, β-glucuronidase and sulfatase enzymes are incubated with urine under conditions of mild pH and temperature (37°C). These mild conditions minimize steroid degradation during hydrolysis, but quantitative release of the aglycone portion of the conjugate is not always achieved if enzyme inhibitors happen to be present in the urine. Additional drawbacks to the enzymatic hydrolysis include limitations on the volume of urine that can be hydrolyzed because of the cost of the enzyme, and long incubation times of 36 to 72 hr. The other approach, hydrolysis by refluxing urine acidified with mineral acid, is rapid and complete, but the harsh conditions have to be optimized to prevent steroid degradation. The published procedures for acid hydrolysis of urinary estrogen conjugates were selected and shown to be compatible with the HPLC determination by Huber. After hydrolysis, estrogens were isolated from the crude reaction mixture by extraction into ether. A typical flow chart for these steps is shown in Fig. 8. The extraction serves two vital purposes. First, it isolates the hydrolyzed steroids from the polar constituents of urine, providing a crude but effective preliminary purification step. Second, it allows concentration before chromatography, by simple evaporation of the extracting solvent. Direct analysis of urinary estrogens is difficult not only because these estrogens occur as conjugates (separations of estrogenic conjugates have been reported, e.g., Fig. 9) but also because their concentrations are so low (<5 mg/liter of urine).

After hydrolysis, extraction, and concentration to 1 ml or less, samples (50 μl) were injected and separated on a ternary stationary phase of water-ethanol-isooctane (229:680:91) coated on diatomaceous earth. The same components in the ratio 19:177:804 made up the mobile phase. Using this system, baseline resolution of estrone, estradiol, and estriol, the major urinary estrogens, was possible in less than 30 min. Chromatograms of estrogens from pregnancy urine contained peaks coincident in retention time to a standard injection of these three estrogens. Besides investigating the feasibility of HPLC analysis of urinary estrogens, Huber clarified its limitations as a general method for steroid analysis in physiological fluids by examining the significance of important instrumental and chromatographic parameters. Among other things, he considered the influence of injection volume on detector response (peak height) and showed that it has a mutually opposing effect on the maximum concentration of steroid reaching the detector, and the resolution of two closely eluting steroids. Since in quantitative analysis the systematic error increases with decreasing resolution and the statistical error increases with decreasing peak height, one must empirically determine the best concentration and volume to inject to achieve a minimum error. In practice the situation is not as bad as it seems, especially since the introduction of improved chromatographic columns capable of generating several thousand plates per meter of available resolving power. The widespread use of superficially porous columns at the time this paper was published made it a significant problem because of their limited capacity.

I	20 ml of urine + 3 ml of 12 N HCl. Heat in boiling water bath 1 hr.
II	Cool and extract 3 times with 20 ml of ether.
III	Combine ether extracts and wash once with 8 ml of saturated carbonate buffer pH 10.5; once with 6 ml of 8% $NaHCO_3$; and twice with 5 ml H_2O.
IV	Evaporate ether to dryness.
V	Dissolve residue in 8 ml benzene + 8 ml petroleum ether.
VI	Extract phenolic steroids once with 8 ml, once with 6 ml, and once with 4 ml of 1% NaOH. Combine extracts.
VII	Neutralize the alkaline phase containing estrogens with concentrated HCl.
VIII	Extract once with 20 ml and twice with 10 ml of ether.
IX	Wash ether once with 6 ml 8% $NaHCO_3$, twice with 4 ml H_2O. Discard washings.
X	Evaporate ether to dryness in a conical tube. Reconstitute residue in solvent for HPLC.

FIG. 8. Flow chart for hydrolysis and extraction of estrogenic steroids.

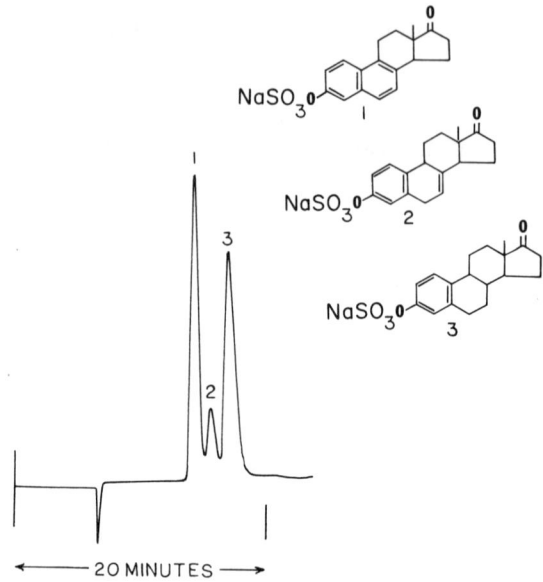

FIG. 9. Separation of some estrogen conjugates. Flow rate 1 ml/min; uv detector 0.08 AUFS, 254 nm; anion-exchange column.

Several factors contributed to the success of HPLC for urinary estrogen analysis. Fortuitously, the estrogens are phenolic compounds with four distinct advantages over other urinary steroids. First, being phenolic, they are somewhat more stable to acid hydrolysis than neutral or polyoxygenated steroids. Second, they can be fractionated rather selectively from both more polar and less polar nonphenolic elements which might have interfered in the chromatographic separation. Third, their unsaturated A ring makes detection of nanogram quantities possible with typical spectrophotometric HPLC monitoring systems. Finally and most importantly, in the late stages of pregnancy (third trimester), urinary estrogen levels rise to 2 to 3 mg/liter, so the 50-ml samples of urine (third trimester pregnancy patients) contained 100 to 150 µg of estrogens. When this amount was extracted and concentrated, several micrograms, an amount easily detected, was deposited onto the column.

Other reports on urinary estrogen analysis have subsequently appeared. Butterfield and co-workers [21] examined four different partition systems for separating up to nine equine estrogens and concluded that a 3-m Permaphase ETH column was best. Quantitative utility was not explored. Dolphin [22] has described an adsorption chromatographic system capable of separating estrone, estradiol, and estriol in less than 6 min and applied his

approach for their quantitation in pregnancy urine. The procedure used
prior to the HPLC step was virtually identical to that reported by Huber.
The chromatographic systems of these two later papers were an improvement over the ternary phase system at least in a practical sense. With
suitable precautions, such as careful temperature, flow rate, and humidity
control, ternary phase systems are quite useful, as evidenced by the separation of the estrogens. However, for adaptation to routine analysis by less
qualified personnel, the simpler systems based on adsorption chromatography
or permanently bonded partition phases would be favorable. Separations
based on reversed-phase systems have also been reported (see Fig. 10), but
have not been applied to urinary assays. Conceivably, these might be even
better, since adsorption columns are prone to change with cumulative deposition of strongly retained materials.

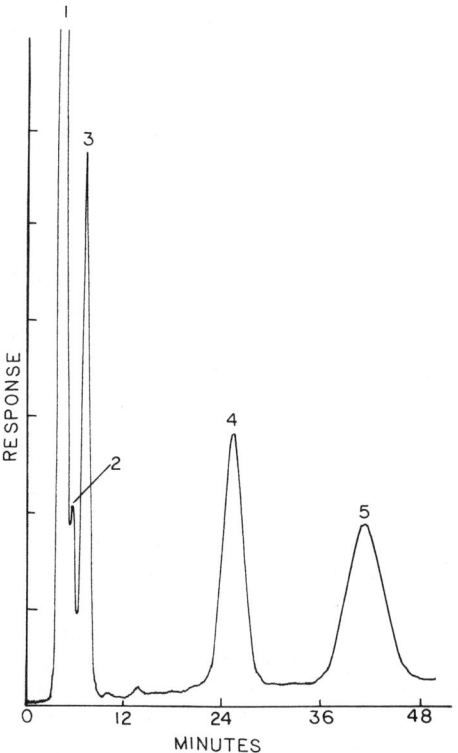

FIG. 10. Reversed-phase separation of estrogens. Column 50 cm, 20%
undecane nitrile on CTFE 2300. 100% H_2O mobile phase, 0.1 ml/min, uv
0.04 AUFS, 254 nm. 1 = estriol; 2 = 16-ketoestrone; 3 = 16-ketoestradiol;
4 = estradiol; 5 = estrone.

The extension of the HPLC methodology described to the analysis of normal or subnormal estrogen levels is conspicuous by its absence. This is simply explained: current HPLC detectors are not sensitive enough. Only Dolphin [22] has reported the extension of his method to the analysis of normal levels of estrogens; however, unpublished work by Schleicher [24] has shown that at least 250 ml of normal urine (estrogen content ~10 μg/liter) must be hydrolyzed and extracted to obtain enough estriol for detection. Besides being impractical from a logistic standpoint, the extraction of this much urine yielded a larger peak eluting with the solvent front, which compromised the quantitative accuracy. Thus, the practical detection and HPLC analysis of urinary estrogens at normal levels remains an elusive problem.

B. Corticosteroid Assays

As was the case with the estrogens, the clinical importance of the corticosteroids and the difficulties associated with their analysis spurred investigations into the use of HPLC. Specific urinary assays for corticosteroids are relatively uncommon, because corticosteroids are metabolized so extensively. However, as an index of adrenal function in both sexes, plasma levels of aldosterone, cortisone, and cortisol are especially significant. Separations of the corticosteroids were described in the first paper on steroid HPLC but a viable, quantitative analysis based on this methodology was not reported until late in 1972. Touchstone and Wortmann [23] separated corticosteroids on high-performance adsorption columns (Sil-X) and were able to determine the cortisol content in 5 ml of human plasma by HPLC. The identity of the cortisol peak was confirmed by isolation and an additional thin-layer chromatographic step, and the incorporation of radiolabeled cortisol into the assay. Cortisol circulates in its free form so hydrolysis was unnecessary. After extraction and concentration of the extract, the sample was ready for separation and analysis. Since cortisol contains an α,β-unsaturated ketone which absorbs strongly at 254 nm, the detector wavelength, and the concentrations found in plasma (~1 μg/ml) are sufficiently high, detection was straightforward. Because of their lower concentrations in normal patients, the method was not amenable to other plasma corticosteroids. Except for occasional variations in retention times no problems were encountered with this method. Two significant advantages pointed out by the authors were that chromatography at room temperature resulted in less sample loss during separation (presumably relative to gas chromatography although this was not explicitly stated) and that a notable decrease in analysis time accompanied the HPLC method. As before, other separation systems, especially bonded reversed-phase systems, have been reported for corticosteroid separations and could be adapted to this analysis.

C. Androgen Assays

The androgens contain several members designated 17-ketosteroids (17-KS) which are of diagnostic significance in several pathological states (Table 3). In normal males about 70% of the urinary 17-KS content is derived from the adrenal cortical hormones and the remainder from the metabolism of testosterone. In normal females essentially all the urinary 17-KS is of adrenal corticol origin, with traces being secreted by the ovaries. In males the urinary 17-KS levels reflect the composite function of both the adrenal cortex and the testes, whereas levels in females reflect adrenal activity alone. The major 17-ketosteroids appearing in the urine as glucuronide or sulfate conjugates are androsterone, etiocholanolone, and dehydroepiandrosterone. Some unconjugated dehydroepiandrosterone and the 11β-hydroxy and 11-keto derivatives of androsterone and etiocholanolone, resulting from hepatic metabolism of cortisol, are also found in urine. Fig. 11 shows that epimeric differences at the 3 and 4 positions are the distinguishing factor among the 17-KS.

A colorimetric group estimation procedure for the 17-ketosteroids based on the formation of a chromophore with m-dinitrobenzene in alkaline solution is the most commonly used analytical method for their determination [25]. Being nonspecific, the method is subject to interferences from both ketonic and nonketonic artifacts as well as steroids with carbonyl groups at positions other than C_{17}. Again, accuracy is improved with preliminary TLC or column-chromatographic stages, but assay time is lengthened.

TABLE 3

Diseases Associated with Altered 17-Ketosteroid Excretion

Decreased 17-KS	Elevated 17-KS
Addison's disease	Adrenal cortical tumor
Cirrhosis of the liver	Adrenal hyperplasia
Myxedema	Interstitial tumor of the testes
Hypogonadism	Androgen-secreting ovarian tumor
Hypopituitarism	Severe stress

Source: From Specialized Diagnostic Laboratory Tests, 9th ed., January 1972, Bio-Science Laboratories, Richmond, California.

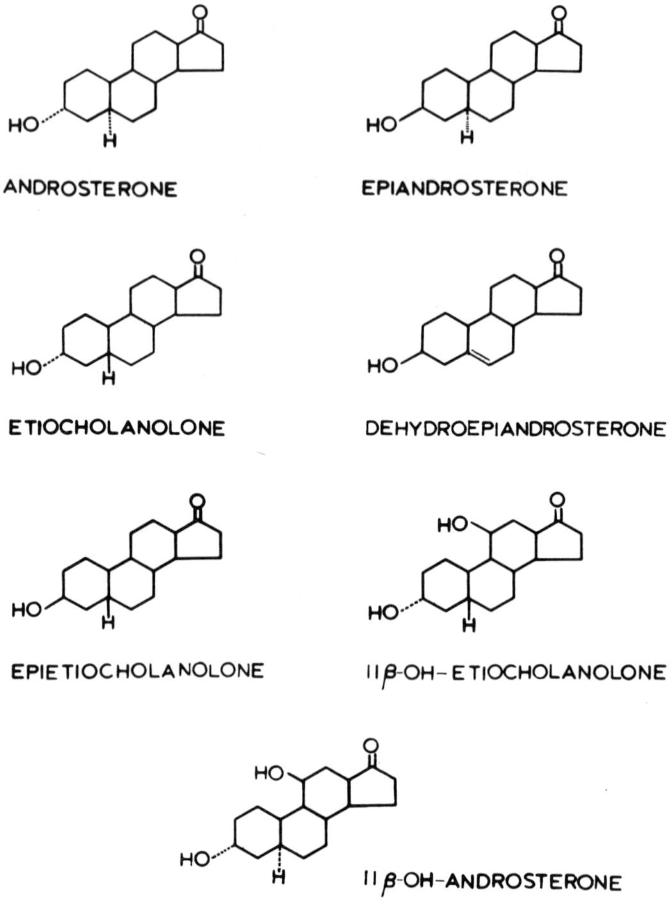

FIG. 11. Structure of typical 17-ketosteroids.

Gas-chromatographic [26, 27] methods have been reported for 17-KS determinations, but the most successful results have been obtained when additional chromatographic cleanup steps (TLC or column chromatography) have preceded the final measurement. Despite their drawbacks, chromatographic methods are more desirable than the group estimation method because of their specificity. In adrenal carcinoma, for example, elevated total urinary 17-KS levels would be insufficient for diagnostic purposes, and a specific dehydroepiandrosterone assay or an assay discriminating the total 3α 17-KS from the total 3β 17-KS would be necessary. For this reason HPLC appeared promising for specific urinary 17-KS analysis.

Like the estrogen assay, a hydrolysis was required to release the free steroids. The neutral 17-ketosteroids are prone to degradation under conditions used for the urinary estrogen hydrolysis, so the modified procedure of Vestergaard and Clausen [28] was used. Acidified urine refluxed concurrently with benzene allowed rapid transfer of the free steroid to the organic phase. Once hydrolyzed, steroids were effectively isolated from the acid phase and sample degradation was minimized. The neutral steroid fraction in the benzene phase was isolated from acidic and phenolic compounds (e.g., estrogens) by successive washes with alkali and water. Since the 17-ketosteroids occur at concentrations of 1 to 2 μg/ml in the urine of normal individuals, detection would not ordinarily have been a problem. Unlike the estrogens and free cortisol, however, the neutral 17-KS had no appreciable absorbance at 254 nm, the operating wavelength of the uv detector. This problem was circumvented by forming uv-absorbing 2,4-dinitrophenylhydrazone derivatives of the 17-KS present in the benzene after hydrolysis and washing. Microgram levels of the different 17-KS epimers were derivatized by standard procedures and separated by liquid-liquid partition chromatography [29] (1%, β,β-oxydipropionitrile on Zipax 1 m × 2 mm). Typical chromatograms of a standard mixture and derivatized urine extracts are shown in Fig. 12. Peak identities were validated indirectly, by using solvolytic cleavage and extraction conditions selective for the 17-KS sulfates dehydroepiandrosterone and epiandrosterone, and also by comparing standard and urinary hydrolysate chromatograms. Enzymatically hydrolyzed samples showed results equivalent to acid-hydrolyzed samples, and the precision of the HPLC method at the microgram level was ±10%. Except for the common problem of maintaining the stationary phase coating at a constant level, no insurmountable difficulties were encountered. Surprisingly, there was no evidence of separate syn and anti peaks for any of the steroid hydrazones formed.

D. Summary

The application of HPLC to the analysis of steroids from physiological fluids has been limited by several problems. The metabolic fate of the steroids, namely excretion into the urine as glucuronide or sulfate conjugates, complicates the assay by necessitating a chemical or enzymatic hydrolysis. Appropriate conditions for this step have been examined thoroughly, but it still remains a potential source of interferences or compound destruction. This difficulty is present for the assay of steroids in physiological fluids regardless of the measurement technique ultimately chosen. A problem particular to HPLC is detection. Besides being conjugated most steroids undergo metabolic transformations at their α,β-unsaturated ketone sites, the sites which provide them with desirable absorption properties for monitoring with a single-wavelength, 254-nm spectrophotometer, the most sensitive HPLC detector. Without strong uv extinction, detection of trace quantities (<50 ng on column is nearly impossible. Even when this deficiency is absent, as in the case of the urinary estrogens or plasma free cortisol, or when it can be amended by the formation of derivatives, as with the 17-ketosteroids, only steroids in the concentration range of 0.5 to 1 μg/ml are readily analyzed by HPLC. Detection of levels as low as 50 to 500 ng/ml is probably possible within the boundaries of current HPLC technology if

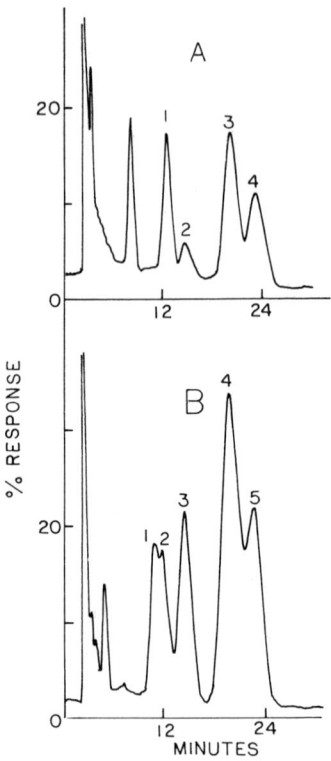

FIG. 12. One meter, 1.5% β,β-oxydipropionitrile on Zipax, 2,2,4-trimethylpentane eluent, 0.50 ml/min, 0.08 AUFS, 254-nm uv detector. (A) Enzymatically hydrolyzed urine: 1 = androsterone; 2 = epiandrosterone; 3 = etiocholanolone; 4 = dehydroepiandrosterone. (B) Standard mixture: 1 = epietiocholanolone; 2 = androsterone; 3 = epiandrosterone; 4 = etiocholanolone; 5 = dehydroepiandrosterone.

improved sample concentration steps such as XAD-2 chromatography are incorporated into the assay either before or after hydrolytic cleavage of the conjugates. Attempts to determine lower concentrations imply that larger sample volumes be processed. For urine this is only a minor difficulty, but for blood or plasma analysis it is a severe one.

On the positive side, improved separations and analytical procedures for several steroids have been reported. The use of chemically bonded, microparticulate packings can enhance separations by avoiding stationary phase losses in liquid-liquid partition chromatography, and by providing higher sample capacity and resolution. HPLC analysis of steroids from physiological fluids is still in its infancy, and advantages of rapid, specific chromatographic assays compared to group estimation methods will focus continued attention on this area.

VI. HIGH-PERFORMANCE LIQUID CHROMATOGRAPHY OF STEROIDS IN PHARMACEUTICALS

The therapeutic effects of steroids have generated and sustained their appeal to the pharmaceutical industry. Consequently, problems associated with their separation and analysis have arisen at all stages in the research and development of new steroidal drug products. Typically, these problems are different but no less challenging than those encountered in physiological fluid analysis. In addition to the analytical aspects of the HPLC separation of steroids in blood and urine, there are important biochemical aspects. HPLC separation of steroidal pharmaceuticals, however, is more of an analytical problem. Close attention to chromatographic considerations is necessary to devise a method which meets strict precision and accuracy requirements and yields reproducible and reliable results over long periods of time. These requirements are eased by less severe limitations in sample size and in many cases prior knowledge about the composition (or decomposition) of the mixture to be separated and analyzed.

A. Development of Separations

Discussion of the selection of the chromatographic system has been intentionally reserved for insertion here because it is the key factor in the analysis of steroids in pharmaceuticals. The guidelines given are equally true for the separation of steroids from physiological fluids, but isolation of intact steroid from the biological matrix remains the critical step in those assays.

All four modes of chromatography, adsorption, partition, ion exchange, and gel permeation, have been used for steroid separations. With enough perseverance, nearly any problem can be solved. In HPLC, like most endeavors, the first step is often the most difficult. Two schools of thought exist on the sequence one should follow to develop an HPLC separation. The first, designated the systematic school, prepares a battery of columns, typically one silica gel adsorption with 100% hexane as the mobile phase; one bonded, reversed-phase column with 100% water as the mobile phase; one cation exchange with a low-pH, low-ionic-strength eluent; and one anion exchange with a high-pH, low-ionic-strength eluent. Samples are injected onto each under a given set of conditions and elution patterns monitored. (For simplicity we must assume that a suitable detector is being used.) If separations under the initial conditions are inadequate, the eluent is varied in such a way as to optimize retention. After equilibration, samples are chromatographed again. In a hypothetical case one might inject a prednisone-prednisolone mixture onto each of the columns just described. For superficially porous columns 1 m long × 2 mm I.D. at flow rates of 1 ml/min, these two steroids would elute as discernible peaks in a reasonable time

(< 1 hr) only on the bonded, reversed-phase column. Knowing this, one could then dilute the 100% aqueous eluent with methanol or acetonitrile to achieve the minimum retention time consistent with the desired resolution. The behavior of the steroids on the reversed-phase column indicates what changes in eluent composition of the adsorption system would be beneficial. In this example, the steroids eluted in a reasonable time with a purely aqueous eluent in the reversed-phase mode, so one might suspect that they were strongly retained by the adsorption column and dilution of the hexane eluent with a more polar solvent (isopropanol or methylene chloride) would resolve them in a reasonable time. This in fact is true. Occasionally, one encounters situations in which one or more of the components of a steroid mixture have eluted in the chromatographic void volume. If the initial eluents have been selected to guarantee maximum retention on the column (e.g., 100% hydrocarbon eluents produce maximum retention on adsorption columns and 100% aqueous eluents produce maximum retention on reversed-phase columns), the steroid in the void volume of the adsorption column will be the last compound eluted from the reversed-phase column and vice versa. This rule of thumb should not be invoked too dogmatically, since subtle alterations in steroid structure can produce drastic and unpredictable changes in chromatographic retention times. Ion-exchange columns are most useful only for steroid conjugate separations. These columns can apparently operate by simple liquid-liquid partitioning in some cases, however, and they may be considered as a "last resort" attempt for separations that were not possible by conventional means.

The second approach to developing HPLC separations is designated the "juggling" school. Having diligently applied the theoretical principles of chromatography in a logical fashion and having usually been faced with confusion, overlapping peaks, and drifting baselines, the author must in all good conscience endorse the juggling school. Here, one column, which experience has shown to be most useful, is selected and the eluent conditions are "juggled" until the steroids elute in an orderly pattern. In my experience these columns have been bonded, reversed-phase columns, either Corasil C18, Permaphase ODS, Phenylcorasil, or most preferably the new bonded, reversed-phase microparticulate columns, for example, μ-Bondapak C18, which show chromatographic efficiencies of several thousand plates per meter. Reversed-phase chromatography is particularly suitable for steroids and other lipophilic compounds, and probably more than half of the separation problems faced can be solved using these columns. The remaining ones can almost always be solved by using silica adsorption columns or permanently bonded ether phase columns. Microparticulate silica gel columns in particular are gaining tremendous acceptance as the standard for many difficult separations. Although the selection process for steroid HPLC systems is basically intuitive and experience provides the best guideline, a compendium of steroid separations reported and the HPLC systems used is shown in the Appendix for reference.

B. Corticosteroid Pharmaceuticals

We have seen that early reports describing the high-performance liquid chromatography of steroids contained limited information on its use for routine quantitative analysis. Concern among chemists in pharmaceutical laboratories for reliable quantitative techniques to ensure the purity and efficacy of steroidal products quickly led to expanded application in this area.

Corticosteroid determinations were a problem which had plagued analysts for years. Because corticosteroids are prone to decomposition, specific assays were needed, usually meaning chromatographic steps prior to measurement. Thin-layer, paper, and column chromatography were all used, but results could be deemed only adequate. Gas chromatography itself provokes degradation and was not useful for their analysis. HPLC possessed none of the disadvantages of the other chromatographic techniques and because the corticosteroids and their decomposition products absorb radiation strongly at 254 nm, highly sensitive detection was possible. Capitalizing on these features, several reports on quantitative HPLC analysis of corticosteroids in pharmaceuticals appeared. The first, by Mollica and Strusz [30], described a partition chromatographic system (1%, β,β-oxydipropionitrile on Zipax, 94.8:5:0.2 hexane-ethyl acetate-acetonitrile mobile phase) to separate and quantitate fluomethasone pivolate ($6\alpha,9\alpha$-difluoro-$11\beta,17\alpha$-dihydroxy-16-methyl-21-trimethylacetoxy-1,4-pregnadiene-3,20-dione) from other ingredients in a topical cream. Placebo creams spiked with the steroid gave 100% recovery when extracted with ethyl acetate, and replicate assays showed a precision of under 3%. Because of its specificity and reproducibility, the method was adapted as a stability-indicating assay. A similar article by Landgraf and Jennings [31] stressed the construction of a modular HPLC from commercially available components, and the advantages of silane-modified supports with ternary phase partition systems. Using a ternary phase system, an analysis for fluocinonide was devised and retention data for 13 additional steroids were reported. Although they have some advocates, ternary phase columns have not been widely adopted. Although they provide flexibility in the selection of operating parameters as the authors assert, they also provide more opportunities for variable retention data if the mobile phase composition is not reproduced exactly from day to day. Column temperature and humidity must be closely controlled for operation near the phase separation region, and ternary phase systems are more susceptible to variation from stationary phase "stripping" losses by shear forces as the eluent is pumped through the column.

The evolution of HPLC column technology is apparent by the increased reliance on stable, commercially available columns in most articles published since 1973. Hydrocortisone, cortisone, and their respective acetate esters in a topical ointment, were separated by Olson [32] on a reversed-phase cyanoethylsilicone column in less than 25 min without interference

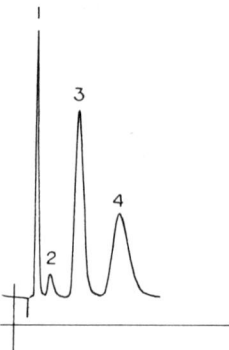

FIG. 13. Separation of hydrocortisone acetate from hydrocortisone and parabens. Bondapak C18 column, 1 m × 2 mm, acetonitrile-water eluent. 1 = methylparaben; 2 = hydrocortisone; 3 = propylparaben; 4 = hydrocortisone acetate.

from excipients (cholesterol, parabens, iodochlorohydroxyquin, and oxytetracycline). Decomposition of the steroid acetates and foreign steroid content were determined in creams, ointments, suppositories, and lotions. A typical chromatogram of such a separation on a similar column is shown in Fig. 13. The determination of foreign steroid content by official compendial methods (U.S.P. or N.F.) requires an extraction of the total steroids followed by spectrophotometric analysis. After thin-layer chromatographic separation of a second aliquot, the zones containing the individual steroids are identified and their contents analyzed by colorimetry after reaction with tetrazolium blue. The foreign steroid content is defined as the difference between the two values. Results of quantitative HPLC analysis of 25 commercial samples correlated well with results by compendial methods. The simplicity of the HPLC assay, however, coupled with its speed and resolution promise its eventual incorporation into the National Formulary and United States Pharmacopoeia.

C. Estrogen Pharmaceuticals

Compendial methods for synthetic estrogens suffer from the same basic disadvantages (discrete chromatographic and analysis steps) as those for the corticosteroids. Although gas chromatography is more compatible with the estrogens, derivatization is still necessary. To avoid these difficulties, Roos [33] developed a reversed-phase HPLC procedure using a 254-nm uv detector for several synthetic estrogens containing a 4,4'-stilbenediol moiety. The isolation and preparation steps prior to the HPLC determination are worth noting because of their general applicability. Tablets proved easiest to prepare, and dissolution by mechanical shaking or ultrasonic vibration promoted the quantitative release of steroid into the solvent. In ascending order of difficulty, creams and lotions were next, requiring longer shaking or vibration and solvents with more affinity for the steroid. Suppositories proved most difficult. Melting followed by vigorous shaking was necessary to free the steroid.* Several dosage forms with estrogen levels ranging from 0.1 mg per tablet to 72 mg per capsule were analyzed. Results by the proposed HPLC method and compendial methods agreed within 2%.

D. Summary

Advantages of HPLC for pharmaceutical quality control and analysis include high specificity, high sensitivity for uv-absorbing steroids, short analysis times, and accuracy and precision of 3% or less under normal operating conditions. Using stable, high-performance columns (ordinarily permanently bonded columns), reliable and reproducible operation over a span of several months is common. The nondestructive nature of HPLC permits the analyst to collect steroids or their decomposition products in a small volume as they elute from the column, and to characterize them by physical chemical methods. It is expected that HPLC analyses of other steroids including progestins and androgens will eventually be developed.

VII. CHEMISTRY BEFORE, DURING, AND AFTER HPLC OF STEROIDS

A. Derivatization to Enhance Detectability

The concept of derivatization to expand the boundaries of common HPLC spectrophotometric detectors was first applied to steroid separations [16, 34]. In principle, an almost unlimited choice of derivatives for enhanced detection exists. Considerations involved in derivative selection are as follows:

1. Derivatization should be quantitative and should occur under conditions which avoid unknown or undesirable alteration of the steroid (e.g., dehydration).

---

*An excellent list of guidelines for preparing compounds for pharmaceutical analysis is available from Waters Associates, Maple Street, Milford, Massachusetts.

2. The derivative should not adversely affect the chromatographic properties of the compound being separated. For example, formation of derivatives with more hydroxyl groups than the parent compound might result in excessive peak tailing.

3. The derivatization reaction should be amenable at levels at which the actual analysis will be carried out. For example, the conditions for forming phenylhydrazones of steroids in the range of 100 to 500 mg are considerably different from reaction conditions on steroids at 10 to 50 µg. The suitability of the procedure should be evaluated at appropriate concentrations.

4. The derivative selected should substantially enhance the properties of the molecule which are necessary to achieve detectability. For example, incorporation of substituted phenyl rings into steroids virtually always lowers their detection limits with a uv absorption detector. Different properties might perhaps be more desirable for the enhancement of the response of an electrochemical detector, or a conductance detector. Excess reagent should not interfere with the detector.

5. The identity of the derivative should always be independently verified.

Because of their diverse structural characteristics, steroids allow considerable ingenuity to be exercised when derivatization is being considered. The 2,4-dinitrophenylhydrazones of steroids containing carbonyl groups have been described by different investigators [16, 29]. For steroids containing hydroxyl groups, benzoates and nitrosubstituted benzoates have been employed [34]. In both cases, reaction was achieved under mild conditions (temperatures under 60°C) and excess reagent was easily eliminated by a simple partition step in the case of the steroid benzoates or it was separated from the reaction products on the HPLC column as in the case of the steroid hydrazones. Generally, the substitution of a nitrosubstituted phenyl ring into the steroid nucleus improves the detection limit by 100- to 1000-fold. For example, about 1 µg deposited onto the column is the detection limit of androsterone with commercially available uv detectors operating at 254 nm. The 2,4-dinitrophenylhydrazone of this same steroid can be detected at the nanogram level, a 1000-fold sensitivity enhancement. Similar improvements were noted for steroid nitrobenzoate esters. Clearly, other possibilities exist for derivatization of steroids, although little work has been done. For example, all 17-ketosteroids have an activated methylene group at the C_{16} position (recall that the group estimation of the 17-KS is based on the formation of a nitrosubstituted adduct at this position), and the formation of benzylidene analogs was explored briefly by this author [35]. Although successful on model ring carbonyl compounds, positive evidence of steroid derivative formation was not obtained. Besides enhancing the uv detection limits of steroids, suitable derivatives often allow resolution of compounds that are difficult to chromatograph in their underivatized forms, or as a certain other derivative. Some steroids, for example, poorly resolved as acetates, can be separated as benzoates [36].

Despite its advantages, derivatization is a mixed blessing. Its primary drawback is that it introduces an additional step into the assay. Curiously, one of the strongest appeals of HPLC initially was that it avoided the formation of derivatives required for the gas-chromatographic separation of many steroids. The use of derivatives with selective detectors can also be a source of confusion in some cases. A dramatic example of this would be the isolation and identification of a peak suspected to contain a steroid benzoate. Although the uv detector might respond as if a single component were present, it is quite possible that other components, not derivatized and thus not responding (e.g., carbonyl-containing steroids), were eluted at the same time. The unexpected presence of an additional compound could thwart any attempted identification. This type of complication is possible with any specific detectors (polarographic, electron capture, or uv), so caveat usor!

More exotic forms of chemical alteration before HPLC have been used with steroids. The irradiative formation of phenanthrenedione followed by reduction to phenanthrenediol, then HPLC on an ETH Permaphase column has been proposed as the basis for the quantitative analysis of synthetic estrogens [33] (diethylstilbestrol) at submicrogram levels. It seems unlikely that the problems facing analytical chemists working with steroids can be solved by purely instrumental approaches. One is then forced to admit that classical organic functional group chemistry will have a role to play in high-performance liquid chromatography for some time to come, as it has in other areas of chromatography since their inception.

B. Chemistry during Chromatography

The very basis of liquid-liquid partition chromatography is the differential chemical interaction of the compounds to be separated with the liquid separating phase. The liberty to select virtually any stationary phase provides the imaginative worker with another opportunity to exploit functional group chemistry. Examples of specific chromatographic interactions are known. Cis-oriented hydroxy groups bind selectively to borate stationary phases allowing facile separation of many cis-trans isomers. Silver-impregnated stationary phases have been widely used to separate compounds with varying degrees of unsaturation. In steroid HPLC specifically, nitrile-terminated hydrocarbons were more effective than other types of functionally terminated hydrocarbons for difficult separations [37]. A specific interaction between the nitrile group and the polar substituents on the steroid has been postulated to account for this effect. Bonded phase supports with different alkyl or aryl groups polymerized on the surface of a silica gel particle also display different chromatographic characteristics with many compounds.

A novel approach [38] for the prediction of steroid partition coefficients relies on the specific chemical interactions involved in liquid-liquid partition chromatography. By assuming a linear relationship between retention time and partition coefficients (PC), the PC value of an unknown steroid can be obtained from its retention time and those of two steroids with known PC values. Errors in this method arise from subtle variations in the mobile and stationary phases from run to run, and deviations from the assumption that only partitioning is operative in affecting the retention time (even deactivated supports showed some adsorptive contributions). By applying linear regression analysis for many steroids, statistical errors were reduced to an acceptable level, allowing PC values to be useful identification parameters. In addition, the prediction of the partition coefficients was accurate enough to allow a rational selection of appropriate liquid-liquid phase compositions to separate specific mixtures.

C. Chemistry after Chromatography

No work has been reported on postcolumn reactions of steroids for detection analogous to amino acid systems. Problems in the design of low-volume reaction chambers which do not distort the resolution achieved on the column have probably stifled efforts in this area.

VIII. THE FUTURE OF HIGH-PERFORMANCE LIQUID CHROMATOGRAPHY OF STEROID HORMONES

A. Instrumental Developments

Generally, improvements in instrumental capabilities spur progress in all areas of a new analytical technique. For high-performance liquid chromatography, detector limitations are currently the most severe obstacle to its expanded use. Multiwavelength spectrophotometric detectors have not replaced 254-nm uv detectors in any sense. Reasons for this are the limited power output of most continuous-wavelength uv sources (deuterium lamps), and the further diminution in incident beam intensity caused by grating monochromators in most commercial instruments. Beam condensing optics have helped to boost the incident power, but basically all multiwavelength detectors are expanded scale spectrophotometers and are not well adapted to the particular requirements of modern HPLC detectors. Because of their sensitivity in the low-uv range they had been heralded as the long-awaited universal HPLC detector, but often the eluting solvent is incompatible with operation at those very wavelengths which might be useful, particularly if one is trying to avoid derivative formation. Limitations in other, so-called, universal HPLC detectors have been described [5], and the introduction of a detector equivalent to the flame ionization detector in gas chromatography will probably come only after some thus far unknown conceptual breakthrough in physics. For most purposes, a commercially available 254-nm uv detector is suitable for steroid applications. At this stage the best ancillary equipment to augment this detector is copies of organic functional group analysis books [39, 40].

B. Column Technology

This chapter has not emphasized the various HPLC systems reported for steroid separations, because many have been rendered obsolete by microparticulate, chemically bonded packings. Researchers using HPLC for the first time should concentrate on the general HPLC literature published since 1973 when selecting appropriate columns. Problems of mobile phase saturation, column stability, and packing uniformity are virtually eliminated by the newer packings. This practical advantage coupled with the improvement in separation capacity and efficiency (e.g., see Fig. 14)

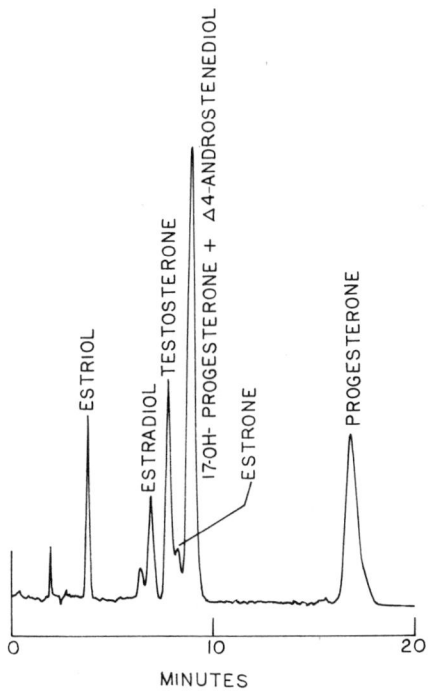

FIG. 14. μ-Bondapak C18, eluent: CH_3CN/H_2O, 55:45. Uv 0.04 AUFS, 254 nm.

minimizes the variability of column selection. Improvements will probably be in the introduction of columns adapted for specific applications since the asymptotic limit of efficiency appears to have been reached.

C. Other Equipment

Improvements in other HPLC equipment appear to have plateaued also. Pumping systems of every conceivable type have been described and their flaws and advantages outlined [5]. It is the opinion of the author that pumping systems capable of generating pressures in excess of 1500 psig are a status symbol rather than a useful asset in HPLC. There could be considerable disagreement over this, however. Temperature-controlled columns are another case in which simplicity is elegance. Generally the improved peak shape attributed to operation at high temperatures ($\sim 60°C$) is largely illusory and reflects decreased eluent viscosity more than anything else.

A piece of ancillary equipment which would be favorably received would be an automatic sample injector capable of unattended overnight operation. DuPont has recently introduced such a device operating on pneumatic valve activation principles which avoids the significant problem of septum rupture. Its main disadvantage is a required sample of 100 μl per injection.

IX. CONCLUSIONS

High-performance chromatography has developed rapidly in areas in which other forms of chromatographic separation have been deficient, notably steroid separations. Enough work has now been done that overly optimistic expectations have been stifled and a more realistic appraisal of the potential of HPLC is possible. Despite drawbacks it remains a powerful aid to the chemist. The ultimate test of its utility will be whether it provides quicker solutions to the problems of health and overpopulation which continue to plague us all.

ACKNOWLEDGMENTS

Thanks are due to Dr. Sidney Siggia, University of Massachusetts, for permission to use Figs. 5 to 7, and to Dr. Jim Little, Waters Associates, for Figs, 9, 13, and 14.

APPENDIX

Steroids separated	Chromatographic system	Ref.
All classes	Reversed-phase partition	15
All classes	Normal phase partition (ODPN on Zipax); reversed-phase partition (Permaphase ODS)	16
Estrogens	Normal phase partition	18
Estrogens	Bonded phase partition	21
Estrogens	Adsorption	22
Corticosteroids	Adsorption; bonded, reversed-phase partition	23
Androgens (derivatized)	Normal phase partition; bonded, reversed-phase partition	29
Corticosteroid analogs	Normal phase partition	30
Corticosteroid analogs	Normal phase partition	31
Corticosteroids	Reversed-phase partition	32
Synthetic estrogens	Reversed-phase partition	33
Androgens (derivatized)	Reversed-phase partition	34
All classes	Reversed-phase partition	37
All classes	Normal phase partition	38
Androgens, corticosteroids	Reversed-phase partition	14

REFERENCES

1. P. R. Brown, High Pressure Liquid Chromatography, Academic Press, New York, 1973.
2. J. J. Kirkland (ed.), Modern Practice of Liquid Chromatography, Wiley-Interscience, New York, 1971.
3. J. J. Kirkland, Anal. Chem., 43, 37A (1971).

4. G. J. Fallick, Advances in Chromatography, Vol. 12 (J. C. Giddings, ed.), Marcel Dekker, New York, 1967.
5. I. Berry and B. Karger, Anal. Chem., 45, 819A (1973).
6. L. Fieser and M. Fieser, Steroids, Van Nostrand-Reinhold, Princeton, New Jersey, 1959.
7. C. W. Shoppee, Chemistry of the Steroids, Butterworth, London, 1964.
8. W. Klyne, The Chemistry of the Steroids, Wiley, New York, 1957.
9. E. Heftman, Biochemistry of Steroids, Van Nostrand-Reinhold, Princeton, New Jersey, 1960.
10. H. Carstensen, Steroid Hormone Analysis, Marcel Dekker, New York, 1967.
11. J. E. Bush, The Chromatography of Steroids, Pergamon, Oxford, 1961.
12. R. Neher, Steroid Chromatography, Elsevier, Amsterdam, 1964.
13. R. Neher, Advances in Chemistry, Vol. 4 (J. C. Giddings and R. A. Keller, eds.), Marcel Dekker, New York, 1967.
14. S. A. Slocum and J. F. Studebaker, Anal. Biochem., 68, 242 (1975).
15. S. Siggia and R. Dishman, Anal. Chem., 42, 1223 (1970).
16. R. A. Henry, J. A. Schmit, and J. F. Dieckman, J. Chromatogr. Sci., 9, 513 (1971).
17. S. Kober, Biochem. Z., 239, 209 (1931).
18. J. F. K. Huber, J. A. R. J. Hulsman, and C. A. Meijers, J. Chromatogr., 62, 79 (1971).
19. A. E. Schindler, V. Ratanasopa, and W. Hermann, Clin. Chem., 13, 186 (1967).
20. J. Brown and H. Blair, J. Endocrinol., 17, 411 (1958).
21. A. Butterfield, B. Lodge, and N. Pound, J. Chromatogr. Sci., 11, 401 (1973).
22. R. J. Dolphin, J. Chromatogr., 83, 421 (1973).
23. J. Touchstone and W. Wortmann, J. Chromatogr., 76, 214 (1973).
24. R. G. Schleicher, unpublished results.
25. W. Zimmerman, Z. Physiol. Chem., 233, 257 (1939).
26. H. Sparagna, E. Kruetman, and W. Mason, Anal. Chem., 35, 1231 (1963).

27. W. Van den Heuvel, E. Hornig, and B. Greech, Anal. Biochem., 4, 191 (1962).
28. P. Vestergaard and B. Clausen, Acta Endocrinol. Suppl., 64, 35 (1962).
29. F. Fitzpatrick, S. Siggia, and J. Dingman, Sr., Anal. Chem., 13, 2211 (1972).
30. J. A. Mollica and R. F. Strusz, J. Pharm. Sci., 61, 444 (1972).
31. W. C. Landgraf and E. C. Jennings, J. Pharm. Sci., 62, 278 (1973).
32. M. C. Olson, J. Pharm. Sci., 62, 2001 (1973).
33. R. Roos, J. Pharm. Sci., 63, 594 (1974).
34. F. Fitzpatrick and S. Siggia, Anal. Chem., 45, 2310 (1973).
35. F. A. Fitzpatrick, Thesis, University of Massachusetts, 1972.
36. R. V. Brooks, W. Klyne, and E. Miller, Biochem. J., 54, 212 (1953).
37. F. Fitzpatrick, Clin. Chem., 19, 1293 (1973).
38. J. F. K. Huber, E. Alderlieste, H. Harren, and H. Poppe, Anal. Chem., 45, 1337 (1973).
39. R. Shriner, R. Fuson, and D. Curtin, The Systematic Identification of Organic Compounds, 5th ed., Wiley, New York, 1964.
40. S. Siggia, Quantitative Organic Analysis via Functional Groups, 3rd ed., Wiley, New York, 1967.

Chapter 3

NUMERICAL TAXONOMY IN CHROMATOGRAPHY

Desire L. Massart and Henri L. O. De Clercq[*]

Farmaceutisch Instituut
Vrije Universiteit Brussel
Bosstraat, Jette, Belgium

I. INTRODUCTION 76
II. NUMERICAL TAXONOMIC TECHNIQUES 77
    A. General Considerations 77
    B. Data 78
    C. Measures of Resemblance 79
    D. Linkage Methods 82
    E. Graph Theoretical Procedures 86
III. APPLICATIONS 87
    A. The Selection of Optimal Sets in TLC 87
    B. The Classification of Stationary
       Phases in GLC 94
    C. The Classification of Substances According
       to Chromatographic Behavior 105
    D. Numerical Taxonomy as a Pattern
       Recognition Procedure 109
ACKNOWLEDGMENTS 110
REFERENCES 110

---

[*]Current affiliation: Academisch Ziekenhuis, Vrije Universiteit Brussel, Brussel, Belgium

I. INTRODUCTION

Chromatography is one of the methods of choice of scientific workers confronted with the necessity of identifying or quantifying substances. This has led to large amounts of research aimed at developing new and better separation systems, with the result that some interesting new stationary phases, solvent combinations, etc., have been obtained, but also that an enormous quantity of often closely similar separation systems has been published. One of the difficulties with which the chromatographer is currently confronted is how to make a choice among all these possibilities. To help him make these decisions he has at his disposal large amounts of data in the form of R_f values, retention indices, distribution coefficients, and so on. The difficulty is then converted into the problem of creating some order out of this mass of data so as to be able to select a few significantly different systems with which to achieve the separation needed. This means that the investigator classifies (usually implicitly) the proposed separation procedures and eliminates those that are redundant.

When one wants to study the chromatographic behavior of a large set of substances (e.g., antibiotics) one often selects some that are thought to be typical of the groups composing the set (e.g., one macrolide, one polypeptide, one tetracycline) and subjects these to a more thorough investigation. This implies a classification of the compounds. Another and more notable example in chromatography is the Rohrschneider index, which can be described as a device to characterize the behavior of all stationary phases toward all chromatographed solutes in gas-liquid chromatography (GLC). Here, the set of compounds comprises all substances that can be subjected to GLC, and these substances are classified according to their interactions with GLC phases, e.g., dipole orientation, proton donors. The selection of five probes is the result of a nonformal classification procedure.

Since chromatographic methods are used to obtain information about the composition of complex mixtures, they themselves generate large quantities of data. In many cases these data are used to group the samples into classes or to assign the sample to one of a number of known classes. A striking example is the determination of the origin of milk by GLC. The fatty acid spectrum is used to decide the animal (cow, goat, sheep) from which the sample originates.

In each of the problems just sketched, one therefore has to classify a number of systems, substances, or types of samples. In general, one finds that no explicit classification is made, and even when this has been done, it is more or less subjective. It is clear that in many cases a formal and explicit classification would be an advantage.

This situation resembles greatly biological or bacteriological systems as they were in the early to mid-1960s. A very important branch of the biological sciences is concerned with taxonomy, i.e., the ordering of individual species into groups on the basis of their mutual resemblances. These groups are then incorporated into larger groups. In this way, a hierarchical classification is elaborated.

This type of classification dates back several centuries when Linnaeus published the first systematic taxonomy and nomenclature in his Species plantarum (1753). Until the early 1960s, plants and animals had their places assigned in the classification usually on the basis of morphological characteristics. However, this ordering remained more or less subjective, which led to many disputes and controversies, such as the placing of some insect into one genus or another. This was due to the absence of objective, i.e., numerical, methods.

The advent of numerical taxonomic methods has revolutionized biological taxonomy by placing numerical classification techniques at the disposition of taxonomists. It is the purpose of this chapter to explain how these methods can be applied in chromatography and to describe the results which have been obtained or can be expected.

II. NUMERICAL TAXONOMIC TECHNIQUES

A. General Considerations

Numerical taxonomy is one of many methods that allows the ordering of data sets. Other examples are factor analysis and principal component analysis. We will discuss the latter techniques here only insofar as necessary to situate numerical taxonomy among the other methods and to discuss the results obtained with this technique.

In fact, the techniques mentioned can be considered to be part of numerical taxonomy, which itself is part of a large body of numerical techniques called cluster analysis. A classification can be considered as the recognition of clusters of points. To make this clear let us consider as an example a number of substances which we want to classify according to their behavior in two chromatographic systems. Although we do not know this at the onset of the classification procedure, the set consists of two very distinct groups of substances, for example, hydrocarbons and fatty acids. The distribution coefficients of the substances in the two chromatographic systems are known and are depicted in Fig. 1. One observes that it is possible to distinguish immediately two clusters (i.e., to separate the set into two taxa) of points, A and B, and, on more careful inspection, two smaller clusters, B' and B". B' and B" are related much more closely to each other than to A (B' and B" are, for example, saturated and unsaturated fatty acids). This conclusion is drawn from Fig. 1 because B" and B' are less distant from each other than from A. In the same way the acid represented by point

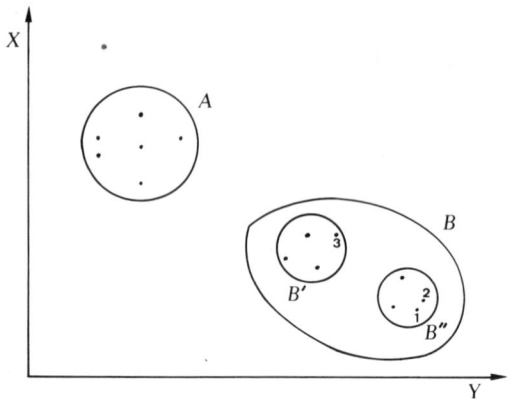

FIG. 1. Spatial representation of OTUs and clusters characterized by two parameters X and Y.

1 more closely resembles the acid of point 2 than the one represented by point 3. One notes the following:

1. Objects with characteristics that are closely alike form clusters.
2. The distance between points constitutes a numerical value of the resemblance of the objects depicted by these points.

This reasoning can be generalized for more than two chromatographic systems. When one has to take n systems into account, then the clusters must be viewed as being in an n-dimensional space.

All the techniques just mentioned enable us to find out something about the clusters present in a data set. However, we will reserve the name numerical taxonomy for those techniques that permit a formal, complete, and hierarchical classification.

Numerical taxonomy has now become a science in itself. It is out of the question to discuss all the possibilities of scaling variables, measuring resemblances, and so on. Instead, we will restrict the discussion to those points that have been used in chromatographic applications. Readers who wish to study the subject in a more detailed way are referred to a book written by the pioneers of numerical taxonomy, Sneath and Sokal [1].

B. Data

The object of numerical taxonomy is to classify operational taxonomic units (OTUs) according to the values of a set of (taxonomic) characters. If one wants to classify TLC solvents according to their migration patterns for synthetic dyes [2], then the TLC solvents are the OTUs and the migration of the dyes, the characters, the values of which are given by their R_f values in the respective solvent systems. These values are arranged in a matrix such as the one represented in Table 1, where OTU 1 is, for example, trisodium citrate (2 g), water (85 ml), and 0.88 ammonia (15 ml).

In general, the data can be used as such or they can undergo some transformation. One of the most common rearrangements is the scaling of variables. It is clear that if one character has extreme values of 1 and 2 and another those of 1 and 1000, that the latter could make larger contributions in the determination of the resemblance or similarity (see Sec. II.C) than the former. Therefore, it is necessary in some cases to transform the data so that the overall variation of the data is about the same. This is not the case in TLC because there all characters have values ranging between 0 and 1, but it is in GLC where the range of retention indices in the classification of McReynolds' set of stationary phases (see Sec. III.B) is much smaller for the character cis-hydrindane (about 300 units) than for the character nitropropane (1100 units). To give an equal weight to each of the characters, the retention indices were standardized in this instance [3]. The mean retention indices and standard deviation for each substance were computed. The individual retention indices were then expressed as deviations from the mean in standard deviation units.

In our first published application of numerical taxonomy in TLC system selection [2] the R_f data were not used as such but replaced by a ΔR_f, the difference between the R_f value of the individual dye in a given chromatographic system and the mean R_f value for all dyes in that system. It was thought that these values would be better able to show variations in the behavior of individual dyes compared to the average one. When using the correlation coefficient as a measure of resemblance, as we have done in our preceding publication on the selection of TLC system combinations by numerical taxonomy (see Sec. III.A), this is no longer necessary.

C. Measures of Resemblance

The next step is to compare each OTU with all other OTUs and to evaluate their resemblance. Taxonomic resemblance or similarity between pairs of OTUs, i.e., between the elements in two rows or two columns of a data matrix, is generally stated by means of a single calculated value, the similarity coefficient. Various similarity coefficients have been proposed in many scientific fields, mostly developed or adapted to meet the needs of a particular classification problem. In chromatography two of these have been used, i.e., distance coefficients and correlation coefficients.

TABLE 1
Example of Data Matrix

Characters	\multicolumn{10}{c}{OTUs (solvent systems)}									
	1	2	3	4	5	6	7	8	9	10
Amaranth	0.6	0.3	0.5	0.6	0.4	0.2	0.6	0.4	0.1	0.2
Bordeaux B	0.2	0.6	0.4	0.6	0.5	0.6	0.7	0.4	0.1	0.1
Carmoisine	0.3	0.7	0.6	0.5	0.7	0.6	0.8	0.4	0.3	0.4
Eosine	0.2	1.0	0.7	0.8	1.0	1.0	0.9	0.6	1.0	1.0
Erytrosine	0.1	1.0	0.7	0.9	1.0	1.0	0.9	0.7	1.0	0.9
Fast Red E	0.4	0.7	0.6	0.7	0.6	0.5	0.8	0.4	0.3	0.4
Ponceau 4R	0.7	0.5	0.9	0.6	0.4	0.3	0.7	0.2	0.1	0.3
Ponceau 6R	0.8	0.2	0.8	0.4	0.3	0.1	0.6	0.1	0.1	0.2
Ponceau MX	0.2	0.7	0.5	0.6	0.5	0.5	0.8	0.4	0.2	0.3
Ponceau SX	0.4	0.7	0.5	0.6	0.5	0.5	0.8	0.4	0.2	0.4
Red 2G	0.6	0.6	0.7	0.6	0.4	0.5	0.7	0.4	0.2	0.3
Red 6B	0.4	0.3	0.4	0.5	0.4	0.2	0.6	0.4	0.2	0.2
Red 10B	0.2	0.5	0.6	0.6	0.4	0.3	0.7	0.4	0.2	0.3
Red FB	0.0	0.3	0.1	0.2	0.4	0.2	0.7	0.6	0.2	0.2
Rhodamine B	0.5	1.0	0.9	1.0	1.0	1.0	0.9	0.8	0.7	0.7
Scarlet GN	0.9	0.7	0.9	0.8	0.8	0.6	0.8	0.5	0.1	0.3
Auramine	0.3	1.0	0.9	1.0	0.9	1.0	0.8	0.8	0.8	0.8
Acid yellow	0.7	0.6	0.9	0.6	0.6	0.5	0.7	0.4	0.3	0.3
Chrysoidine	0.1	0.8	0.7	0.9	0.8	0.9	0.9	0.8	1.0	1.0
Chrisoine S	0.5	0.8	0.8	0.6	0.9	0.7	0.8	0.5	0.8	0.5
Naphthol yellow S	0.6	0.7	0.8	0.7	0.7	0.6	0.7	0.5	0.2	0.3
Orange G	0.8	0.7	0.9	0.6	0.7	0.5	0.7	0.4	0.2	0.3
Orange GGN	0.6	0.7	0.8	0.6	0.6	0.5	0.7	0.4	0.3	0.3
Orange I	0.4	0.8	0.8	0.7	0.9	0.7	0.8	0.5	0.8	0.8
Sunset yellow	0.6	0.7	0.8	0.6	0.6	0.5	0.7	0.4	0.3	0.3
Tartrazine	0.8	0.4	0.8	0.4	0.5	0.3	0.5	0.3	0.1	0.1

Source: From Ref. 2. Reprinted with permission. Copyright by the American Chemical Society.

1. Distance Coefficients

The distance coefficients provide a measure of the taxonomic distance between OTUs in a sample space: the greater the distance between the characters of the OTUs (see Sec. II.B), the greater the lack of similarity between them. In this way, the distance ought to be seen as a quantification of dissimilarity. If two OTUs display identical characters, the distance between them will obviously be zero.

Various mathematical formulations have been introduced to express the distance between OTUs. Because of its simplicity, the mean character difference (MCD) is a rather popular one. It is based on the average difference between any two OTUs in an n-dimensional hyperspace (i.e., between any two OTUs presenting n characteristics). The mathematical expression for the MCD between two OTUs j and k (presenting n characteristics each) equals

$$\frac{1}{n} \sum_{i=1}^{n} |X_{ij} - X_{ik}| \tag{1}$$

where X_{ij} is the value of character i for OTU j.

Unfortunately, the MCD suffers from the drawback of underestimating the distance between OTUs in a hyperspace. For this reason, an ameliorated quantity was introduced by Sneath and Sokal [1], the taxonomic or Euclidean distance Δ_{jk}. Between OTUs j and k, the taxonomic distance is

$$\Delta_{jk} = \left[\sum_{i=1}^{n} (X_{ij} - X_{ik})^2 \right]^{1/2} \tag{2}$$

Since Δ_{jk} depends on n, the number of characters, Eq. (2) cannot be used when some of the values X_{ij} are missing. To avoid this difficulty, it is preferable to compute the average distance:

$$d_{jk} = \frac{\Delta_{jk}}{\sqrt{n}} \tag{3}$$

2. Correlation Coefficients

Correlation coefficients are to be regarded as mathematical expressions of the dependence and/or proportionality between pairs of OTUs. Among the various similarity coefficients employed in numerical taxonomy, the Pearson product moment is undoubtedly the most familiar one. Calculated between a pair of OTUs j and k, this correlation coefficient is defined as

$$r_{jk} = \frac{\sum_{i=1}^{n} (X_{ij} - \overline{X}_j)(X_{ik} - \overline{X}_k)}{\sum_{i=1}^{n} (X_{ij} - \overline{X}_j)^2 \sum_{i=1}^{n} (X_{ik} - \overline{X}_k)^2} \tag{4}$$

where X_{ij} is the value of character i in OTU j, $\overline{X}_j$ the mean of all values for OTU j, and n the total number of characters. Correlation coefficients range from -1 to +1. Negative correlation coefficients between OTUs can have a physical and a mathematical significance, but do not bear any meaning for classification purposes. Consequently, it is necessary when employing correlation coefficients as the measure of similarity, to take only absolute values in consideration.

D. Linkage Methods

After the similarity coefficients have been computed, a similarity matrix $i_{max} \times i_{max}$ (where i_{max} is the number of OTUs) can be constructed. This will serve as the basis for the actual classification step. The purpose of a classification is to create groups or clusters of OTUs for which "group distance or variance within is minimized and, by definition, between group variance maximized" [4]. To achieve this, a variety of grouping, linkage, or clustering techniques have more or less frequently been employed in numerical taxonomy. To aid comprehension, we will first give a numerical example of one linkage technique and will then proceed to a more general discussion of the available methods.

Consider the following similarity matrix:

	A	B	C	D	E
A	0				
B	24	0			
C	30	10	0		
D	5	36	20	0	
E	40	30	50	50	0

where A,..., E are OTUs to be classified and the numerical values are similarity coefficients (e.g., distances). The closest relationship is between liquid phases A and D, since they have the smallest distance. They are

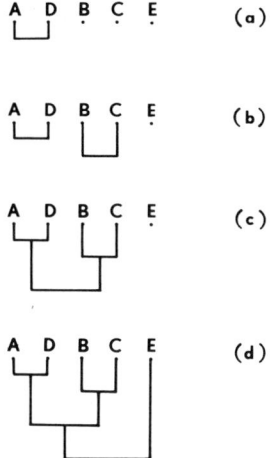

FIG. 2. Successive links in average linkage pair-group methods (dendrograms).

considered to form one OTU A', which is depicted by a link, in what will eventually be the dendrogram (Fig. 2a). There now remain four OTUs to be classified (A', B, C, E) and a new matrix is constructed by reduction of one column and one row of the original matrix. Distances between OTUs which do not include A' are not affected and remain unchanged. Distances including A' as one of the two OTUs which are compared are the mean of the corresponding distances including A and D. For example, the distance between A and B is 24, and between D and B 36; therefore, the distance in the new reduced matrix between A' and B is 30. In the new matrix the closest

	A'	B	C	E
A'	0			
B	30	0		
C	25	10	0	
E	45	30	50	0

relationship is between B and C. These are considered to form class B'. They are linked in the dendrogram shown in Fig. 2b and the new matrix

	A'	B'	E
A'	0		
B'	37,5	0	
E	45	40	0

is formed. Here A' and B' are the most closely related. They form a new OTU, A", as depicted in the dendrogram of Fig. 2c. In the new reduced matrix there remain only two OTUs, namely E and A", the latter grouping

	A"	E
A"	0	
E	42,5	0

the original OTUs A, B, C, and D. By joining these two together, the dendrogram is now completed (Fig. 2d) since all the OTUs are linked to another OTU or group of OTUs. This particular technique, the average linkage pair group method, is part of the so-called SAHN techniques [1], by which is meant a group of techniques that are sequential, agglomerative, hierarchical, and nonoverlapping.

Most clustering techniques employed in numerical taxonomy and in particular the techniques reviewed here, are <u>sequential</u>; that is, the OTUs are classified by a stepwise repetition of the clustering algorithm. Most clustering techniques are also agglomerative (and not divisive). In <u>agglomerative</u> techniques, a number i_{max} of individual OTUs is grouped into successively fewer sets leading eventually to one single set containing all the original OTUs. Divisive methods, on the contrary, start with one set containing all the OTUs and proceed by subdividing this into one or more subsets, which themselves can be further subdivided into still smaller subsets, and so on. In <u>hierarchical</u> methods the classification obtained consists in fact of one set of OTUs which is divided into two subsets, which can be further divided into subsubsets, and so on. Classifications depicted by a dendrogram are typically hierarchical whereas the graph theoretical procedure as employed by us is not. Finally, in a <u>nonoverlapping</u> clustering method, OTUs cannot be member of more than one taxon at the same rank. The combination of hierarchical and nonoverlapping clustering is conceptually the most familiar. It is, for example, the one obtained in classical biological taxonomy: A whelk is an OTU of the taxon of the gastropods which is a subset of the hierarchically higher taxon of the molluscs. This excludes the possibility that the whelk should be considered also as part of the lamellibranchiae (bivalves), a taxon of the same rank as the gastropods. For SAHN clustering methods,

NUMERICAL TAXONOMY IN CHROMATOGRAPHY

numerous alternative algorithms have been proposed, among which single linkage and average linkage (weighted or unweighted) are the most favored.

The technique described in the example is, as just stated, an average linkage method. This is so because one obtains the similarity coefficient U between a newly formed cluster (J, K) and other OTUs, for example L, by using

$$U_{(J,K),L} = \frac{1}{2} U_{JL} + \frac{1}{2} U_{KL} \tag{5}$$

i.e., by making the arithmetic average between the similarity coefficients of the two members of the cluster and OTU L. When J and/or K are taxa already consisting of a cluster of OTUs (such as A" in the example), then the method is weighted when the number of OTUs in J and K is not the same. If the number of OTUs in J and K is t_J and t_K, respectively, and if t_J t_K, then by using Eq. (5) one confers more importance to the OTUs from K than to those from J.

In an unweighted average linkage method Eq. (5) changes to

$$U_{(J,K)L} = \frac{t_J}{t_{(J,K)}} \cdot U_{JL} + \frac{t_K}{t_{(J,K)}} \cdot U_{KL} \tag{6}$$

Two other possibilities are single and complete linkage clustering. In single clustering, the similarity between an OTU and a taxon containing more than one OTU is expressed as the similarity between the OTU and the nearest or most similar OTU of the taxon. In complete linkage, on the other hand, the similarity between an OTU and a group of OTUs is considered equal to the similarity between the OTU in question and the least similar OTU of the taxon. To illustrate this, let us again consider the example. The distance between A' and B would now be 24 using single linkage and 36 with complete linkage. Whereas single linkage clustering will lead to widely diffuse clusters, complete linkage clustering will lead to tight, compact clusters. Both methods have disadvantages compared to average clustering techniques, so that the latter methods are used most often.

It should be noted here that we linked only two OTUs or clusters at each step. This procedure is called the pair-group method, and results in bifurcation dendrograms. In contrast, when several OTUs or clusters can fuse simultaneously, the procedure is called a variable group method, and results in multiple-furcation dendrograms. However, since pair-group methods are by far easier to program, pair-group methods are the most popular.

E. Graph Theoretical Procedures

Several graph theoretical procedures, such as Kruskal's algorithm [5] for the calculation of a minimum spanning tree in a network, can be used as linkage methods for nonhierarchical classifications [6]. This algorithm can be introduced by the following example, in which a production unit is to be connected with a number of clients through a pipeline. One wants to know in which way to interconnect the clients and the production unit so that the pipeline has a minimal length. The distances between the clients (B, C, D, E, F, and G) and the production unit A, and between the clients themselves are given in Table 2. By drawing all the possible interconnections between the points, a graph is obtained where clients and production unit are the nodes, and the interconnections the edges whose value is given in Table 2. All points must be linked directly or indirectly to each other, and contain no cycle. This is called a tree, and the tree for which the sum of the values of the edges is minimal is called the minimum spanning tree. Kruskal's [5] algorithm for finding the minimum spanning tree, i.e., the optimal, shortest distribution, can be stated as follows: "Add to the tree the edge with the smallest value that does not build up a cycle with the edges already chosen." Applying this algorithm to the values of Table 2, one starts by selecting edge DF (value 17), followed by AB(21), BC, and EF(22). In a next step, edge DE (value 31) should be chosen; however, since this edge forms a cycle with the already chosen edges DF and EF, edge DE is of no use in the construction of the minimum spanning tree. In a series of ensuing steps, edge FG(36) is chosen, but not edges GD(41) and AC(42) because they also form a cycle. Eventually edge GA(51) completes the construction of the minimum spanning tree.

TABLE 2

Distance between the Nodes in Fig. 3.

	A	B	C	D	E	F	G
A	0						
B	21	0					
C	42	22	0				
D	90	87	78	0			
E	108	110	104	31	0		
F	87	88	83	17	22	0	
G	51	53	54	41	52	36	0

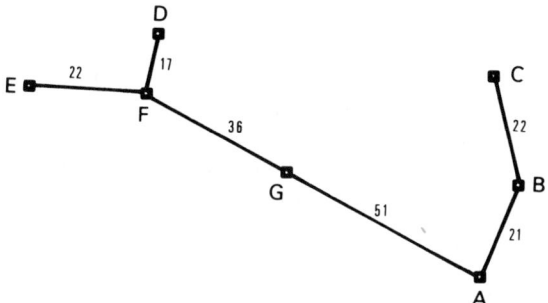

FIG. 3. Minimum spanning tree for a production unit (A) and six clients.

The optimum distribution network is therefore as given in Fig. 3. By careful inspection of this figure, which is drawn to scale, two clusters can be distinguished, namely (A, B, C) and (D, E, F, and G). These clusters can be obtained formally by breaking up the longest edge (AG) in the tree. This distribution problem can be used in TLC for the classification of systems according to similarities in their chromatographic behavior. The resemblance of the chromatographic systems can be pictured as a distance, for instance, a Euclidian distance [Eq. (2)], or one minus the correlation coefficient. The chromatographic systems can then be considered as constituting nodes in a graph. The values of the edges are given by the distance between the nodes: the smaller the distance, the larger the similarity between the nodes. If A...G were chromatographic systems, one should first break the longest edge, AG. This yields two classes A, B, C and D, E, F, G. Then one should break the second largest link, so that one obtains the classes A, B, C/D, E, F/G.

III. APPLICATIONS

A. The Selection of Optimal Sets in TLC

Thin-layer chromatography is one of the chromatographic techniques in which there is the largest number of alternative separation systems (by which a particular combination of solvents and stationary phase is meant) available. During a literature search preliminary to the selection of an optimal scheme for the identification of sulfonamides not less than 63 such systems were found. The question that should then be asked is, what is the combination of n systems that allows the largest possible number of the investigated group of compounds to be identified?

One possibility is, of course, to evaluate each of the possible combinations. This can be done using a particular concept, such as the information content which has been described for the evaluation of single systems by one of us [7] and which can be easily generalized for combinations of two or more systems. Another evaluation criterion is the so-called discriminating power described by Moffat and co-workers [8-11]. These authors consider two compounds to be unresolved in a particular chromatographic system if the difference between their R_f values does not exceed a certain value, the error factor E. To compute the discriminating power of a system or combination of systems in which N compounds are investigated, the total number, M, of unresolved pairs of compounds (within the limits of E) is counted. The discriminating power is then given by

$$DP = 1 - \frac{2M}{N(N-1)}$$

To test out all the possible combinations of chromatographic systems, specially written computer programs are necessary. Moreover, one has to make a choice using one rigid criterion, such as two substances are always separated when their R_f values are larger than 0.1. In practice, a chromatographer may want to include other criteria in his choice, such as availability of stationary phases, reproducibility, and personal preferences (most people will exclude pyridine as a possible solvent, except when it is really much better than other solvents). Numerical taxonomy allows the selection of combinations of systems in TLC on a rational basis. The basic philosophy of this selection is as follows:

1. Each system in the combination must be as good as possible: i.e., each chosen system must yield a good separation. In the rest of this section, we will accept that this means that the individual systems must yield as much information (as defined in Ref. 7) as possible. However, as stated earlier, other criteria can be used to do this.

2. The systems must yield different information. It is a well-known fact that many systems give rise to very analogous separations and it is clear that it is therefore impossible simply to select the n best individual systems. If one classifies the different systems according to their similarity and chooses from each of the n most different classes the best system, then one obtains a combination of n systems which is, at least, near optimal.

To illustrate the preceding, and at the same time, to prove that the best separation is indeed obtained, we will consider here in detail the selection of the optimal combination of 4 systems from 14 described for 100 basic drugs by Moffat and Clare [12]. The data set consists therefore of 1400 R_f values. As similarity coefficients, the correlation coefficients ρ were used. This means that the correlation coefficient between each pair of systems has to be calculated. This offers no difficulties since computer centers have this kind of program available, and if necessary, the coefficients can be computed on table or pocket computers with standard programs. The matrix of ρ values is shown in Table 3. The dendrogram of Fig. 4 is constructed by weighted linkage. If one wants a combination of four systems, one should use one system each of the groups 4-5-6, 7-8, 2-3-9-11-12-14, and 1-10-13.

The selection was carried out with the information content of the individual systems as a criterion, which yielded the combination of systems 6, 7, 13, 14. It was shown by Moffat and Clare that this combination separates 99.4% of all the pairs of basic drugs which can be formed from the original set. It is also interesting to note that the classification makes chemical sense. For example, systems 4 and 5 are the only two to use silica gel dipped in $KHSO_4$.

An application, developed by us and which has not been published until now, is the selection of the optimal sequence of TLC methods for the identification of toxicologically important organic bases. Sunshine et al. [13] proposed seven possible systems for the qualitative analysis of 139 such substances. Since one system does not allow the unambiguous identification of all these substances, the authors propose to carry out first a TLC run with their system 1; all substances with an R_f within ±0.05 R_f units of the unknown are considered as possibly present. Then system 2 is used to make a further discrimination. The substances with an R_f within ±0.05 of the unknown are retained as possible. This is continued with further TLC systems until unambiguous identification is achieved.

Sunshine et al. state that the order in which they give the systems is arbitrary and does not necessarily constitute the best sequence to achieve identification with the smallest possible number of TLC systems. To determine what the optimal sequence is, we determined the correlation coefficients between the seven systems. The resulting similarity matrix is given in Table 4. By carrying out a weighted pair-group reduction on this matrix the dendrogram of Fig. 5 is obtained.

The following groups of systems are isolated from this dendrogram:

2/1, 3-7

2/1, 5, 7/3, 4, 6

2/1, 5/3, 4, 6/7

TABLE 3

Similarity Matrix (Correlation Coefficient) between 14 Systems for TLC Identification of Basic Drugs

	1	2	3	4	5	6	7	8	9	10	11	12	13	14
1	1													
2	0.007	1												
3	0.256	0.804	1											
4	0.265	0.396	0.502	1										
5	0.001	0.352	0.465	0.902	1									
6	0.277	0.387	0.449	0.583	0.596	1								
7	0.288	0.462	0.434	0.226	0.177	0.525	1							
8	0.324	0.368	0.279	0.001	0.068	0.285	0.544	1						
9	0.555	0.483	0.532	0.230	0.179	0.493	0.449	0.519	1					
10	0.594	0.155	0.312	0.025	0.073	0.310	0.405	0.437	0.642	1				
11	0.474	0.390	0.396	0.159	0.096	0.418	0.371	0.568	0.883	0.679	1			
12	0.010	0.792	0.583	0.251	0.188	0.237	0.297	0.331	0.643	0.301	0.634	1		
13	0.910	0.041	0.218	0.036	0.078	0.221	0.271	0.316	0.626	0.695	0.564	0.090	1	
14	0.293	0.629	0.586	0.281	0.186	0.407	0.512	0.474	0.724	0.647	0.708	0.719	0.329	1

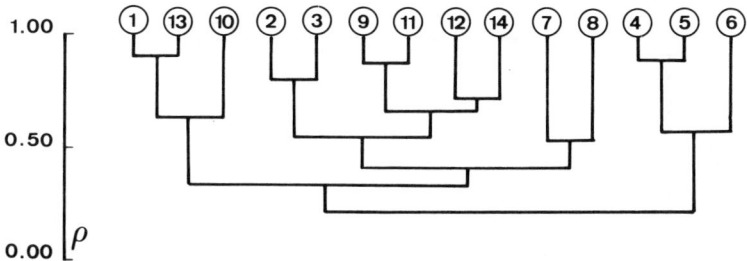

FIG. 4. Weighted pair-group linkage (correlation coefficients) of 14 systems for TLC identification of basic drugs.

TABLE 4

Similarity Matrix (Correlation Coefficient) between Seven Systems for TLC Identification of Toxicologically Important Organic Bases

	1	2	3	4	5	6	7
1	1						
2	0.095	1					
3	0.240	0.171	1				
4	0.520	0.315	0.796	1			
5	0.889	0.096	0.258	0.538	1		
6	0.636	0.189	0.631	0.871	0.592	1	
7	0.528	0.405	0.422	0.504	0.540	0.456	1

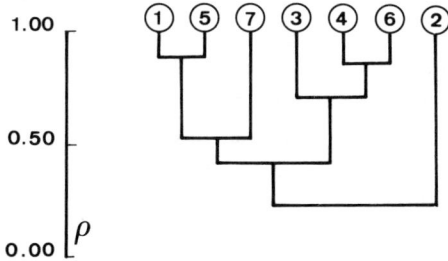

FIG. 5. Weighted pair-group dendrogram of seven systems for TLC identification of toxicologically important organic bases (similarity matrix in Table 4).

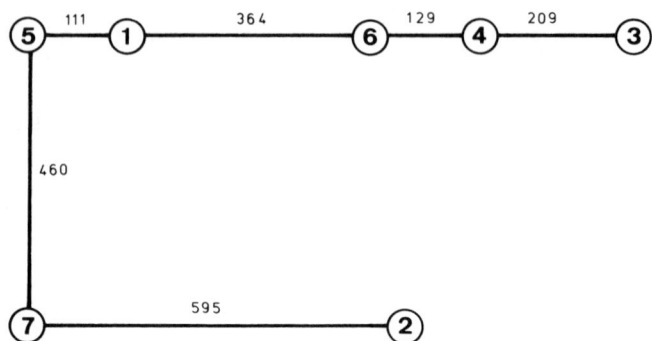

FIG. 6. Minimal spanning tree $[(1 - \rho) \times 1000$ as a distance] between seven systems for TLC identification of toxicologically important organic bases (similarity matrix in Table 4).

The minimal spanning tree $[(1 - \rho) \times 1000$ as a distance] is given in Fig. 6. By successively cutting through the longest edges, one arrives at the groups

    2/1, 3-7

    2/1, 3-6/7

    2/1, 5/3, 4, 6/7

By selecting in both classifications the best system in each group, i.e., the system with the highest information content (Table 5), one arrives at the conclusion that if one chooses to use four systems, these should be systems 7, 4, 2, and 1. In Table 6 the number of unambiguously identified organic bases is given for the original sequence and for the sequence of systems determined by us. One finds, for example, that using the latter 98 substances (87% of the substances that can be identified) are determined with four systems compared to 81 (72%) with the former sequence.

    Other applications of numerical taxonomy for the selection of TLC systems have been published for food dyes [2] and for a set of systems for basic drugs [14] that is smaller than the one just discussed.

    All the examples given until now illustrate the effectiveness of the method for selection from a relatively restricted set of systems. We have also studied [31] a more important set of systems for the separation of sulfonamides. In this instance, we used R_f values for 22 sulfonamides in 41 systems on SiO_2 and 10 systems on Al_2O_3. Some of the systems were selected from the literature, others were devised by us. These 51 systems were preselected using the method described in Sec. III.C from more than

TABLE 5

Information Content of Seven Systems for TLC Identification of Toxicologically Important Organic Bases*

System	Information content (bit)
1	3.61
2	3.76
3	3.65
4	3.99
5	3.58
6	3.63
7	4.09

*R_f classes of 0.05 R_f units.

TABLE 6

Number of Identified Organic Bases as a Function of the Combination of Chromatographic Systems

Original sequence		Numerical taxonomy	
1/2	11	7/4	17
1/2/3	57	7/4/2	75
1/2/3/4	81	7/4/2/1	98
etc.	88	7/4/2/1/3	105
	92	7/4/2/1/3/6	109
	113	7/4/2/1/3/6/5	113

70 systems and all the data were obtained by chromatography on commercially available precoated plates in a Vario S chamber. Numerical taxonomy with correlation coefficients as the similarity coefficients yielded two highly separated classes ($\rho = 0.25$), one of them containing all the Al_2O_3 and only

2 out of the 41 SiO_2 systems. From each class the best systems according to information content were chosen, namely n-hexanol/SiO_2 and acetone NH_4OH 25% (80-15)/Al_2O_3, and it was found that these two together allow the separation of all but 2 pairs of the 22 sulfonamides. By breaking the lowest remaining links in the dendrogram ($\rho = 0.50$), one isolates first one system on SiO_2 from 38 others. This allows the separation of one of the unseparated pairs. Then ($\rho = 0.59$) one separates a group of four Al_2O_3 systems from six others and two SiO_2 systems. One of these, namely butanol saturated with water/Al_2O_3, permits the separation of both the unresolved pairs of sulfonamides. In our opinion, these applications show that numerical taxonomy, combined with information theory, constitutes an excellent tool for the systematic optimization of TLC separations and identification schemes. The application of this method allows an objective and, almost certainly, nearly optimal combination of systems. However, one should be warned against overbelief in such formal procedures. Let us consider, for example, the extreme situation in which a system with only R_f values of zero is present among the systems from which one is to choose. Because this system shows special retention characteristics, it will almost certainly be present in a group by itself and linked to all other systems in the dendrogram at a very low level. If one now applies the rule of breaking the lowest links, one will isolate this class and be obliged to select this obviously uninteresting separation system. This shows that these methods should not be considered as absolute and infallible automata, but rather as tools to help the chromatographer make better decisions.

B. The Classification of Stationary Phases in GLC

The classification of GLC phases is a problem which has interested many chromatographers in the last decades. It is not our intention to describe in detail all this work, but rather to limit the discussion to a few milestones in the development of the research in this domain, the outline of some clustering or ordination techniques related to numerical taxonomy, and the results obtained by the latter technique. In the early stages, research workers are concerned mostly with the development of methods to describe the phases. For example, Brown [15] classified stationary phases according to three values depending on electron donor and acceptor properties. This work has culminated in the development in 1961 by Rohrschneider [16] of what is now known as the Rohrschneider system of functional probes. This consists of a set of five substances (benzene, ethanol, ethyl methyl ketone, nitromethane, and pyridine), each representative of some type of interaction. Much additional research concerning the choice of these probes has since then be carried out. Some of this work and a recent method proposed by the authors of this chapter are discussed in Sec. III.C. In 1970,

McReynolds [17] applied the same ideas to characterize 226 (i.e., most) liquid phases then in use, with 10 probes. One of the larger manufacturers of GLC phases lists the retention indices of seven of these in his latest catalog [18] and uses the sum of the first five as a polarity index. The phases are then ordered according to this sum. This sequencing is in fact a simple attempt at classification. The very large number of phases which have been proposed until now and frequent pleas in the literature for a smaller set of so-called preferred phases have also stimulated attempts to arrive at better and more complete classifications. Several research teams used different approaches to arrive at this result. However, each of these methods used numerical methods to achieve the classification. This clearly illustrates that there is a need for such procedures in chromatography.

The first attempt is due to Leary et al. [19], who used a so-called nearest neighbor technique. The first step consists in the selection of 12 preferred phases from the 226 phases in McReynolds data set in such a way that these 12 span the complete polarity range. These function as the nucleus of a group. The Euclidian distances [Eq. (2)] between each of the other phases and the 12 preferred phases are then calculated. The nearest of these 12 is called the nearest neighbor and each of the phases is assigned to the group of their nearest neighbor. This classification has the merit of being the first numerically obtained classification. It also has some flaws, the most serious of these being the inherent subjectivity due to the preliminary choice of 12 preferred phases and its unflexibility (it is not possible to derive in an easy way from the obtained 12-group classification an 11- or 13-group classification). This classification also has a limited physico-chemical meaning since phases with special interactions with some of the probes are not detected.

A principal components analysis was performed on the same data by Wold and Andersson [20]. This is a method which allows the determination of the number of factors (components) necessary to reproduce the retention indices within specified limits. Three components allow a 30 retention index (RI) limit to be obtained. The retention of the i-th solute on the k-th liquid phase is then represented by

$$RI = z_{k1} v_{1i} + (z_{k2} + 3.15583) v_{2i} + x_{k3} u_{3i} + e_{ki}$$

where v_{1i}, v_{2i}, and u_{3i} are weights given to the probes, z_{k1}, z_{k2}, and x_{k3} are the components whose values characterize the liquid phases, and e_{ki} are the residuals, i.e., the discrepancy or error between the calculated and observed value.

Although a characterization of the phases is achieved, this was not used to obtain a completely formal and hierarchical classification. Nevertheless, interesting conclusions were obtained. Since these fall outside the scope of this chapter, we will relate only one of these. Phases with a large residual e_{ki} were classified as "abnormal" in the sense that they show some specific strong interaction with one or another solute.

McCloskey and Hawkes [21] performed a very similar principal components analysis on the same data. They found that two components describe 98.5% of the total variance, which allows a two-dimensional representation of the stationary phases. The two axes, representing the two factors or compounds, were identified as a general polarity axis and an "alcohol-attracting ability" axis. The results obtained can serve as the basis of a classification, but do not allow a complete formal classification.

The same set of data was classified by us using numerical taxonomy. In a first publication [3] we proposed a classification based on distances as the similarity coefficients and the average linkage-weighted pair clustering technique. In a later publication [22], we used the unweighted pair clustering method. The results obtained with the latter method are in general very analogous to those obtained using the weighted pair-group method, but are slightly superior since the method corrects some minor flaws in the first published classification. This was to be expected since a weighted pair-group method has more chance of introducing a bias than an unweighted pair-group method. Since both classifications have been published, we will not reproduce these (very long) tables herein, nor will we discuss them in detail, but only point out some remarkable results.

The classification of the apolar phases shows that the results obtained agree with what could have been predicted from the chemical structures of these phases. This is because one observes first the formation of a group of saturated hydrocarbons, then a group of Apiezons, two groups of silicones, one of them containing the methylsilicones and the other the methyl, phenyl, vinyl silicones, a small group of three esters, and a group of three fluorinated hydrocarbons. This shows that the method used yields a logical classification and that the results are significant. It should also be noted that the method allows the selection of "abnormal" phases, which are found as very small classes (usually containing only one member). These abnormal phases are the same as those found by both Wold and McCloskey (see Table 7), which confirms the fact that meaningful results are obtained.

One of the criticisms which can be leveled against the combined work of Leary, Wold, McCloskey, and ourselves, is that these investigations deal with a data set containing many phases that are outdated. Therefore, we have carried out the same work on the phases listed in the latest Applied Science catalog [18] using the method described in Ref. 3. In Table 8 and Fig. 7 are given the classes emerging after reduction to 32 × 32 of the original 121 × 121 similarity matrix. At this level 19 groups containing more than one phase are obtained. The remaining classes consist of only one (abnormal) phases. These are listed in Table 9. The phases Amine

TABLE 7

Comparison of "Abnormal" Phases

Abnormal according to McCloskey and Hawkes [21], Wold and Andersson [20], and numerical taxonomy [3, 22]

   N,N,N',N'-Tetrakis(2-hydroxyethyl)ethylenediamine (THEED)
   Octakis(2-hydroxypropyl)sucrose (Hyprose SP-80)
   Diglycerol
   Sodium dodecylbenzene sulfonate (Siponate DS-10)
   Zinc stearate
   Trimer acid
   Fluoroalkyl camphorate (Zonyl E91)
   Trifluoropropyl methyl silicone (QF1, OV-210)
   Trifluoropropyl methyl vinyl silicone (LSX-30295)
   Tetra(fluoroalkyl)-1,2,4,5-tetracarboxybenzene (Zonyl E-7)
   Stepan DS-60

Abnormal according to McCloskey and Hawkes [21] and numerical taxonomy [3, 22]

   N,N,N',N'-Tetrakis(2-hydroxypropyl)ethylenediamine (Quadrol)
   Tributoxyethyl phosphate

Abnormal according to Wold and Andersson [20] and numerical taxonomy [3, 22]

   N,N-Bis(2-cyanoethyl)formamide

Abnormal according to McCloskey and Hawkes [21]

   Triethyl hexyl phosphate
   Versamid 930
   Versamid 940

Abnormal according to Wold and Andersson [20]

   Tergitol NPX (due to an error in the data: see Ref. 3)
   Epon 1001
   Ethylene glycol phthalate

Abnormal according to numerical taxonomy

   Silicone XF-1150
   Apiezon H

TABLE 8

Classification of "Normal" Phases in Applied Science [18] Catalog Listing

Group 1	Group 3
Squalane	Di(2-ethylhexyl) sebacate
Nujol	Diisodecyl adipate
Apiezon M	Octyl decyl adipate
Apiezon L	Dilauryl phthalate
SF 96	Diisodecyl phthalate
Apiezon J	Dinonyl phthalate
Apiezon N	DC 710
SE 30	Dioctyl phthalate
OV-1	POLY-L 110
M and B silicone oil	Diisooctyl phthalate
DC 200	OV-17
OV-101	[Bis(2-ethylhexyl) tetrachloro-phthalate]
DC-410	
Versilube F-50	Group 4
DC 11	UCON LB-550-X
Group 2	Span 80
SE 52	Castorwax
SE 54	POLY-A-103
DC 560	Polypropylene glycol
OV-3	POLY-A 101A
Fluorolube HG 1200	UCON LB-1715
Kel F Wax	Group 5
Apiezon H	OV-22
OV-7	OV-25
DC 550	Polyphenyl ether OS-124 (five rings)
[Dexsil 300]*	Polyphenyl ether OS-138 (six rings)

TABLE 8 (Cont.)

Group 6
Acetyl tributyl citrate
Didecyl phthalate
Tributyl citrate

Group 7
Neopentyl glycol sebacate (HI-EFF-3CP)
Squalene
UCON 50-HB-280X
Tricresyl phosphate
Sucrose acetate isobutyrate

Group 8
QF-1
OV-210
DC LSX-3-0295

Group 9
Ethofat 60/25
Igepal CO-630
UCON 50-HB-2000
Triton X-100
UCON 50-HB-5100
[Tween 80]

Group 10
XE 60
OV-225

Group 11
Neopentyl glycol adipate (HI-EFF-3AP)

Group 11 (Cont.)
UCON 75-H-90000
Igepal CO 880
Triton X-305
HI-EFF-8BP

Group 12
Neopentyl glycol succinate (HI-EFF-3BP)
Igepal CO 990
Carbowax 20M
Carbowax 20M (TPA)
Carbowax 6000
Carbowax 4000

Group 13
EGSP-Z
Epon 1001

Group 14
Silicone ASI 50phenyl150cyanopropyl
Ethylene glycol isophthalate (HI-EFF-2EP)
XF-1150

Group 15
Ethylene glycol adipate (HI-EFF-2AP)
Butane-1,4-diol succinate (HI-EFF-4BP)
Phenyldiethanolamine succinate (HI-EFF-10BP)
Reoplex 400
LAC-1-R-296

TABLE 8 (Cont.)

Group 15 (Cont.)	Group 17 (Cont.)
Diethylene glycol adipate (HI-EFF-1AP)	Diethylene glycol succinate (HI-EFF-1BP)
Carbowax 1540	LAC-3-R-728
LAC-2-R-446	**Group 18**
EGSS-Y	Silicone APOLAR-9CP (formerly SILAR-9CP)
Group 16	
Silicone APOLAR-7CP (formerly SILAR-7CP)	Silicone APOLAR-10C (formerly SILAR-10C)
ECNSS-M	**Group 19**
ECNSS-S	Tetracyanoethylated pentaerythritol
Group 17	Ethylene glycol succinate (HI-EFF-2BP)
EGSS-X	
Ethylene glycol phthalate (HI-EFF-2GP)	

*Phases in brackets were not included in the classification (one retention index missing), but should be classified with the groups under which they are named on the basis of the six other retention indices.

220, etc., were not included in the classification because of missing RI values. On the basis of the other RI values they are, however, clearly "abnormal." It should be noted that in this classification only 7, instead of 10, probes were used, so that some interactions or effects will not be noted.

In the Applied Science catalog the phases are characterized by and ordered according to $\sum_1^5$, the sum of the first five retention indices. This is considered to be a general selectivity guide. In fact, it would be better called a polarity guide, since no specific interactions are considered. The tables obtained by numerical taxonomy are true selectivity guides since they isolate groups with similar characteristics and single phases with special characteristics. One typical example is the group QF-1, OV-210, DC LSX-3-0295 which, according to $\sum_1^5$, should resemble closely a phase such as Ethofat 60/25. In fact, as is well known, these phases show very special interactions with several groups of solutes, such as the alcohols.

NUMERICAL TAXONOMY IN CHROMATOGRAPHY 101

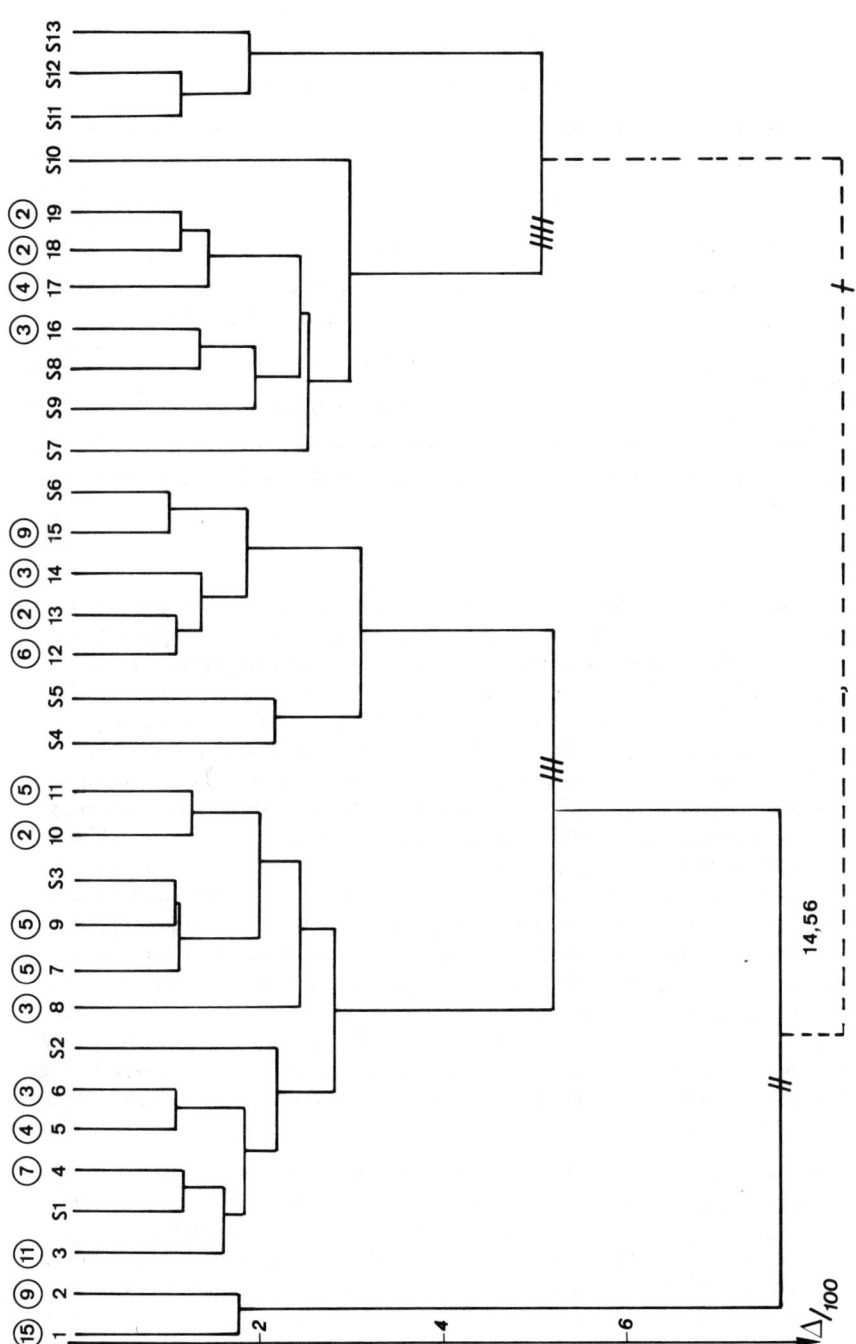

FIG. 7. Weighted pair-group dendrogram (Δ = Euclidian distances) of the GLC phases listed in the Applied Science [18] catalog. The numbers refer to the groups or special phases in tables 8 and 9. The numbers in circles indicate the number of phases in a group.

TABLE 9

Abnormal Phases in the Applied Science Catalog [18] Listing*

S1	Hallcomid M-18	S8	EGSP-A
S2	Trimer acid	S9	Diglycerol
S3	Tergitol NPX	S10	THEED
S4	Siponate DS-10	S11	1,2,3-Tris(2-cyanoethoxyl)propane [1,2,3,4-tetrakis(CYANO-B)
S5	Quadrol		
S6	Carbowax 1000	S12	Cyanoethylsucrose
S7	Hyprose SP-80	S13	N,N-Bis(2-cyanoethyl)formamide

*Amine 220, POLY A-135, and CYCLO-N should also be considered special phases.

In a recent publication, a list of preferred phases is proposed by a committee of experts [23]. This list is meant to replace the very large quantity of often redundant phases now available. This means that the proposed set should consist of phases that are as similar as possible. The Applied Science listing contains 23 of the 24 phases of Table II of Ref. 23, i.e., the larger of the two proposed preferred phases lists. We observe that there are several groups in the numerical taxonomic classification that contain more than one of these 23 phases. Since phases in the same group are very similar one can speculate whether some of these phases are not redundant, for example, in group 1, squalane, dimethylsilicone, and Apiezon L; in group 11, nonylphenoxypoly(ethylene oxy)ethanol (Igepal CO-880) and cyclohexane dimethanol succinate (HI-EFF-8BP); and in group 18, 90% 3-cyanopropyl 10% phenylmethylsilicone (SILAR-9 CP) and 3-cyanopropylsilicone, 100% (SILAR 10 C). On the other hand, some of the other groups (the most important being group g) in the numerical taxonomic classification are not represented. It seems, therefore, that from the viewpoints of dissimilarity, this list is not optimal. It must be remembered, however, that dissimilarity is not the sole criterion that has to be taken into account for the choice of a list of preferred phases.

Numerical taxonomy and the methods of Wold, Leary, and others as just discussed are classification methods. If one wants to compose a preferred set, one can proceed in the same way as in the TLC problem, i.e., one can select the best liquid phase from a number of classes. By the best, one can mean the cheapest, the most stable, and so on. We have never proposed such a preferred set because it is our opinion that such a proposal should not be the work of an individual but rather of a group of specialists, patronized by an international organization such as IUPAC. However, we have investigated one of the more fundamental aspects of such a selection, namely how to use numerical taxonomy to optimize the information obtained with a restricted set of phases selected from a larger group.

Dupuis and Dijkstra [24] used an information theoretical approach to select an optimal set of stationary phases for qualitative analysis (retrieval). It is impossible to give the mathematical details in this chapter, but the two main points can be summarized as follows:

1. The information given by one phase is calculated using the distribution of the retention indices of the substances considered. The wider the distribution is, the more information is obtained.

2. To quantify information from a set of columns one must also take into account the fact, which we have discussed before, that different information must be obtained from the selected columns. In information theory, this is carried out by taking into account the covariance matrix, which is a measure of the correlation between the considered liquid phases.

It was found [25] that, by using correlation coefficients as the similarity coefficient, and by selecting from each class in the resulting numerical taxonomic classification the phase with the individually highest information, sets of preferred phases are obtained that are very analogous to those obtained using the information theoretical approach of Dupuis and Dijkstra. This was investigated more systematically by applying both numerical taxonomy and information theory to select a few stationary phases from a set of 16 for several sets of retention indices of related compounds (alcohols, esters, and aldehydes/ketones). A typical result is shown in Table 10, which gives the number of the phases in the order in which they are selected by information theory and the classification obtained by successively breaking the lowest link in the dendrogram. By applying the same principle as for TLC, namely the selection of the best phase from each group, one selects the same sets of phases as obtained from information theory.

In general, the more important conclusions of comparison between information theory and numerical taxonomy are as follows:

1. In all cases the agreement between selections according to both methods is excellent. Since the information theoretical approach always yields the set of phases that displays the highest possible information, this proves that numerical taxonomy is capable of selecting optimal and at least near-optimal sets in this fundamentally important respect. The main advantage of numerical taxonomy is that it also allows the consideration of such factors as stability and temperature range, if these are considered more important than information yield.

TABLE 10

Combination of Columns Which Yield a Maximum Amount of Information and the Classification of Columns, for the Subset Esters

Number of columns	Obtained combination of columns	Classification of columns with	
		Correlation coefficients	Taxonomic distances
1	12		
2	1, 8	8, 10/1-7, 9, 11-16	1-7, 9, 12, 14, 16/8, 10, 11, 13, 15
3	2, 8, 11	1-4/8, 10/5-7, 9, 11-16	1-4/5-7, 9, 12, 14, 16/8, 10, 11, 13, 15
4	2, 8, 10, 11	1-4/5-7, 9, 11-16/8/10	1-4/5-7, 9, 12, 14, 16/8/10, 11, 13, 15
5	2, 8, 10, 11, 15	1-4/5, 6, 9, 11, 12, 16/7, 13-15/8/10	1-4/5, 9/6, 7, 12, 14, 16/8/10, 11, 13, 15

Source: From Ref. 25.

2. It is found, both by information theory and numerical taxonomy, that different sets of phases are selected for alcohols, esters, and aldehydes/ketones. This proves, in a mathematical way, the thesis defended by the antagonists of the selection of a restricted set of preferred phases for general use, namely that in each specialized area separation problems occur that cannot be solved with such a general set but require a specialized set.

C. The Classification of Substances According to Chromatographic Behavior

To be able to obtain a classification of chromatographic systems, one must use a number of characters. When the object of the classification is simply to obtain a separation scheme for a certain class of compounds, there is no difficulty: All the compounds must be considered characters. However, when the purpose is to develop a general classification, such as in the classification of McReynolds' or Applied Science's sets, the complete set of characters consists of all the substances which could be chromatographed by GLC. It is clear that is is impossible to obtain values for all the characters and OTUs. Therefore, one has to select a smaller number of characters, and in GLC these are called functional probes. A different number of probes is used by various workers. The best known and most generally accepted character sets at this moment are those proposed by Rohrschneider [16], containing 5, and by McReynolds [17] containing 10 such probes. It is beyond the scope of this chapter to explain how these functional probes were obtained. Let us state simply that each probe is thought to be typical of a certain interaction between solutes and GLC phases. For example, in a recently proposed set [26] consisting of benzene, nitroethane, n-propanol, and dioxane, these probes are considered to be representative of dispersion forces, dipole orientation, proton donors, and proton acceptors, respectively. In fact, to arrive at a set of functional probes, one classifies implicitly the complex interactions observed between chromatographed substances and stationary phases to arrive at a restricted number of main interactions. This type of classification is the reverse of the classification problems we have discussed up to now. Indeed, if one wants to classify solutes according to their chromatographic behavior, which is the case here, then the solutes are the OTUs and the GLC phases the characters, the values of which are given by retention indices.

An explanation of nomenclature should be inserted at this point. In biological numerical taxonomy a difference is established between Q and R techniques (the equivalent of which is sometimes called the first and the second space). The association of pairs of characters is called an R technique, whereas procedures for the association of OTUs are called Q techniques. Although it is clear that in biology the OTUs are usually species of plants,

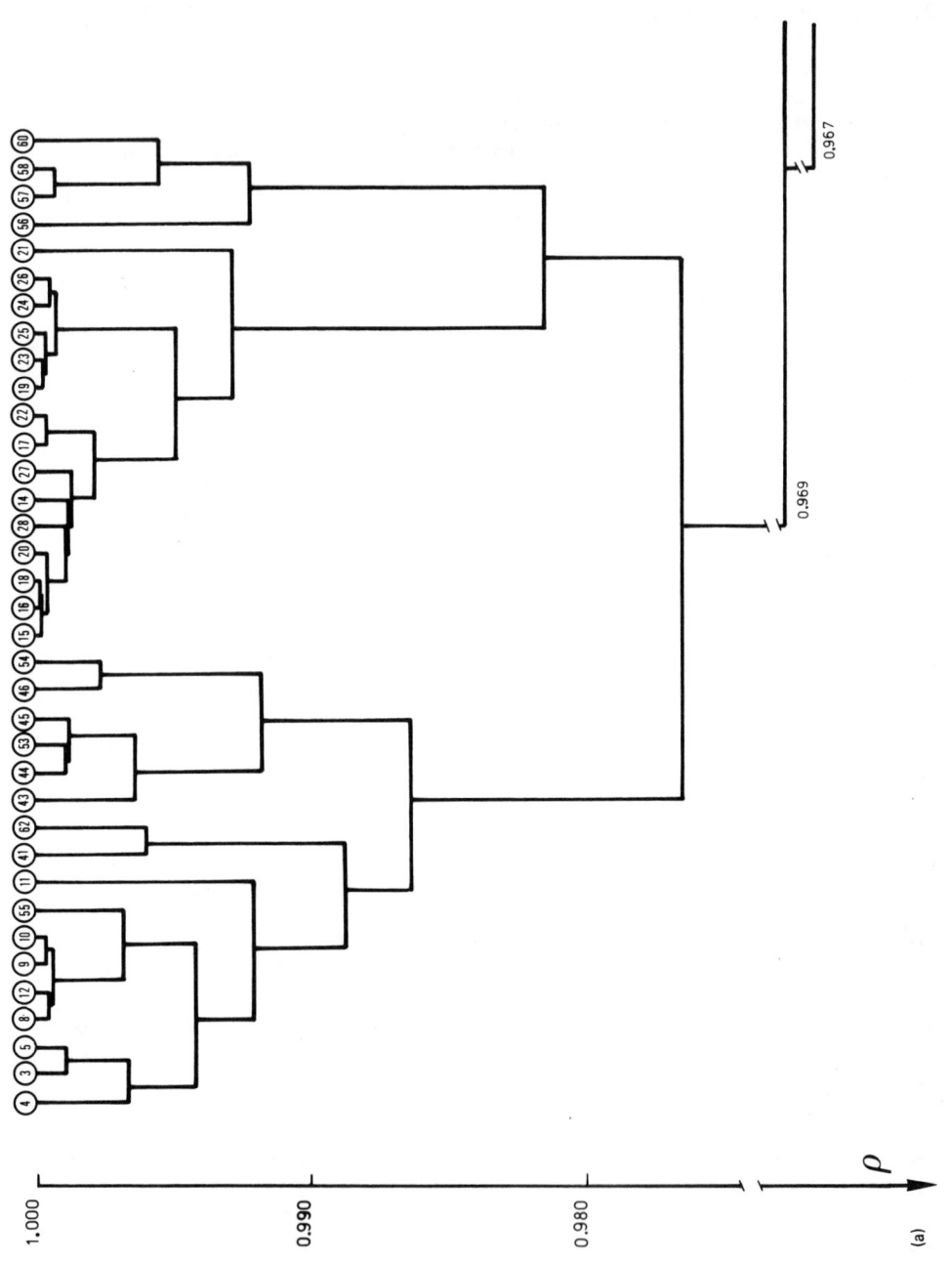

FIG. 8. Classification of 62 substances (for an explanation of the numbers, see the text) according to GLC behavior.

NUMERICAL TAXONOMY IN CHROMATOGRAPHY

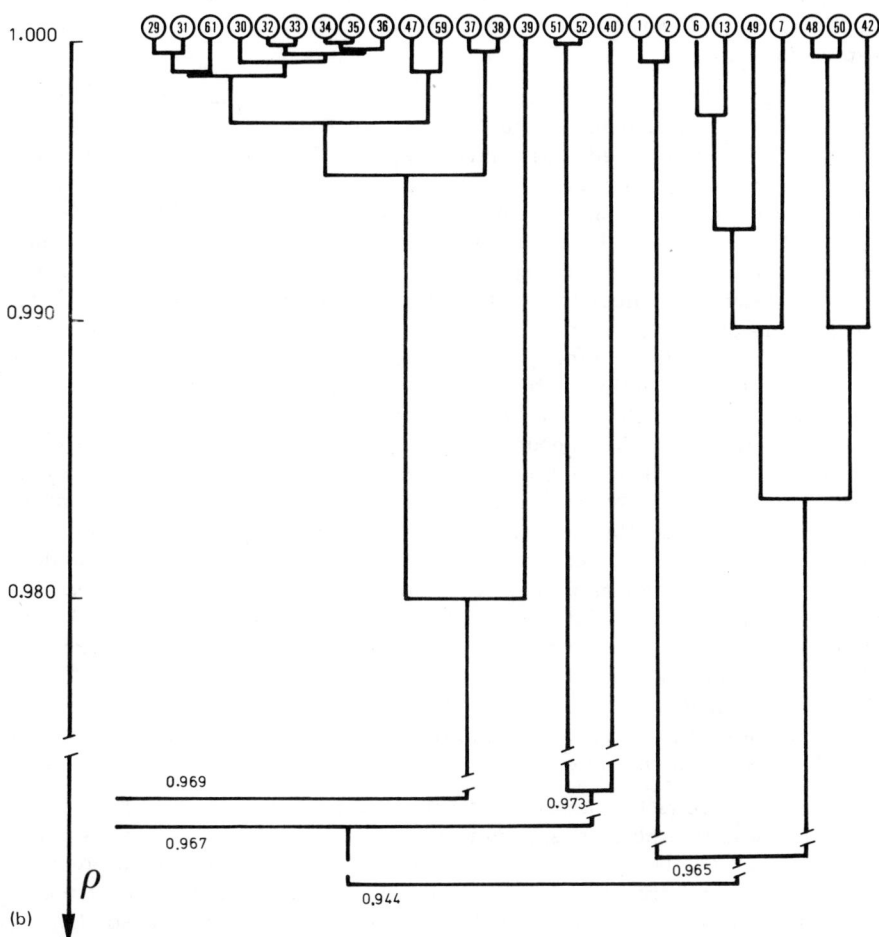

FIG. 8 (Cont.)

and so on, and the emphasis must be on the classification of these, it is not so evident whether in chromatographic research the priority should be given to the classification of chromatographic systems or of chromatographic behavior of solutes, so there is no fundamental reason to make a distinction between Q and R techniques. However, for ease of explanation such a distinction is useful and therefore we will call the classification of compounds R techniques.

In Fig. 8 the numerical taxonomic classification of 62 substances according to their GLC behavior on 25 phases is given, using the correlation coefficient ρ to measure the similarity between the substances and an average weighted pair-group method as linkage technique. The data come from a data set composed by McReynolds (personal communication), and also used by Hartkopf et al. [26]. The same numbering of substances as in the latter publication is used, except for the first five alkanes, which show a range of retention indices that is too small, resulting in values of correlation coefficients of no significance because these are determined by the random error on the retention indices, and for 2-iodobutane, since it was assumed that some of the retention indices for this substance were in error. These six substances were therefore left out of the classification (i.e., compound nr 1 = cis-hydrindane, compound nr 2 = decalin, ..., compound nr 62 = pyridine). Before discussing the selection of probes, it is necessary to make some remarks about the composition of the data set. This set and a smaller set by Rohrschneider [16] have been used as the basis of most of the work on the selection of functional probes. Ideally, such work should be carried out on the complete universum of GLC data (i.e., the retention indices of all possible solutes on all the phases) and the selection of the substances and phases in this set implies a first classification, the soundness of which is impossible to check completely. However, one observes from the figures that it is certainly not ideal. In developing a numerical taxonomic classification, it is necessary to avoid biased character sets, by eliminating closely correlated characters.

Figure 8 shows that narrow clusters of compounds are present. One may wonder if this clustering pattern is really representative for the universum of all compounds and whether, for example, the number of alcohols present is not relatively large. Although there is some criticism on the composition of the original data set, one may hope with good reason that functional probes representing this set in an optimal way, will be sufficiently representative for general GLC use. One way of selecting representative probes is to group the compounds in p classes by carrying out an R classification and to choose the most representative of each class. If the most representative member of a class is considered to be the one nearest to the center of the cluster constituting this class, then by choosing the solute nearest to the center for the p classes, one has developed an objective strategy of selecting probes. The sets selected in this way from McReynolds' were found to be less able, but only slightly, to predict the retention index values of the complete set of phases and compounds. The numerical taxonomy method might, however, prove useful to classify compounds in more restricted GLC areas (e.g., lipids) and to select probes in a rapid way. It should be added that this can also be done using an integer programming method [6] that does allow the selection of better probe sets than those published so far. However, this method uses much more sophisticated mathematics. A situation in which such probes were developed de facto is a systematic study of the optimization of TLC of basic drugs.

One of the favorite occupations of many research workers in the TLC field is the development of a few more solvent/stationary phase systems for restricted sets of compounds (e.g., basic drugs) in the hope of discovering a system with properties different from those already existing. In Sec. II.A we discussed how to make use of published systems to find optimal combinations. One can also ask the following question: How should one explore the many tried and untried possible combinations of solvent mixtures and stationary phases for separating a group of substances? For a large group of substances such as basic drugs it is out of the question simply to try out all the possible solvent systems with all the drugs. Detaevernier [27] solved this difficulty by carrying out the numerical taxonomic classification of 100 basic drugs chromatographed in eight systems [9-11]. From the 10 final classes in the classification, one compound near the center of the class was chosen. In this way, it was found that the TLC behavior on SiO_2 of this group of compounds can be explored best using a preliminary screening procedure with the following 10 probes: chlorpromazine, carbinoxamine, isocarboxazide, iproniazide, diamorphine, yohimbine, tetracaine, atropine, diphenhydramine, piperidolate.

D. Numerical Taxonomy as a Pattern Recognition Procedure

Chromatography is used mainly to identify or determine the concentration of certain substances or groups of substances in a known sample. Less often, although by no means seldom, the purpose of the analysis is to identify the nature of the sample.

The pattern constituting the chromatogram is often more or less typical for a certain kind of sample. Very often, too, the information present is used only to a small degree. A typical example is the work done on the use of the fatty acid distribution in milk fat. This distribution, which is determined by GLC of the methyl esters of the acids, differs according to the animal from which the milk originates, and therefore the chromatogram yields a pattern which can be used for the determination of the origin of the milk sample. Although this has been stated very often by the authors, who have worked in this field, none of them really uses the pattern concept.

An example is the work of Gattuso and Fazio [28] who, after obtaining the complete GLC chromatograms for 20 goats, 20 sheep, and 20 cows, decided to use the ratio of capric acid to caproic acid to distinguish between the animals, thereby leaving the information present in the rest of the chromatogram unused.

There are a number of techniques that allow a more complete use of the relevant information. These are termed pattern recognition techniques. Numerical taxonomy is a method which allows the recognition of clusters of related patterns and is therefore a pattern recognition technique. In the terms used by Kowalski [29], it is an unsupervised learning procedure. Until now, numerical taxonomy has not been used in the field of chromatography for the identification of samples and, in fact, we know of only one analytical application, namely the classification of pottery shards according to trace element patterns determined by activation analysis [30]. It would seem that methods of supervised learning such as the learning machine and statistical isolinear multicomponent analysis have a greater future in this field than numerical taxonomy. Nevertheless, because of its extremely simple mathematics, this technique seems eminently suited for the preliminary investigation of data sets consisting of chromatographic or other analytical results of a number of samples to be classified into categories.

ACKNOWLEDGMENTS

The authors thank Fonds voor Kollektief en Fundamenteel Onderzoek and Fonds voor Geneeskundig Wetenschappelÿk Onderzoek for financial assistance.

REFERENCES

1. P. H. A. Sneath and R. R. Sokal, Numerical Taxonomy, Freeman, San Francisco, 1973.
2. D. L. Massart and H. De Clercq, Anal. Chem., $\underline{46}$, 1988 (1974).
3. D. L. Massart, M. Lauwereys, and P. Lenders, J. Chromatogr. Sci., $\underline{12}$, 617 (1974).
4. R. J. Johnston, Ann. Assoc. Am. Geogr., $\underline{58}$, 575 (1968).
5. J. B. Kruskal, Proc. Am. Math. Soc., $\underline{7}$, 48 (1956).
6. D. L. Massart and L. Kaufman, Anal. Chem., $\underline{47}$, 1244A (1975).
7. D. L. Massart, J. Chromatogr., $\underline{79}$, 157 (1973).

8. K. W. Smalldon and A. C. Moffat, J. Forensic Sci. Soc., 13, 291 (1973).

9. A. C. Moffat, K. W. Smalldon, and C. Brown, J. Chromatogr., 90, 1 (1974).

10. A. C. Moffat and K. W. Smalldon, J. Chromatogr., 90, 9 (1974).

11. A. C. Moffat, A. H. Stead, and K. W. Smalldon, J. Chromatogr., 90, 19 (1974).

12. A. C. Moffat and B. Clare, J. Pharm. Pharmacol., 26, 665 (1974).

13. I. Sunshine, W. Fike, and H. Landesman, J. Forensic Sci. Soc., 11, 428 (1966).

14. H. De Clercq and D. L. Massart, J. Chromatogr., 115, 1 (1975).

15. J. Brown, J. Chromatogr., 10, 284 (1963).

16. L. Rohrschneider, J. Chromatogr., 22, 6 (1966).

17. W. O. McReynolds, J. Chromatogr. Sci., 8, 685 (1970).

18. Applied Science Laboratories, Inc., Catalog 18, 1975, pp. 18-20.

19. J. J. Leary, J. B. Justice, S. Tsuge, S. R. Lowry, and T. L. Isenhour, J. Chromatogr. Sci., 11, 201 (1973).

20. S. Wold and K. Andersson, J. Chromatogr., 80, 43 (1973).

21. D. H. McCloskey and S. J. Hawkes, J. Chromatogr. Sci., 13, 1 (1975).

22. H. De Clercq, D. Van Oudheusden, and D. L. Massart, Analusis, 3, 500 (1975).

23. S. J. Hawkes, D. Grossman, A. Hartkopf, T. L. Isenhour, S. Leary, J. Parcher, S. Wold, and J. Yancey, J. Chromatogr. Sci., 13, 105 (1975).

24. P. F. Dupuis and A. Dijkstra, Anal. Chem., 47, 379 (1975).

25. A. Eskes, P. F. Dupuis, A. Dijkstra, H. De Clercq, and D. L. Massart, Anal. Chem., 47, 2168 (1975).

26. A. Hartkopf, S. Grunfeld, R. Delumeya, J. Chromatogr. Sci., 12, 124 (1974).

27. M. Detaevernier, personal communication.

28. A. M. Gattuso and G. Fazio, Ind. Agrar., 12, 3 (1974).

29. B. R. Kowalski, Anal. Chem., 47, 1152A (1975).

30. J. Op de Beeck and J. Hoste, Analyst, 99, 973 (1974).

31. H. De Clercq, D. L. Massart, L. Dryon, J. Pharm. Sci., 66, 1269 (1977).

Chapter 4

CHROMATOGRAPHY OF OLIGOSACCHARIDES AND RELATED COMPOUNDS ON ION-EXCHANGE RESINS

Olof Samuelson

Department of Engineering Chemistry
Chalmers University of Technology
Goteborg, Sweden

I. INTRODUCTION 114
II. PARTITION CHROMATOGRAPHY OF OLIGOMERIC
   SUGARS AND ALDITOLS 114
   A. General Considerations 114
   B. Choice of Resin 118
   C. Influence of Eluent Concentration and
      Structure of the Oligomers 119
   D. Aspects of Practical Separations 124
   E. Aspects of Identification 127
III. SEPARATIONS OF OLIGOSACCHARIDES
   AS BORATE COMPLEXES 130
IV. PERMEATION CHROMATOGRAPHY OF OLIGO-
   SACCHARIDES IN AQUEOUS SOLUTION 131
V. ANION-EXCHANGE CHROMATOGRAPHY
   OF OLIGOMERS WITH A TERMINAL
   ALDONIC ACID GROUP 134
   A. Correlation between the Distribution Coefficients
      and the Number of Monomeric Units 134
   B. Influence of Temperature 138
   C. Separations for Analytical Purposes 138
   D. Separations for Preparative Purposes 140

VI. ANION-EXCHANGE CHROMATOGRAPHY
   OF OLIGOMERS WITH ONE
   URONIC ACID MOIETY 143

VII. ANION-EXCHANGE CHROMATOGRAPHY
   OF DICARBOXYLIC OLIGOMERS 144

VIII. ANION-EXCHANGE CHROMATOGRAPHY
   OF OLIGOGALACTURONIC ACIDS 146

   REFERENCES 147

I. INTRODUCTION

Chromatographic separations of oligosaccharides and related compounds such as oligomers containing uronic acid moieties have become increasingly important in studies of the structure of polysaccharides. Structural details of parent polysaccharides are often revealed by characterization of the oligomers obtained following their partial hydrolysis. In addition, oligosaccharides are important constituents in food, beverages, and wastewater from industries which produce compounds containing polysaccharides, e.g., the cellulose industry. Paper chromatography [1] and adsorption chromatography on charcoal-Celite [2] are widely used in separations for preparative and analytical purposes. Efficient separations which depend on differences in the degree of polymerization of the oligosaccharides can be achieved by gel permeation chromatography [3].

Recent investigations have shown that with complex mixtures of oligosaccharides, improved separations can be obtained by partition chromatography on ion-exchange resins. The main purpose of this chapter is to elucidate this technique as well as to review recent developments in separations of oligomeric acids by anion-exchange chromatography.

II. PARTITION CHROMATOGRAPHY OF OLIGOMERIC SUGARS AND ALDITOLS

A. General Considerations

Early investigations by Samuelson and co-workers showed that sugars can be retained effectively both on anion-exchange resins [4] and on cation-exchange resins [5] in aqueous ethanol and that these systems can be used for chromatographic separations [4, 6]. Studies of disaccharides and some higher oligomers showed that the elution order was reversed compared to that observed in aqueous solution, i.e., that with few exceptions the distribution coefficients increase with increasing number of sugar moieties in the oligomers [7].

Figure 1 shows the separation of four monosaccharides and five disaccharides on a cation-exchange resin in the lithium form in 85% (wt/wt) ethanol. The separation of all species required about 5 hr. The pressure drop was 4.0 MPa, implying that it would be possible to speed up the separation without introducing difficulties by using finer resin beads [9].

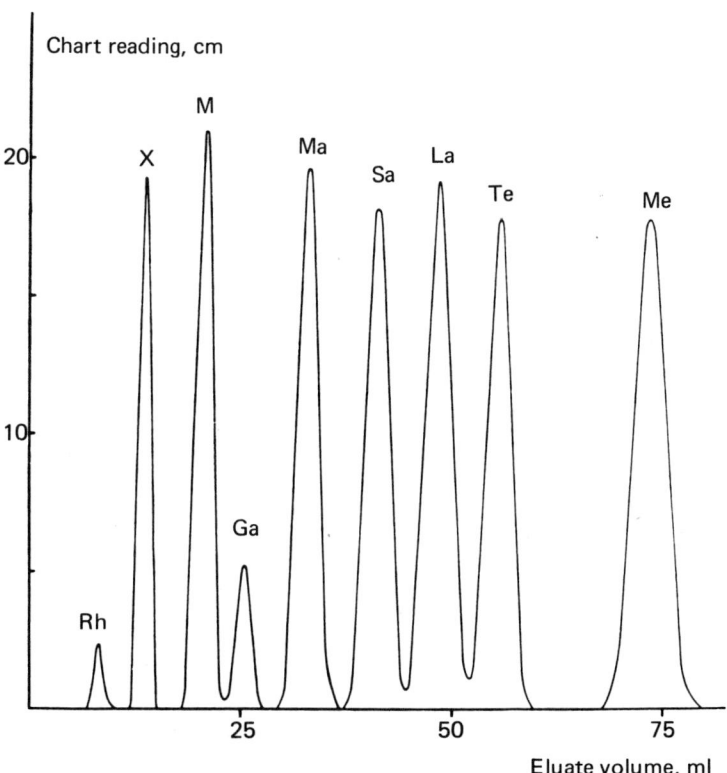

FIG. 1. Separation of various mono- and disaccharides at 75°C in 85% ethanol. Resin bed: 2.6 × 1225 mm, Dowex 50W-X8, Li^+, 14 to 17 μm. Nominal linear flow: 5.2 cm/min (calculated for an empty column). Rh, rhamnose (2 μg); X, xylose (20 μg); M, mannose (70 μg); Ga, galactose (10 μg); Ma, maltose (100 μg); Sa, saccharose (100 μg); La, lactose (100 μg); Te, trehalose (100 μg); Me, melibiose (100 μg). (From Samuelson and Stromberg [8]. Courtesy of Zeitschrift fur analytische Chemie.)

As shown in Fig. 2 members of the same oligomeric series are eluted in order of increasing number of monomeric moieties in the compound (DP). This has been confirmed for several oligomeric series.

In aqueous ethanol the mole fraction of water inside the resin is much higher than that in the external solution [11]. Strongly polar solutes prefer the water-rich phase and therefore their concentration is higher in the resin phase than in the external solution. The distribution coefficients calculated from solubility data and determination of the solvent composition in the two phases differ markedly from the observed values [12, 13]. Hence, the resin cannot be considered as an inert solid support. Under the conditions used in most chromatographic work the observed distribution coefficients are significantly higher than the values calculated from the solubility ratio. This can be explained by the combined effect of solvent-resin interactions [14] and interactions between the ions in the resin and polar solutes. Hence, an exchange of counterions in the resin can affect the elution order of many sugars [15]. In many systems of practical interest these "salting-in"

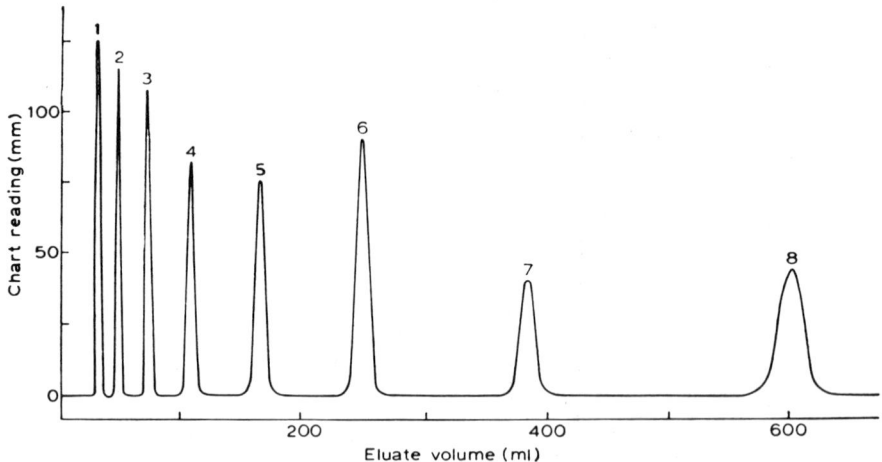

FIG. 2. Partition chromatography of xylan oligosaccharides in 75% ethanol at 75°C. Resin bed: 4 × 600 mm, Technicon T5C, SO_4^{2-}, 14 to 17 μm. Nominal linear flow: 2.8 cm/min. 1, D-xylose (5 μg); 2, di- (5 μg); 3, tri- (6.5 μg); 4, tetra- (9 μg); 5, penta- (13 μg); 6, hexa- (25 μg); 7, hepta- (18 μg); and 8, octa-saccharide (25 μg). (From Havlicek and Samuelson [10]. Courtesy of Carbohydrate Research.)

effects are much more important than the effect of the pressure-volume term in the Gibbs-Donnan equation (see Sec. IV).

It is well known that even anomeric sugars, e.g., α-D-glucose and β-D-glucose, have different solubilities and that an equilibrium solution of glucose contains both anomers. In practice these can be well separated at low temperature, e.g., -10°C [16]. A similar separation of α-maltose and β-maltose is illustrated in Fig. 3.

In most practical separations it is desirable to obtain all forms of the same sugar as a single peak. This can be achieved by carrying out the separation at elevated temperature (75-85°C) so that the different forms of a single sugar are rapidly interconverted during their migration through the resin bed. Another advantage gained by working at higher temperatures is an increased diffusion rate, which is even more important with oligomers than with monomers. Under conditions used in practical separations the rate-determining mass-transfer step is the diffusion inside the resin particles. Rapid separations can therefore only be made at elevated temperature and with fine resin particles. This leads to a beneficial decrease in the viscosity of the mobile phase and permits separations to be made without an excessively high pressure drop in the column.

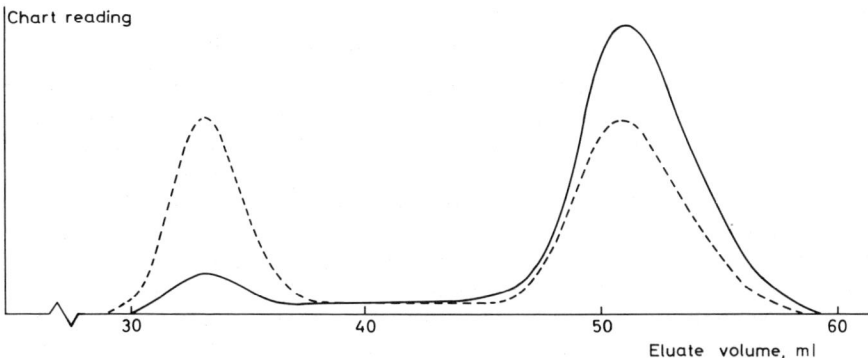

FIG. 3. Chromatography of D-maltose (500 µg) at -10°C in 75% ethanol on an anion exchanger in the sulfate form (Technicon T5C, 10 to 17 µm). Resin bed: 4 × 131 mm. Nominal linear flow: 0.53 cm/min. Solid line: Crystalline D-maltose dissolved in water (+2°C) and chromatographed immediately after addition of ethanol (-20°C). Dashed line: The same sample solution chromatographed after 24 hr at room temperature. (O. Ramnas and O. Samuelson, unpublished chromatogram.)

B. Choice of Resin

In common with other separations on ion-exchange resins, partition chromatography of oligosaccharides is preferably made in columns packed with a slurry of fine resin beads of uniform size. Commercially available resins having a styrene-divinylbenzene matrix and with a nominal cross-linking of 8% (8% divinylbenzene, DVB) have been used in most chromatographic work. Successful separations of the monosaccharides of interest in studies of wood pulps on strongly basic anion exchangers in the sulfate form have made this type of resin the most popular in our practical work. Investigations of other sugars showed that a few monosaccharides which could not be resolved on this type of resin were well separated on a cation-exchange resin of sulfonic acid type in the lithium form [15]. In many applications the choice between a cation exchanger and an anion exchanger is less critical than is the choice of counterions. Evidently, specific interactions between the polar solutes and the counterions exert a great influence on the distribution coefficients.

If all oligomers to be separated belong to the same oligomeric series, the separations can be conveniently made on either an anion exchanger in the sulfate form or a cation exchanger in the lithium form, and probably also on resins with other counterions. In separations of oligomers belonging to different series these two resin types supplement each other. If only the terminal reducing moieties differ, that resin which gives the best separation of the monosaccharides present at the reducing end is to be recommended.

The effect of the nominal cross-linking of the resin on the volume distribution coefficients (D_V) of oligomers of the isomaltose series is shown in Fig. 4. (D_V is defined as the total amount of a solute in the ion exchanger calculated per milliliter of bed volume divided by the analytical concentration in the external solution. It can be calculated from the peak elution volume $\overline{V}$, the bed volume X, and the interstitial fraction ϵ_I by the equation $D_V = \overline{V}/X - \epsilon_I$). All resins were of the benzyltrimethylammonium type and used in their sulfate form. The X4 resins (4% nominal DVB) exhibited higher distribution coefficients for glucose and its oligomers than did resins with a higher degree of cross-linking [17]. This is in agreement with studies of the equilibrium distribution of glucose by Ruckert and Samuelson [12] who ascribed this higher distribution coefficient to the fact that at a given ethanol concentration in the external solution the mole fraction of water was higher in a resin of low cross-linking than in a resin with 8% nominal DVB. It is seen that at constant ethanol concentration in the external solution the incremental increase in log D_V for each successive oligomer (and hence the separation factor) is greater for resins with a lower degree of cross-linking.

As shown in Figs. 5 and 6, a similar effect can be obtained by increasing the ethanol concentration. This means that in separations of small amounts of lower oligomers for analytical purposes, no practical advantages are gained

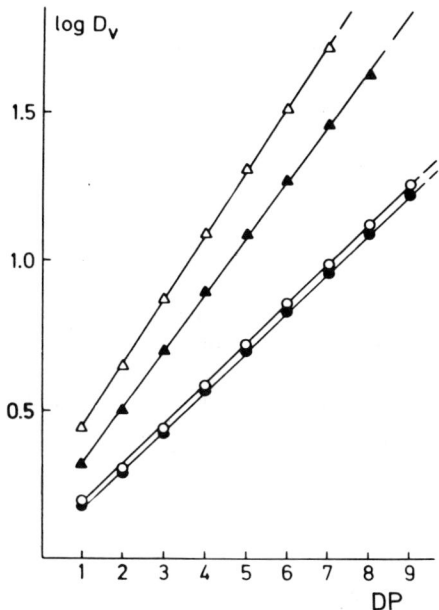

FIG. 4. Effect of the degree of cross-linking on the distribution coefficients of α-(1 → 6)-linked D-glucosides. Anion-exchange resins in the sulfate form, 60% ethanol, 75°C. △, Durrum DA-X4, 15 to 25 μm; ▲, Bio-Rad Ag 1-X4, 15 to 35 μm; ●, Dowex 1-X8, 12 to 18 μm; and ○, Dowex 1-X10, 14 to 17 μm. (From Havlicek and Samuelson [17]. Reprinted with permission from Analytical Chemistry. Copyright by the American Chemical Society.)

with lightly cross-linked resins. With oligomers with a DP of about 10 and higher, precipitation occurs at the ethanol concentration required for efficient separations on resins with 8% nominal DVB. In such cases resins of lower cross-linking can be used to advantage [17].

The application of ion exchangers with other matrices has been briefly studied with monosaccharides [18] but their usefulness in the chromatography of oligomers has not been extensively investigated.

C. Influence of Eluent Concentration and Structure of the Oligomers

As predicted from determinations of the ethanol distribution between stationary and mobile phases and solubility measurements, the distribution coefficients of sugars and alditols increase with an increased ethanol concentration in the eluent [13]. The effect of the ethanol concentration for the oligomers of the β-(1 → 4)-linked D-xylose series is shown in Fig. 5.

At constant eluent composition a linear relationship exists between log D_V and the number of monosaccharide units (DP) in oligomers belonging to the same series. This rule has been confirmed for all oligomeric sugars and alditols studied [10, 17, 19, 20]. Evidently, the slope of these lines is equal to the logarithm of the separation factor (log α = log $D_{V(n+1)}$ − log $D_{V(n)}$) for successive oligomers DP = n and DP = n + 1.

Since D_V is an equilibrium distribution coefficient the results show that each new sugar moiety in an oligomer adds a constant increment [$-\Delta(\Delta G°)$] to the change in free energy of the sorption process. Hence the slope is proportional to $\Delta(\Delta G°)$. It can therefore be anticipated that in oligomeric series with the same type of glycosidic linkages and the same sugar moieties

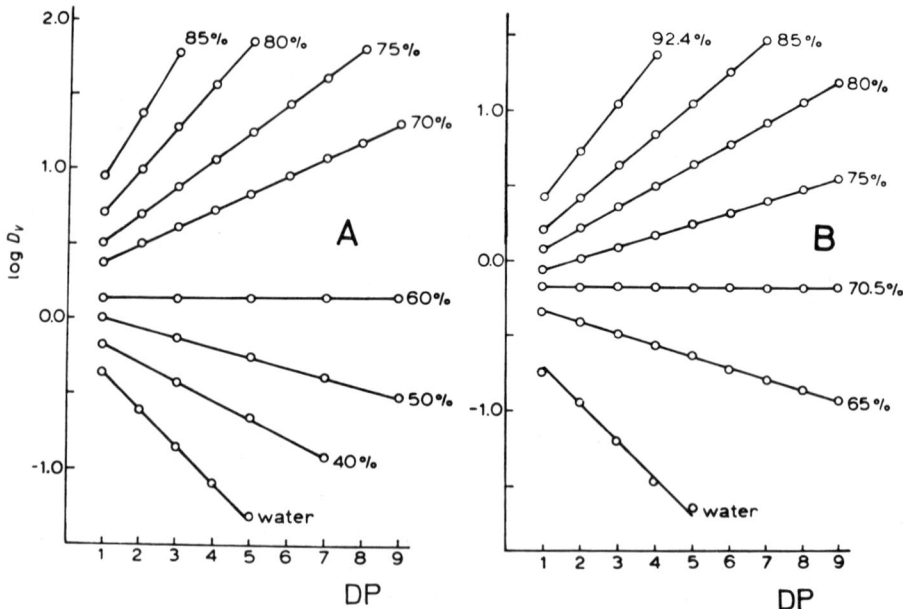

FIG. 5. Relationship between log D_V (dekadic logarithm) at 75°C and number of D-xylose residues in the oligosaccharides of the β-(1 → 4)-linked D-xylose series at various concentrations of ethanol. (A) Sulfate resin (Technicon T5C, 14 to 17 μm); (B) lithium resin (Dowex 50W-X8), 17 to 21 μm). (From Havlicek and Samuelson [10]. Courtesy of Carbohydrate Research.)

in the oligomeric chain, the terminal moiety will not affect the slope of this straight line. This was confirmed in experiments using oligomers of the β-(1 → 4)-linked xylose series in which the terminal moiety was reduced to a xylitol group (Fig. 6). The same result was found in analogous experiments with β-(1 → 3)-, β-(1 → 4)-, and α-(1 → 6)-linked glucose oligomers both before and after the reduction of the terminal moiety to a glucitol end group [21].

Analogously, the D_V values are changed whereas the slopes remain constant when the terminal moiety is isomerized. This was confirmed in experiments with β-(1 → 4)-linked oligomers of xylose and the isomerization products containing lyxose and xylulose end groups (Fig. 7). Among the monosaccharides, xylose fitted into the straight line whereas the D_V values for lyxose and xylulose were significantly lower. The elution order of the

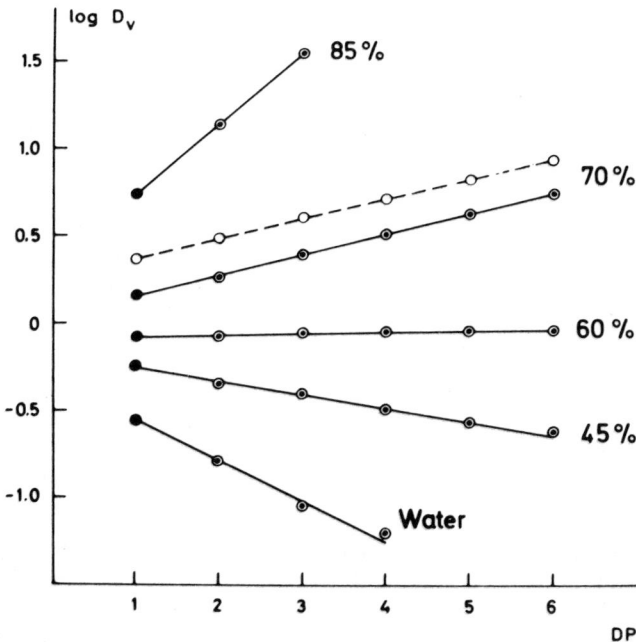

FIG. 6. Relationship between log D_V at 75°C and number of monomeric units in oligomers at various concentrations of ethanol. Sulfate resin (Technicon T5C, 10 to 17 μm). ○, Xylose and oligosaccharides of β-(1 → 4)-linked D-xylose series; ●, xylitol; ◉, oligomeric sugar alcohols of O-β-D-Xylp-(1 → 4)-[O-β-D-Xylp-(1 → 4)]$_{DP-2}$-D-xylitol series. (From Havlicek and Samuelson [21]. Courtesy of Chromatographia.)

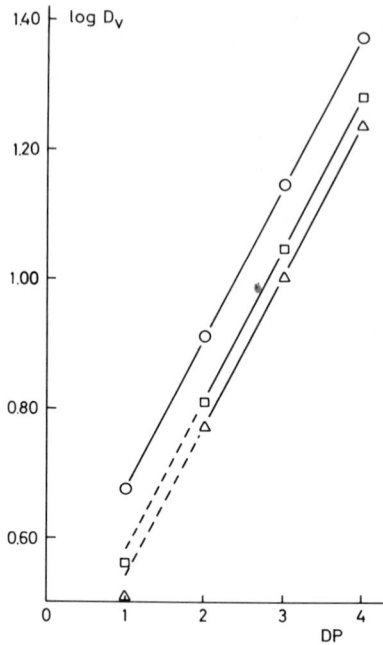

FIG. 7. Relationship between log D_V at 75°C and DP of β-(1 → 4)-linked xylose oligomers with terminal reducing (O) xylose, (□) lyxose, and (△) xylulose moieties. Sulfate resin (Technicon T5C, 10 to 15 μm), 78% ethanol. (M. Johansson and O. Samuelson, unpublished work.)

monosaccharides was, however, the same as that of the corresponding oligosaccharides. Evidently, $-\Delta(\Delta G°)$ for the first glycosyl moiety in reducing oligosaccharides is affected by the reducing end group whereas for glycosyl moieties more distant from this end group $-\Delta(\Delta G°)$ is constant. Similar effects were observed with oligomeric alditols [21].

In contrast, both the type of glycosyl moiety and the type of glycosidic linkage have a great influence on the slope. Experiments with glucose oligomers belonging to different oligomeric series indicate that the slope increases markedly with increasing length of the oligomer compared at a given DP [17]. At low ethanol concentration the differences are small, and in pure water the differences in D_V between oligomers of the same DP are insignificant. Since the partial molar volume in water is virtually unaffected by the type of glycosidic linkage [22], these results are in agreement with those referred to in Sec. IV.

Within the principal range of interest in chromatographic separations of oligomers belonging to the same series, the slope of log D_V versus DP is positive and increases with increasing ethanol concentration. The steeper the slope, the more favorable are the separation factors. On the other hand, an excessively high ethanol concentration leads to very high D_V values and hence slow separations of the highest oligomers. A compromise choice of ethanol concentration is therefore made. Figure 5 shows that at a given ethanol concentration both the slope and the D_V of the sugars are much higher for the sulfate resin than for the lithium resin. A higher ethanol concentration is therefore recommended for separations on a lithium resin than for those on a sulfate resin (Table 1).

This linear relationship is also valid at low ethanol concentrations where the concentration of the sugar inside the resin is lower than that in the external solution ($D_V < 1 - \epsilon_I$, where ϵ_I is the interstitial fraction, approximately 0.39). As a consequence a critical eluent concentration should exist at which all members of one oligomeric series exhibit the same distribution coefficient and hence appear in the same elution peak [10].

For the β-(1 → 4)-linked xylose oligomers the critical ethanol concentration at 75°C was 60% for the sulfate resin. At this concentration all xylose oligomers were eluted at D_V 1.3. For the lithium resin the critical concentration was 70% and the corresponding D_V equal to 0.7. This means that the concentration distribution ratio D_C (amount of solute per milliliter

TABLE 1

Recommended and Critical Ethanol Concentrations (% wt/wt) at 75°C for Separation of Saccharides of the Same Oligomeric Series*

	Recommended concentration		Critical concentration	
	Li^+	SO_4^{2-}	Li^+	SO_4^{2-}
β-(1 → 4)-linked D-xylose	80-85	70-75	70	60
β-(1 → 3)-linked D-glucose	82-88	72-78	77	65
α-(1 → 4)-linked D-glucose	78-83	68-73	69	60
β-(1 → 4)-linked D-glucose	78-83	68-73	70	59
α-(1 → 6)-linked D-glucose	70-75	60-65	62	49

*Li resin: Aminex A-6, 15 to 19 μm; SO_4 resin: Dowex 1-X8, 12 to 18 μm.

of swollen resin divided by the amount per milliliter of external solution) was 2.1 for the sulfate resin and 1.1 for the lithium resin.

As in the case of the slope (log α), it can be anticipated that the critical concentration, i.e., that concentration at which $\Delta(\Delta G°) = 0$, should depend both on the type of glycosyl moiety and the type of glycosidic linkage but be independent of the terminal moiety. The greater the slope at a given ethanol concentration, the lower the critical ethanol concentration should be. For oligomers with the same sugar moieties the critical ethanol concentration should therefore decrease with increasing length of the oligomer compared at the same DP. The results given in Tables 1 and 2 confirm these predictions.

Contrary to the results described for oligomeric sugars the concentration distribution ratio, D_c, of oligomeric alditols at the critical ethanol concentration was significantly higher for the lithium resin than for the sulfate resin. For all species, except for the α-(1 → 6)-linked D-glucose-D-glucitol series on the sulfate resin, the molar concentration inside the resin was significantly higher than that in the external solution.

Studies of the influence of temperature showed that as in the case of monosaccharides and alditols, the oligomers are eluted much earlier at high temperatures [17].

Both the D_v of the disaccharides and the slopes decrease markedly with increasing temperature. As expected the critical concentration is lowered when the temperature is decreased. This can in part be ascribed to an increased mole fraction of water in the resin phase.

D. Aspects of Practical Separations

As discussed earlier, the separation factor for oligomers of the same series is constant and increases with increasing ethanol concentration and decreasing temperature. For reasons already mentioned a high temperature, e.g., 75 to 90°C, is recommended. Certain sugars undergo detectable decomposition at 90°C [8] and for this reason most separations have been made at 75°C. If it is desired to obtain a complete separation of the monosaccharide and disaccharide, the time for eluting the highest oligomers is very long (e.g., 22 hr for the chromatogram reproduced in Fig. 2). At the same time, however, the separation does not require any attendance and can be carried out to advantage during the night. If a slight overlapping of the first two peaks can be tolerated, the time for the elution of the last compound can be reduced to about one-third, e.g., by decreasing the column length. It is also possible to reduce the time by decreasing the particle size and increasing the pressure in the column [9]. When sugars belonging just to one oligomeric series are present the time can often be shortened by

TABLE 2

Critical Ethanol Concentration (% wt/wt) and Distribution Coefficients for Various Oligomeric Alditols at 75°C[a]

Oligomeric series	Critical ethanol concentration (%)		Concentration distribution ratio, D_c, at the critical concentration[b]		Volume distribution coefficient, D_V, of the aldobiitol			
					85% ethanol		70% ethanol	
	Li^+	SO_4^{2-}	Li^+	SO_4^{2-}	Li^+	SO_4^{2-}	Li^+	SO_4^{2-}
β-(1 → 4)-linked D-xylose–D-xylitol	71	59	1.54	1.34	4.70			1.81
β-(1 → 3)-linked D-glucose–D-glucitol	77	64	4.10	3.20	12.1			4.80
β-(1 → 4)-linked D-glucose–D-glucitol	71	59	2.70	2.10	12.0			4.05
α-(1 → 6)-linked D-glucose–D-glucitol	62	49	1.16	0.98	15.6			4.26

[a] Data from Havlicek and Samuelson [21].
[b] $D_c = D_V/(1 - \epsilon_I)$.

using a cation-exchange resin in the lithium form instead of an anion exchanger in the sulfate form [10].

Gradient or stepwise elution with decreasing concentration of ethanol can be applied to speed up the elution of the higher oligomers, but it should be emphasized that only small changes in ethanol concentration can be made without disturbances occurring in the resin bed due to swelling changes [11]. In separations of large amounts for preparative purposes, and in analyses in which several monosaccharides are present, it is recommended to elute the monosaccharides together with the lowest oligomers as a group at low ethanol concentration (e.g., 55-70% on a sulfate resin, depending on the type of oligomers) and subsequently to rechromatograph these lower oligomers at higher eluent concentration (cf. Table 1). If it is desired to isolate several grams of individual oligomers, it can be advantageous to separate crude fractions on charcoal-Celite and to purify these by partition chromatography on an ion-exchange resin [10].

Compared to gel permeation chromatography on cross-linked dextran or polyamides in water, partition chromatography has the disadvantage that it is limited to oligomers soluble in aqueous ethanol. On the other hand, a much better resolution is achieved. This is important both in preparative and accurate analytical work. Its greatest advantage is, however, that it permits the separation of complex mixtures of oligomers which differ in chemical structure.

The linear relationship between log D_V and DP of the oligomers permits the prediction of the positions of all oligomers of a given series on a chromatogram, provided the positions of two members of the series (e.g., of the disaccharide and one higher oligomer) are known. For comparison of retention data it is recommended to tabulate the D_V of the disaccharide and the slope of the straight line. Calculations based on published results [17] show, for instance, that several oligomers of the β-(1 → 4)-linked D-xylose series and those of the β-(1 → 4)-linked D-glucose series (both of interest in studies of wood and other materials of botanical origin) can be well separated in a single run. The same is true for the α-(1 → 4)-linked D-glucose (maltose) series and the α-(1 → 6)-linked D-glucose (isomaltose) series. With complex mixtures overlapping of some compounds will occur, but these can be resolved by rechromatography at a higher or lower ethanol concentration or on an alternative resin.

The situation is simplified when the straight lines are parallel, which as already mentioned occurs for oligomers that differ only in their end groups.

The application of partition chromatography on an anion exchanger in the sulfate form, for the separation of the oligosaccharides referred to in Fig. 7, is illustrated by the chromatogram in Fig. 8. It is taken from a recent study of the kinetics of the endwise degradation of xylotetraose in

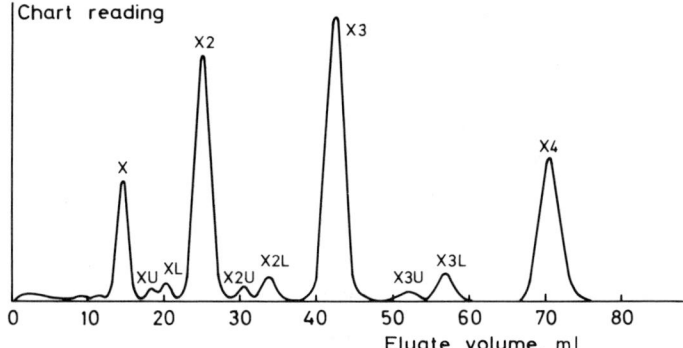

FIG. 8. Separation of isomerization and degradation products formed during alkali treatment of xylotetraose (X4) by partition chromatography in 78% ethanol at 75°C on the sulfate form on an anion exchanger. Resin bed: 2.8 × 476 mm, Technicon T5C, 10 to 15 μm.

alkaline medium [20]. The last peak on the chromatogram corresponds to the remaining xylotetraose (X4). The major reaction products were xylotriose (X3), xylobiose (X2), and xylose (X). The isomerized tetrasaccharide with a xylulose end group (X3U) is an important intermediate which rapidly decomposes in alkaline solution. It could be determined quantitatively, as could the tetrasaccharide with a lyxose end group (X3L) and the isomerized trisaccharides, X2U and X2L, with xylulose and lyxose end groups. The isomerized disaccharides XU and XL were also recorded. A quantitative determination of these species was not necessary for the kinetic study and the conditions were chosen so that these solutes overlapped. The time required to record this chromatogram was 6 hr.

With complex mixtures containing members of different oligomeric series, advantage can be taken of the fact that all members of one series appear in a single peak at the critical ethanol concentration. Figure 9 illustrates the separation of isomaltose (II) and isomaltotriose (III) as separate peaks from a peak (I) containing glucose, maltose, maltotriose, and maltotetraose. Separation was performed on a sulfate resin at the critical concentration for the maltose series. The sugars contained in the first peak were then rechromatographed at a higher ethanol concentration e.g., 70%. A complete separation of all sugars was obtained by this two-stage method.

E. Aspects of Identification

The linear relationship between log D_V and DP for oligomers having the same glycosyl moieties and the same type of glycosidic linkages, together

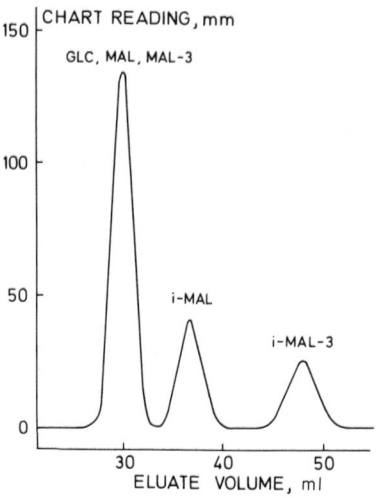

FIG. 9. Separation of oligosaccharides of maltose and isomaltose series in 60% ethanol at 75°C. Resin bed: 4 × 1300 mm, Dowex 1-X8, SO_4^{2-}, 8 to 13 μm. Nominal linear flow: 1.7 cm/min. Glucose, maltose, maltotriose, isomaltose, and isomaltotriose, 10 mg of each. (From Havlicek and Samuelson [17]. Reprinted with permission from Analytical Chemistry. Copyright by the American Chemical Society.)

with the fact that the slope of this line is independent of the terminal (reducing or reduced) moiety, is of great value for identification purposes. The results suggest that it should be possible to predict the D_V of oligomers containing different types of glycosidic linkages [e.g., of oligomers linked by both α-(1 → 4)- and α-(1 → 6)-glucosidic linkages] from the D_V values of the disaccharides and the slopes determined for each series of oligosaccharides with only one type of linkage (Fig. 10). Plots of this type have been used in studies of the oligosaccharides present in beer and wort [23].

In addition the application of multichannel analyzers in recording the eluate composition facilitates the identification of the chromatographic peaks. Hence, all sugars and oligomers can be automatically recorded with the orcinol method [24] or a differential refractometer. Sugars containing reducing ketose end groups [25] give a strong response in the periodate-formaldehyde channel [26] (oxidation with periodate and subsequent colorimetric determination of formaldehyde). The conditions during the periodate oxidation may be advantageously chosen (pH 2) so that aldoses give a negligible response unless present in very large amounts. The simultaneous application of these analytical channels makes it possible to distinguish oligomers with reducing ketose end groups from other oligomeric sugars.

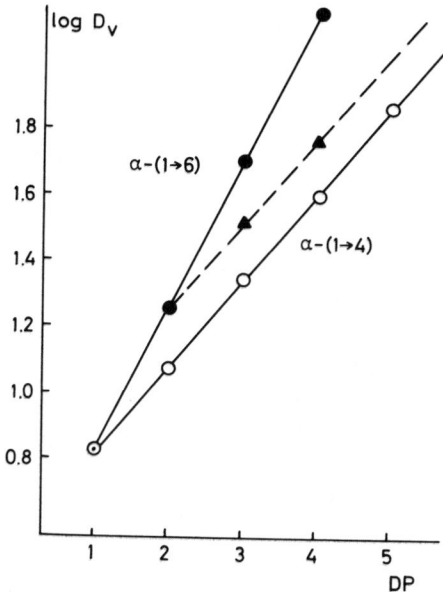

FIG. 10. Relationship between log D_V at 75°C and the number of D-glucose residues for oligomers with different types of glucosidic linkages. Sulfate resin (Durrum DA-X4), 75% ethanol. ●, α-(1 → 6)-linked oligomers; ○, α-(1 → 4)-linked oligomers; ▲, α-(1 → 4)-linked oligomers with one α-(1 → 6) linkage. (J. Havlicek and O. Samuelson, unpublished work.)

An example of the application of the two-channel analyzer in analysis of mixtures of sugars and alditols is given in Fig. 11.

Oligomeric alditols give rise to different amounts of formaldehyde as the position of the glycosidic bond in the alditol moiety is varied. Hence 2 mol of formaldehyde is formed for each mole of 3-O- and 4-O-glycosides of hexitols and 3-O-glycosides of pentitols whereas only 1 mol is formed when another site is blocked by a glycosidic bond. The response in the periodate-formaldehyde channel relative to that in the orcinol (or refractive index) channel gives valuable information about the type of glycosidic linkage [21]. Additional information can be obtained by employing a method in which only reducing sugars are recorded, e.g., the automatic copper-2,2'-bicinchoninate method devised by Mopper and Gindler [27].

In analyses of complex mixtures containing principally unknown compounds, e.g., sewage, it is recommended to isolate the fractions obtained by chromatography on a sulfate resin and to check the purity by rechromatography on a lithium resin. Final identification can then be made by

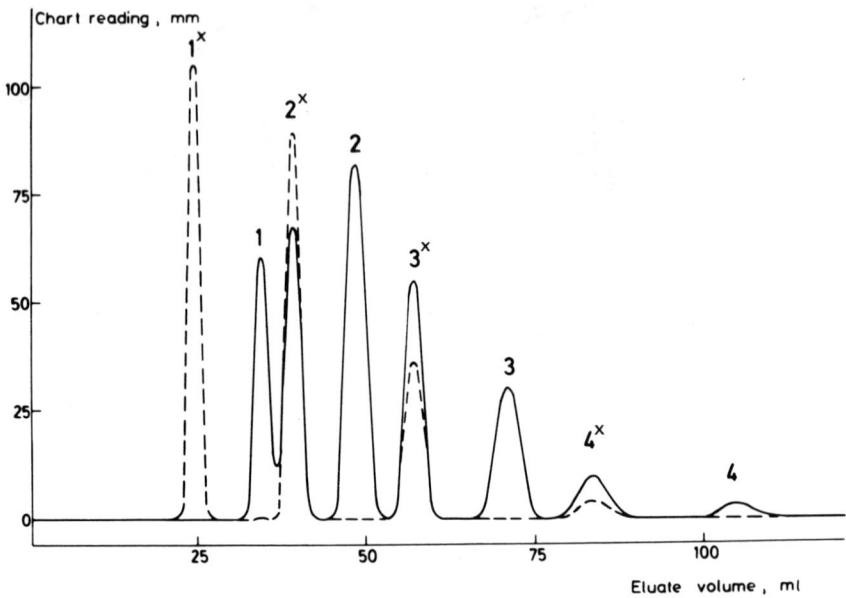

FIG. 11. Separation of a mixture of sugars and alditols at 75°C in 70% ethanol. Resin bed: 4.4 × 720 mm, Technicon T5C, SO_4^{2-}, 10 to 17 μm. Nominal linear flow: 3.2 cm/min. Orcinol method, solid line; periodate-formaldehyde method, dashed line. 1, Glucose (50 μg); 2, maltose (100 μg); 3, maltotriose (50 μg); 1*, glucitol (50 μg); 2*, maltitol (100 μg); 3*, maltotriitol (100 μg); 4, maltotetraose (traces); 4*, maltotetraitol (traces). (From Havlicek and Samuelson [21]. Courtesy of Chromatographia.)

conventional methods such as methylation analysis and analysis of the products formed by acid hydrolysis.

III. SEPARATIONS OF OLIGOSACCHARIDES AS BORATE COMPLEXES

The separation of sugars as borate complexes first suggested by Khym and Zill [28] is widely used for the separation of monosaccharides, alditols [29], and mixtures of monosaccharides with disaccharides and trisaccharides [30-32]. For species of approximately the same molecular size the number and position of vicinal hydroxyl groups have a predominant influence on D_V. The hydroxyl groups in pyranose moieties have little affinity for the borate resin as compared to cis-1,2-diol groupings in a furanose moiety. Hence, maltose, which cannot exist in the furanose form, exhibits a very low affinity for the resin whereas isomaltose is held very strongly. Mutarotation of reducing moieties can enhance the complex formation and contribute to the D_V [29]. This explains the elution order trehalose < cellobiose. For acyclic moieties vicinal hydroxyl groups preferentially in the gauche conformation contribute much more to D_V than comparable moieties with vicinal hydroxyl groups in the anti conformation [33].

Torii and Sakakibara [34] have recently demonstrated the usefulness of the borate method for the separation of glucose and the first six oligomers of the isomaltose [α-(1 → 6)-glucose] series. The compounds which were stepwise eluted at 65°C with three buffers of increasing borate concentration and pH emerged in the order of decreasing DP. The species with DP 1 to 6 were well separated whereas the hexaose and heptaose exhibited serious overlapping in a run which required 5 hr. The observed elution order indicates that only the terminal reducing moiety gives strong complexes with borate.

This and other separations [35] carried out recently at elevated temperature show that in many systems the risk of isomerization is less than reported by Carubelli [36]. A comparison between the results currently available shows that an equilibration of reducing sugars with the eluent should be avoided before application of the sample to the column.

For oligomers of different types, the borate complexing can have a greater effect on the elution order than can molecular size. Hence, sucrose, raffinose, and stachyose are eluted in that order [37, 38].

As in other separations in aqueous medium chromatography in borate solution has the advantage that gradient elution or stepwise elution can be applied without disturbance to column packing. On the other hand, reconditioning of the column is necessary when these elution methods are employed. In analysis of complex mixtures of oligosaccharides the borate method and partition chromatography in aqueous ethanol will supplement each other. With unknown mixtures containing several monosaccharides it is recommended to use partition chromatography to remove the monosaccharides before an attempt is made to separate oligosaccharides in borate medium.

IV. PERMEATION CHROMATOGRAPHY OF OLIGOSACCHARIDES IN AQUEOUS SOLUTION

Permeation chromatography on ion-exchange resins in water or non-complexing salt solutions is less efficient in practical separations of oligomers than the methods referred to previously. In common with gel permeation chromatography on uncharged gels chromatography in water has the advantage of simplicity but the disadvantage that all species are eluted within narrow range. When resins with 8% nominal DVB are used pentaoses and higher oligomers are eluted at a peak elution volume close to the interstitial volume and monomers at a volume much less than the total column volume.

As an example it can be mentioned that on an anion exchanger in the acetate form, the D_V was 0.01 for xylopentaose and 0.23 for xylose [39]. Although the separation factors are very favorable it is difficult to obtain a complete separation on a column of reasonable size unless the conditions are such that very sharp peaks are obtained. Small resin particles, low flow rate, low column loading, and an analytical system that causes a minimum broadening are prerequisites for a complete separation of the recorded peaks. Practical separations of lower oligomers have been reported [40, 41]. Recycling chromatography can be used to increase the efficiency of a given column.

The distribution coefficients increase with decreasing degree of cross-linking. Improved separations can therefore be obtained by chromatography on a lightly cross-linked resin. As shown by Larsson and Samuelson [39] the acetate form of an anion-exchange resin, with nominal 2% DVB, can be used for a complete separation of the oligomers of DP 1 to 5 belonging to the xylobiose series. The first 15 oligomeric acids of the xylobionic acid series [xylobionic acid with an increasing number of β-(1 → 4)-linked xylose moieties] were separated in the same run. With the analytical method employed (automatic colorimetric analysis with orcinol) a baseline separation was achieved only when the flow rate was very low. Equally important is that the amounts of the higher oligomeric sugars are also low. Figure 12 shows a chromatogram from the same investigation. In this run the amounts of the sugars were increased by a factor of 3 while the other conditions were the same as in the experiment in which a complete separation was achieved. It is seen that the first peaks overlapped seriously, and that even the second and the third peaks gave some overlapping.

As shown in Figs. 5, 6, and 13 a linear relationship exists between the logarithm of the distribution coefficient and the number of monosaccharide moieties in the oligomers (DP). If the resin is considered to be a homogeneous phase, the equilibrium distribution can be described by the Gibbs-Donnan equation [42].

$$\ln \frac{m_r}{m} = - \frac{\pi \bar{v}}{RT} - \ln \frac{\gamma_r}{\gamma}$$

where m_r/m is the ratio of the molality in the resin phase to that in the external solution, π the swelling pressure, $\bar{v}$ the partial molar volume of the solute, and γ_r/γ the activity coefficient ratio. When dilute solutions are chromatographed the activity coefficients in the external solution can be neglected.

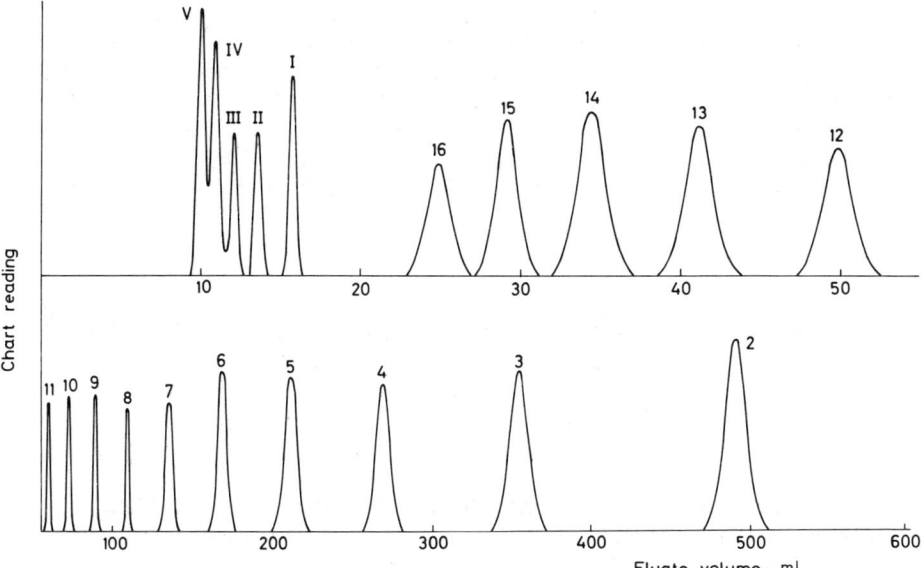

FIG. 12. Separation of xylose (I), β-(1 → 4)-linked xylose oligomers of DP II to V, and aldonic acids of the O-β-D-xylopyranosyl-(1 → 4)-[O-β-D-xylopyranosyl-(1 → 4)]$_{DP-2}$-D-xylonic acid series at 30°C in 0.02 M sodium acetate on a lightly cross-linked anion-exchange resin. Applied amounts of sugars and acids ~3 µg of each compound. Column: 4.5 × 1290 mm, Dowex 1-X2, acetate, 23 to 40 µm. Nominal linear flow: 0.21 cm/min for the elution of sugars; 0.43 cm/min for the elution of aldonic acids of DP 16 to 8; 2.7 cm/min for aldonic acids of DP 7 to 2. (K. Larsson and O. Samuelson, unpublished chromatogram.)

The partial molar volume of oligosaccharides is linearly related to DP [22]. Hence, we have

$$\bar{v} = \bar{v}_2 + \Delta\bar{v}(DP - 2)$$

where $\bar{v}_2$ refers to the disaccharide and $\Delta\bar{v}$ to the incremental change for each additional sugar moiety. Since D_v is proportional to m_r/m the equation of the straight line starting with the disaccharide (subscript 2) can be written

$$\ln D_v - \ln D_{v(2)} = -\frac{\pi}{RT}\Delta\bar{v}(DP - 2) - \ln \frac{\gamma_r}{\gamma_{r(2)}}$$

The straight-line relationship requires that the activity coefficient term should be either very small or proportional to DP - 2. To decide unambiguously if it is permissible to neglect the activity coefficient term, determinations of the swelling pressure should be made for the same batch of resin as used in studies of the distribution coefficients. Unfortunately, no results from such experiments are available. The calculation of the swelling pressure from chromatographic retention data is discussed in the following section.

V. ANION-EXCHANGE CHROMATOGRAPHY OF OLIGOMERS WITH A TERMINAL ALDONIC ACID GROUP

A. Correlation between the Distribution Coefficients and the Number of Monomeric Units

According to the Gibbs-Donnan theory, the selectivity coefficient of the anion-exchange equilibrium between aldonate ions (A) and acetate ions (Ac) is given by the equation

$$\ln k_{A/Ac} = \frac{\pi}{RT} (\bar{v}_{Ac} - \bar{v}_A) + \ln \frac{\gamma_{r(Ac)}}{\gamma_{r(A)}} + \ln \frac{\gamma_A}{\gamma_{Ac}}$$

where

$$k_{A/Ac} = \frac{[A]_r}{[A]} \cdot \frac{[Ac]}{[Ac]_r}$$

and the other symbols are as defined in the preceding section.

Experiments with a large number of aldonic acids on a resin having a normal degree of cross-linking (8% DVB) showed that the distribution coefficients in 0.08 M sodium acetate were in close agreement with those calculated from experiments in 0.02 M solution by dividing by 4. Hence, the ratio γ_A/γ_{Ac} can be taken as unity, which means that the last term in the Gibbs-Donnan equation can be omitted [43].

In chromatography with trace amounts of aldonate ions in the presence of a large excess of acetate D_V is proportional to $k_{A/Ac}$. It is reasonable to assume that starting with aldobionate ions (DP = 2) an increase in DP within a series of oligomeric acids corresponds to the same increase in partial molar volume as observed for the corresponding oligomeric sugar. Hence, the Gibbs-Donnan equation can be rewritten

$$\ln D_v - \ln D_{v(2)} = -\frac{\pi}{RT} \Delta \bar{v}(DP - 2) - \ln \frac{\gamma_{r(A)}}{\gamma_{r(2)}}$$

With the exception that the last term refers to the ratio between the activity coefficients of the anions, this equation is the same as that valid for oligomeric sugars. If this term can be neglected or is proportional to DP - 2, the linear relationship between $\ln D_V$ and DP will also exist for oligomeric acids. Investigations by Havlicek and Samuelson [43] show that this is true for the D_V of oligomeric acids belonging to the cellobionic and xylobionic acid series (DP 2-6) on the acetate form of Dowex 1-X8.

The foregoing has been confirmed in a recent investigation (Fig. 13). In addition the D_V values for oligomeric sugars belonging to the same series were determined under identical conditions [39]. It is seen that at both eluent concentrations the observed D_V values of the oligomeric acids of DP 7-9 were greater than those determined by extrapolation of the straight line. The D_V values of these oligomers are, however, extremely low. As an example it can be mentioned that the observed D_V of xylodecaonic acid was

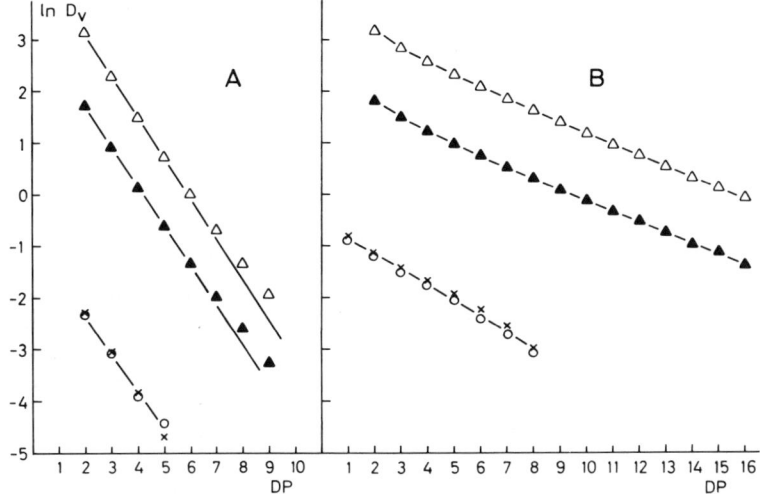

FIG. 13. Relationship between $\ln D_V$ (natural logarithm) at 30°C and the number of monomeric units in oligomers. (A) Dowex 1-X8, acetate, 13 to 20 μm; (B) Dowex 1-X2, acetate, 23 to 40 μm. ○×, Oligomeric sugars of the xylobiose series in 0.02 and 0.08 M sodium acetate; △▲, oligomeric acids of the xylobionic acid series in 0.02 and 0.08 M sodium acetate. (K. Larsson and O. Samuelson, unpublished work.)

0.08 in 0.02 M sodium acetate and 0.02 in 0.08 M solution. A slight error in the determination of the dead volume and the interstitial volume may explain the observed deviation from linearity. The slope of the straight line calculated by the least-squares method was -0.78 for the experiments in 0.02 M solution and -0.77 when the elution was made in 0.08 M sodium acetate.

As expected the D_V of the sugars was virtually unaffected by this change in eluent concentration. Since xylopentaose was almost completely excluded from the resin phase, the slope of the straight line was calculated from the D_V of the three lowest oligomers (DP 2-4). The calculated values (-0.79 in 0.02 M solution and -0.76 when the concentration was 0.08 M) are in close agreement with the slopes determined for the oligomeric acids. Hence, the incremental change in free energy [$\Delta(\Delta G°)$] calculated per monosaccharide moiety in an oligomer (DP > 2) is the same for sugars and aldonic acids with the same type of glycosidic bonds.

In previously published experiments with a lightly cross-linked resin (2% nominal DVB) it was observed that the plot of the logarithm of D_V versus DP for aldonic acids of DP 2 to 16 exhibited slight but significant deviations from linearity [44]. Extended studies confirm these results (Fig. 13).

In comparing the retention data in 0.08 M sodium acetate with those in 0.02 M sodium acetate the shrinkage of the resin bed (2%) at the higher eluent concentration must be taken into consideration. For the oligomeric acids of DP 2 to 8 the observed distribution coefficients (per unit weight of the resin) in 0.08 M sodium acetate differed by less than 2% from those calculated from the determinations in 0.02 M sodium acetate (by dividing by 4). The difference between the observed and calculated values increased with increasing DP of the oligomers and was 7% for the highest oligomer (DP = 16). These deviations can be ascribed at least in part to a decrease in swelling pressure with increasing electrolyte concentration in the external solution.

Another and more serious complication is that it is difficult to determine the interstitial fraction unambiguously. Both the compression of the resin bed following prolonged use and the choice of "nonretarded" test substance affect the determination. These errors are of little importance for the calculation of D_V for the lower oligomeric acids, which exhibit high D_V, but become increasingly important with increasing DP of the acids. Abnormally low values (ϵ_I = 0.31-0.33, depending on the bed compression) were obtained when dextran T 2000 (from Pharmacia, Uppsala, Sweden) was used as the nonretarded substance. Approximately the same values were found when polyethylene glycols with a molecular weight of 9000 and 4000 were employed.

The interstitial volume was therefore determined graphically from the peak elution volumes recorded for the oligomeric sugars. A plot of the peak elution volume of different sugars versus that of the next higher oligomer gives a straight line which is extrapolated to the point at which the ordinate and the abscissa are equal, i.e., to the point at which an oligomer of DP = n has the same retention volume as that of DP = n + 1.

This method gave a value ($\epsilon_I = 0.37$) which was much higher than that obtained with the other test substances. Since no cracks could be observed by microscope in the resin particles it can be concluded that in a column packed with lightly cross-linked resin, there exist regions on the surface of the resin particles which are unaccessible for very large solutes such as dextran whereas smaller molecules such as monosaccharides and lower oligomers can penetrate into these regions. Hence, the interstitial fraction is not a well-defined quantity under these conditions. Although it can be questioned whether it is justifiable to calculate any distribution coefficients from these chromatograms, the values are useful in planning chromatographic separations and for tentative identification of oligomeric acids when authentic samples are lacking.

For the oligomeric sugars the determination of the D_V for oligomers of DP 6 and higher is rather uncertain. A calculation of the slope for a linear relationship between $\ln D_V$ and DP based on the experiments with oligomers of DP 2 to 5 gave the value -0.28 for 0.02 M sodium acetate and -0.27 for 0.08 M sodium acetate. Calculations based on the experiments with the corresponding oligomeric acids gave the values -0.29 and -0.28, respectively. These results, together with those obtained with the more tightly cross-linked resin, show that if the distribution coefficient of an aldobionic acid (or one higher oligomer) is known, the positions of the other oligomers can be approximately predicted from the peak positions of the oligomeric sugars belonging to the same series.

Previous studies of the relative importance of the pressure-volume term and the activity coefficient term in the Gibbs-Donnan equation show that, for strongly polar solutes such as alditols, the pressure-volume term has the greatest effect on the equilibrium distribution [45]. This has been confirmed by experiments with some oligomeric sugars [10, 46]. It is therefore reasonable to assume that in an oligomeric series, the ratio between the activity coefficients will approach unity. This implies that the slope of the straight line ($\ln D_V$ versus DP) should be equal to $-\pi \Delta \bar{v}/RT$, i.e., that it should be possible to predict the slope from the swelling pressure or, conversely, to obtain a rough estimate of the swelling pressure from the slope of the straight lines [43]. Determinations of $\Delta \bar{v}$ are available for various types of oligomeric sugars [22]. For the β-(1 → 4)-linked xylose oligomers the value is 0.085 dm^3/mol.

The swelling pressure calculated with these assumptions from the slopes of straight lines discussed earlier is 23 MPa (227 atm) for the acetate form of Dowex 1-X8. This value is about 10% higher than that reported for another batch of the same type of resin [43]. The values are reasonable when compared to those reported by Boyd and Soldano [47] in their classical paper on determinations of the swelling pressure by isopiestic measurements. For Dowex 1-X2 the corresponding value is 8.5 MPa (84 atm) in 0.02 M sodium acetate and 8.1 MPa (80 atm) in 0.08 M sodium acetate.

The results show that in contrast to ion-exchange equilibria where simple inorganic ions [48] or hydrophobic organic ions [49] are involved, the anion-exchange equilibrium of oligomeric, strongly hydrophilic ions is mainly determined by the pressure-volume term in the Gibbs-Donnan equation.

B. Influence of Temperature

A study of the influence of temperature [43] on separations of oligomeric acids in sodium acetate showed that over a wide temperature range (0-95°C) ln D_V was linearly related to DP. Plots of the logarithm of the selectivity coefficient versus 1/T resulted in curved lines for all species investigated (DP 1-6). At high temperature the enthalpy change ($\Delta H°$) was positive for all species. For the highest oligomers $\Delta H°$ was also positive at low temperature. At a given temperature, $\Delta H°$ increased with increasing DP, i.e., with decreasing ion-exchange affinity. Within a series of oligomeric acids, starting with the aldobionic acid, a linear relationship exists between $\Delta H°$ (at constant temperature) and DP. The results show that $\Delta H°$ for the ion-exchange process is an additive property within an oligomeric series of acids.

As far as the separation factors are concerned advantages are gained when working at very low temperature. However, it is inconvenient to work below room temperature because of the cooling required, the increased viscosity of the eluent, and the broadening of the elution curves due to slow diffusion of higher oligomers inside the resin particles. The separation factors of the oligomeric acids are scarcely affected by an increase in temperature from 30°C to 95°C. As expected, the elution curves become sharper and the counterpressure lower when the elution is carried out at high temperature.

C. Separations for Analytical Purposes

Excellent separations of aldobionic acids can be achieved by chromatography on anion exchangers with 8% nominal DVB with sodium acetate or borate as eluent [50]. The same type of resin is suitable for a complete separation of oligomeric aldonic acids of DP 2 to 8 in sodium acetate and acetic acid media [43].

The separation of oligomeric acids belonging to two oligomeric series in 0.02 M sodium acetate is demonstrated in Fig. 14. The last acid was eluted within 4.5 hr. This chromatogram illustrates the usefulness of a three-channel analyzer for automatic colorimetric analysis of the eluate [51]. All acids are oxidized by chromic acid and recorded in the nonspecific chromic acid channel. As expected all oligomeric species give a response with carbazole while xylonic and gluconic acids are not recorded. On the other hand, these species give a very strong response in the periodate-formaldehyde channel. Cellobionic and cellotrionic acids are also recorded in this channel, whereas the oligomeric acids of the xylonic acid series, which are lacking a terminal diol group, cannot give rise to formaldehyde.

With anion exchangers with 8% nominal DVB, oligomeric acids containing more than eight monomeric units (DP > 8) are almost completely excluded. The application of a resin with lower cross-linking (2% DVB) permits

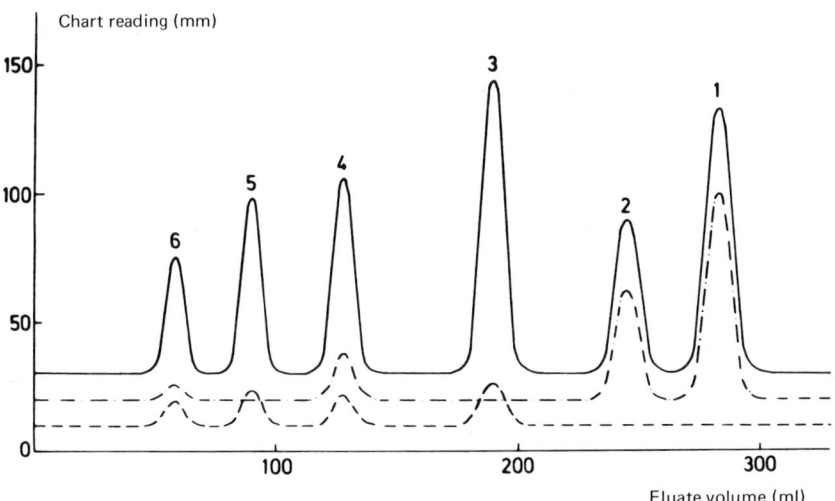

FIG. 14. Separation of oligomeric acids at 30°C in 0.02 M sodium acetate (pH 5.9). Resin bed: 4 × 670 mm Dowex 1-X8, acetate, 13 to 18 μm. Nominal linear flow: 8.5 cm/min. 1, Xylonic (1.0 mg); 2, gluconic (0.6 mg); 3, xylobionic (1.0 mg); 4, cellobionic (0.6 mg); 5, xylotrionic (0.5 mg); 6, cellotrionic acid (0.3 mg). Analysis channels: solid line, chromic acid; dashed line, carbazole; dash-dot line, periodate-formaldehyde. (From Havlicek and Samuelson [43]. Courtesy of the Journal of Chromatography.)

the complete separation of oligomers of a DP up to 16 (Fig. 12). On a longer column, even higher oligomers were resolved by elution with very dilute (0.005 M) sodium acetate solution. A broadening of the elution peaks was observed for the highest oligomeric acids when a high flow rate was employed. Evidently, the height of a theoretical plate for these compounds is seriously affected by their slow diffusion inside the resin phase. Separate experiments showed that the time can be shortened by two-thirds by increasing the eluent concentration during the run. A further improvement can be obtained by chromatography at elevated temperature [43].

In acetic acid, as in sodium acetate, the oligomers are eluted in the order of decreasing DP, thereby indicating that the pressure-volume term in the Gibbs-Donnan equation has a predominant influence in this medium also. Plots of log D_V versus DP exhibit significant deviations from linearity. The results show that the acid strength has a significant influence in acid media. Hence, the separation factor xylonic acid:gluconic acid is higher in 0.5 M acetic acid than in sodium acetate. The same holds true for the oligomers of equal DP. The separation of the four lowest members of each series is therefore more favorable in acetic acid than in sodium acetate. On the other hand, the D_V in 0.5 M acetic acid of cellopentaonic acid (0.72) is fairly close to that of xylohexaonic acid (0.80). This means that sodium acetate of low concentration (0.01 or 0.02 M) is preferable in chromatography of mixtures of higher oligomers belonging to both series [43].

When tetraborate solution is used as eluent the formation of borate complexes with vicinal hydroxyl groups has a predominant influence on D_V. Hence xylonic acid is eluted much earlier than gluconic acid [52] while in sodium acetate the acids appear in the order of decreasing molar volume. Within the oligomeric series studied the acids are eluted in the order of decreasing DP (Fig. 15). This means that the complexing with the (1 → 4)-linked glycosyl moieties contributes less to D_V than the factors which tend to increase $\Delta G°$. At flow rates comparable to that referred to in Fig. 14, symmetrical and comparatively sharp elution peaks are recorded for the three lowest oligomers whereas the higher oligomers give broad nonsymmetric peaks. Experiments at different flow rates show that equilibrium is established slowly when higher oligomers are involved [43].

An application of a similar technique for the determination of oligomeric sugars (DP 2-5) has been proposed recently [53]. The method involves oxidation of the reducing sugar moiety to an aldonic acid end group followed by chromatography of the oligomeric acids in borate-acetate solution with increasing concentration of sodium acetate.

D. Separations for Preparative Purposes

In agreement with the data reported by Mattisson and Samuelson [45] and confirmed by Ginzburg and Cohen [46] the results just presented clearly

CHROMATOGRAPHY OF OLIGOSACCHARIDES 141

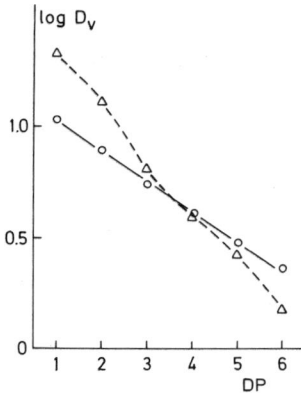

FIG. 15. Relationship between log D_V at 30°C in 0.15 M potassium tetraborate solution and the number of monomeric units in oligomeric acids (Dowex 1-X8, 13 to 18 μm). △, Cellobionic acid series; ○, xylobionic acid series. (Data from Havlicek and Samuelson [43].)

show that the partial exclusion of sugars from the swollen gel cannot be explained by a model of mechanical screening because of the network structure of the stationary phase. The most striking observation is that an oligosaccharide with a DP of 8 is almost completely excluded from a lightly crosslinked resin (Dowex 1-X2), whereas the oligomeric acids of DP 15 and 16 exhibit such high distribution coefficients that they are completely separated (Fig. 12). The high distribution coefficients of oligomeric sugars in aqueous ethanol compared to those in water can also be recalled in this connection. All these results can only be explained by the assumption that, from a macroscopic point of view, all parts of the swollen resin are available for the solutes.

The absence of mechanical barriers to the penetration of trace amounts of oligomers of fairly high molecular weight does not mean that the space and amount of water inside the swollen resin are sufficient to permit the accommodation of the same number of moles of an oligomeric acid as monomeric acids. This is confirmed by the results given in Table 3. It is seen that the practical specific exchange capacity for gluconic and xylonic acids is almost the same as the total capacity determined for chloride ions. Only 89% of the exchange sites can, however, be occupied by xylobionate ions. This corresponds to an uptake of 0.90 g of xylobionic acid/g of resin, which is a higher value than observed for the monomer (0.56) and for xylotrionic (0.71) and higher oligomeric acids. For xylopentaonic acid the practical specific capacity was 17% of the total capacity. This corresponds to 0.41 g of acid. If all exchange sites were occupied by xylopentaonate anions, 1 g of resin should accommodate 2.4 g of acid, which would be impossible considering space and water requirements.

TABLE 3

Practical Specific Exchange Capacity Calculated per Gram of Dry Dowex 1-X8 in the Chloride Form[a]

	Observed exchange capacity		
Acids	mmol/g	Percent of total capacity[b]	g acid/g
Gluconic	3.36	98	0.66
Xylonic	3.36	98	0.56
Xylobionic	3.03	89	0.90
Xylotrionic	1.65	48	0.71
Xylotetraonic	1.03	30	0.58
Xylopentaonic	0.59	17	0.41
Xylohexaonic	0.29	8	0.24
Xyloheptaonic	0.14	4	0.13
Xylooctaonic	0.05	1.5	0.06
Xylononaonic	<0.01	0	<0.01

[a] Experimental data from Havlicek and Samuelson [44].
[b] Determined for chloride ions (3.42 mmol/g).

Lightly cross-linked resins that contain a larger amount of water can accommodate a larger number of oligomeric ions per gram of resin. As an example it can be mentioned that for xylopentaonic acid the practical specific capacity of Dowex 1-X2 (total capacity 3.58 mmol/g) was 32% of the total capacity. This is almost twice the value observed for Dowex 1-X8.

The drastic decrease in the practical exchange capacity for oligomers of increasing DP, together with the fact that in batch experiments several hours are required to reach equilibrium, shows that care must be exercised in separations of large amounts of oligomeric acids. Although it is very convenient to separate small amounts of oligomeric acids of DP 2 to 8 for analytical purposes on resins with 8% DVB, resins of this type are unsuitable for separations of these oligomers for preparative purposes or for analyses of unknown mixtures where it may be necessary to isolate several milligrams for additional identification. For such separations it is strongly recommended to use a resin with a lower degree of cross-linking (2% DVB) even if only oligomers of DP 2 to 8 are present. In separations of higher oligomers it is necessary to apply a lightly cross-linked resin. Satisfactory separations of oligomers of DP 11 to 17 for preparative purposes are achieved on a resin with 2% DVB by elution with 0.05 M sodium acetate. The lower oligomers and the monomeric acid are then eluted by increasing the concentration to 0.1 M either stepwise or by gradient elution [44].

The method has been applied in analysis of spent liquors from cellulose bleaching which in addition contained other organic acids and oligomeric sugars [54].

VI. ANION-EXCHANGE CHROMATOGRAPHY OF OLIGOMERS WITH ONE URONIC ACID MOIETY

Uronic acids are important constituents in several types of hemicellulose and pectic substances. Anion-exchange chromatography of the hexuronic and aldobiouronic acids formed after hydrolysis has been used in numerous investigations. Recently higher oligomers obtained by mild hydrolysis of xylan have been studied. The predominant acidic products contain one 4-O-methylglucuronic acid moiety and one or more xylose moieties. Both in sodium acetate and in acetic acid [55] these oligomers are eluted in the order of decreasing DP. The first peak on the chromatogram reproduced in Fig. 16 contained at least two isomeric 2-O-(4-O-methyl-α-D-glucopyranosyluronic acid) xylotrioses. The second peak contained O-(4-O-methyl-α-D-glucopyranosyluronic acid)-(1 → 2)-O-β-D-xylopyranosyl-(1 → 4)-D-xylose, and the predominant constituents in peak 4 were 2-O-(4-O-methyl-α-D-glucopyranosyluronic acid)-D-xylose and 4-O-(α-D-galactopyranosyluronic acid)-D-xylose. The simple hexuronic acids in peaks 6 to 8 appeared in the order 4-O-methylglucuronic < galacturonic < glucuronic acid.

A calculation of the D_V in sodium acetate of the oligomers with three xylose moieties based on the D_V of that containing two moieties was made using the same assumptions as used for other xylose oligomers. The calculated D_V was in good agreement with the observed value. A linear relationship between log D_V and DP of the oligomers of the DP 3 to 10 has been reported for experiments in 0.5 M acetic acid on Dowex 1-X8 [56]. Studies in the author's laboratory show that like the higher oligomers with an aldonic acid end group the oligomers containing one uronic acid moiety can be better separated on a lightly cross-linked resin (Dowex 1-X2).

Experiments with these acids in borate medium show that the elution order is the same as in sodium acetate and acetic acid [57].

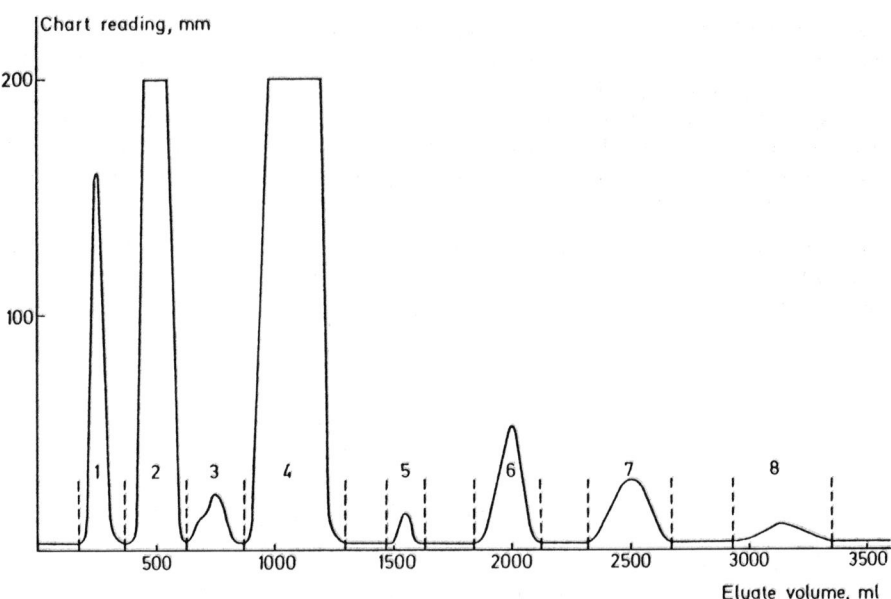

FIG. 16. Separation on a preparative scale of uronic acids isolated after hydrolysis of birch xylan. Resin bed: 30 × 800 mm, Dowex 1-X8, acetate, 25 to 32 µm. Eluent: 0.08 M sodium acetate (pH 5.9), 30°C. Nominal linear flow: 0.3 cm/min. (From Shimizu and Samuelson [55]. Courtesy of Svensk Papperstidning.)

VII. ANION-EXCHANGE CHROMATOGRAPHY OF DICARBOXYLIC OLIGOMERS

Anion-exchange chromatography in acetate media was recently applied in a study of dicarboxylic acids present in spent liquor from the bleaching of wood pulp [58]. The D_V values of the relevant acids which, with the exception of 4-O-methylglucaric acid (included for comparison), contained one uronic acid and one aldonic acid moiety are listed in Table 4. In agreement with the results discussed for monocarboxylic acids the D_V values of the dicarboxylic acids in sodium acetate medium decrease markedly with increasing molecular size. The results indicate that the pressure-volume term in the Gibbs-Donnan equation has a predominant influence on the ion-exchange equilibrium. This was confirmed by a calculation of the D_V of compound VII from that of its lower oligomer (IV) by the method described for oligomeric aldonic acids.

When sodium acetate is used as eluent the isomers containing one xylose moiety (III and IV) exhibit small differences in D_V. This is also true for the

TABLE 4

Volume Distribution Coefficients (D_v) of Dicarboxylic Oligomers in Acetate Media at 30°C

Acids		Sodium acetate solution, pH 5.9				Acetic acid	
		0.08 M	0.2 M	0.3 M	3 M	3 M	5 M
I	4-O-Methylglycaric (3-O-Methylgulcaric)	—	61.4	29.0	35.1	35.1	11.6
II	2-O-(4-O-Methyl-α-D-glucopyranosyluronic acid)-D-xylonic	—	23.4	9.7	73	73	22.2
III	O-(4-O-Methyl-α-D-glucopyranosyluronic acid)-(1 → 2)-O-β-D-xylopyranosyl-(1 → 4)-D-xylonic	81	12.6	5.5	8.0	8.0	2.2
IV	O-β-D-Xylopyranosyl-(1 → 4)-[O-(4-O-methyl-α-D-glucopyranosyluronic acid)-(1 → 2)]-D-xylonic	88	13.5	5.9	33.2	33.2	8.7
V	O-(4-O-Methyl-α-D-glucopyranosyluronic acid)-(1 → 2)-O-β-D-xylopyranosyl-(1 → 4)-O-β-D-xylopyranosyl-(1 → 4)-D-xylonic	32.9	5.6	2.6	3.2	3.2	0.73
VI	O-β-D-Xylopyranosyl-(1 → 4)-[O-(4-O-methyl-α-D-glucopyranosyluronic acid)-(1 → 2)]-O-β-D-xylopyranosyl-(1 → 4)-D-xylonic	—	6.2	2.9	4.7	4.7	1.05
VII	O-β-D-Xylopyranosyl-(1 → 4)-O-β-D-xylopyranosyl-(1 → 4)-[O-(4-O-methyl-α-D-glucopyranosyluronic acid)-(1 → 2)]-D-xylonic	38.2	6.6	3.1	12.2	12.2	3.0

*Data from Ericsson and Samuelson [58].

isomers with two xylose moieties (V, VI, VII). For both types of oligomers the distribution coefficients increase with decreasing distance between the carboxylic acid groups.

In acetic acid the elution order of the oligomeric acids VII < IV < II, which must differ only slightly in acid strength, is the same as in sodium acetate. The large differences in D_V indicate that the molar volume also has a great influence in acetic acid. As expected from their predicted acid strengths, the isomeric acids are eluted in the order of decreasing distance between the carboxylic acid groups. Interestingly, VII is retained more strongly than III, and II more strongly than I. This shows that for these species an increase in acid strength exerts a greater influence on D_V than the decrease in molar volume. The higher D_V of II compared to that of the aldaric acid is explained by the strongly acid properties of the uronic acid moiety.

No systematic investigations have been made with dicarboxylic acids containing two uronic acid moieties and two or more sugar moieties. As can be predicted from its size, the only acid studied, O-(4-O-methyl-α-D-glucopyranosyluronic acid)-(1 → 2)-O-β-D-xylopyranosyl-(1 → 4)-O-[4-O-methyl-α-D-glucopyranosyluronic acid-(1 → 2)]-D-xylose, exhibited a somewhat lower D_V than IV in sodium acetate medium. Its D_V in acetic acid was significantly higher than that of IV. This is explained by the higher acid strength of the uronic acid group [59].

VIII. ANION-EXCHANGE CHROMATOGRAPHY OF OLIGOGALACTURONIC ACIDS

In contrast to the oligomeric uronic acids discussed earlier, oligogalacturonic acids lack sugar moieties. The galacturonic acid moieties are directly linked together by glycosidic bonds. The number of carboxylic acid groups is therefore equal to the DP. As expected the oligomers are held very strongly by anion-exchange resins.

Derungs [60, 61] demonstrated a complete separation of a mixture of galacturonic, digalacturonic, trigalacturonic, and tetragalacturonic acids by stepwise elution in 0.1 to 2.5 M formic acid on a weakly basic anion-exchange resin (Dowex 3). The risk of acid hydrolysis during the separation can be avoided by elution with sodium formate. Nagel and Wilson recommend gradient elution with this eluent on a strongly basic resin [62]. The separation of oligomers of DP 1 to 8 was reported. Unsaturated oligogalacturonic acids were also separated by this technique. These were held more strongly by the resin than the saturated compounds. In separations of mixtures of both types of oligomers the fractions were rechromatographed to ensure purity of each compound.

The method has subsequently been applied for preparative purposes by other authors [63]. As expected with regard to the increasing number of carboxylic acid groups, the oligomers are eluted in the order of increasing DP, i.e., the reverse of the order observed with the oligomeric monocarboxylic acids. On the other hand, polygalacturonic (pectic) acid is not retained by anion-exchange resins [64]. The exclusion of this solute can therefore be explained as a mechanical screening effect.

REFERENCES

1. C. T. Bishop, Can. J. Chem., 33, 1073 (1955).
2. R. L. Whistler and C. C. Tu, J. Am. Chem. Soc., 74, 3609 (1952).
3. W. Brown and O. Andersson, J. Chromatogr., 57, 255 (1971).
4. O. Samuelson and E. Sjostrom, Sv. Kem. Tidskr., 64, 305 (1952).
5. H. Ruckert and O. Samuelson, Sv. Kem. Tidskr., 66, 337 (1954).
6. O. Samuelson, 9. Nordiske Kemikermode i Aarhus, 1956, 105.
7. O. Samuelson and B. Swenson, Anal. Chim. Acta, 28, 426 (1963).
8. O. Samuelson and H. Stromberg, Z. Anal. Chem., 236, 506 (1968).
9. P. Jonsson and O. Samuelson, Science Tools, 13:2, 17 (1966).
10. J. Havlicek and O. Samuelson, Carbohydr. Res., 22, 307 (1972).
11. H. Ruckert and O. Samuelson, Acta Chem. Scand., 11, 303 (1957).
12. H. Ruckert and O. Samuelson, Acta Chem. Scand., 11, 315 (1957).
13. E. Paart and O. Samuelson, J. Chromatogr., 85, 93 (1973).
14. O. Samuelson, in Ion Exchange and Solvent Extraction, Vol. II (J. A. Marinsky, ed.), Marcel Dekker, New York, 1969.
15. P. Jonsson and O. Samuelson, Anal. Chem., 39, 1156 (1967).
16. O. Ramnas and O. Samuelson, Acta Chem. Scand., B28, 955 (1974).
17. J. Havlicek and O. Samuelson, Anal. Chem., 47, 1854 (1975).
18. P. Jonsson and O. Samuelson, J. Chromatogr., 26, 194 (1967).
19. E. Martinsson and O. Samuelson, J. Chromatogr., 50, 429 (1970).
20. M. Johansson and O. Samuelson, Chem. Scr., 9, 151 (1976).
21. J. Havlicek and O. Samuelson, Chromatographia, 7, 361 (1974).
22. W. Brown and K. Chitombo, Chem. Scr., 2, 88 (1972).

23. J. Havlicek and O. Samuelson, J. Inst. Brew., 81, 466 (1975).
24. L.-I. Larsson and O. Samuelson, Mikrochim. Acta, 1967:2, 328.
25. O. Samuelson and H. Stromberg, J. Food Sci., 33, 308 (1968).
26. O. Samuelson and H. Stromberg, Carbohydr. Res., 3, 89 (1966).
27. K. Mopper and E. M. Gindler, Anal. Biochim., 56, 440 (1973).
28. J. X. Khym and L. P. Zill, J. Am. Chem. Soc., 74, 2090 (1952).
29. J. X. Khym, Analytical Ion-Exchange Procedures in Chemistry and Biology, Prentice-Hall, Englewood Cliffs, New Jersey, 1974.
30. J. I. Ohms, J. Zec, J. V. Benson, Jr., and J. A. Patterson, Anal. Biochem., 20, 51 (1967).
31. R. L. Jolley and M. L. Freeman, Clin. Chem., 14, 538 (1968).
32. E. F. Walborg, Jr., and L. E. Kondo, Anal. Biochem., 37, 320 (1970).
33. K. Larsson and O. Samuelson, Carbohydr. Res., 50, 1 (1976).
34. M. Torii and K. Sakakibara, J. Chromatogr., 96, 255 (1974).
35. J. X. Khym, R. L. Jolley, and C. D. Scott, Cereal Sci. Today, 15, 44 (1970).
36. R. Carubelli, Carbohydr. Res., 2, 480 (1966).
37. G. R. Noggle and L. P. Zill, Arch. Biochem. Biophys., 41, 21 (1952).
38. A. Floridi, J. Chromatogr., 59, 61 (1971).
39. K. Larsson and O. Samuelson, J. Chromatogr., 134, 195 (1977).
40. J. K. N. Jones, R. A. Wall, and A. O. Pittet, Chem. Ind. (London), 1959, 1196; Can. J. Chem., 38, 2285 (1960).
41. H. G. Walker and R. M. Saunders, Cereal Sci. Today, 15, 140 (1970).
42. F. G. Donnan and E. A. Guggenheim, Z. Phys. Chem., Abt. A., 162, 346 (1932).
43. J. Havlicek and O. Samuelson, J. Chromatogr., 83, 45 (1973).
44. J. Havlicek and O. Samuelson, J. Chromatogr., 114, 383 (1975).
45. M. Mattisson and O. Samuelson, Acta Chem. Scand., 12, 1386 (1958).
46. B. Z. Ginzburg and D. Cohen, Trans. Faraday Soc., 60, 185 (1964).
47. G. E. Boyd and B. A. Soldano, Z. Elektrochem., 57, 162 (1953).
48. G. E. Boyd, S. Lindenbaum, and G. E. Myers, J. Phys. Chem., 65, 577 (1961).

49. L. Bengtsson and O. Samuelson, J. Chromatogr., 61, 101 (1971).
50. O. Samuelson and L.-O. Wallenius, J. Chromatogr., 12, 236 (1963).
51. B. Carlsson, T. Isaksson, and O. Samuelson, Anal. Chim. Acta, 43, 47 (1968).
52. K. Larsson, L. Olsson, and O. Samuelson, Carbohydr. Res., 38, 1 (1974).
53. Y. Sakai, K. Okawa, and Y. Kamiyama, Agr. Biol. Chem., 39, 545 (1975).
54. S.-I. Andersson and O. Samuelson, Cellulose Chem. Technol., 10, 209 (1976).
55. K. Shimizu and O. Samuelson, Sven. Papperstidn., 76, 150 (1973).
56. M. H. Simatupang, M. Sinner, and H. H. Dietrichs, Technicon Symposium 74, No. 1802.
57. M. Sinner, M. H. Simatupang, and H. H. Dietrichs, Wood Sci. Technol., 9, 307 (1975).
58. T. Ericsson and O. Samuelson, J. Chromatogr., 134, 337 (1977).
59. K. Shimizu and O. Samuelson, Sven. Papperstidn., 76, 156 (1973).
60. R. Derungs, Trennung von Oligogalakturonsauren an Anionenaustauschern, Thesis Eidgenossische Technische Hochschule, Zurich, 1958.
61. R. Derungs and H. Deuel, Helv. Chim. Acta, 37, 657 (1954).
62. C. W. Nagel and T. M. Wilson, J. Chromatogr., 41, 410 (1969).
63. B. A. Dave, R. H. Vaughn, and I. B. Patel, J. Chromatogr., 116, 395 (1976).
64. K. T. Williams and C. N. Johnson, Ind. Eng. Chem., 16, 23 (1944).

Chapter 5

APPLICATIONS AND THEORY OF FINITE
CONCENTRATION FRONTAL CHROMATOGRAPHY

Jon F. Parcher

Chemistry Department
University of Mississippi
University, Mississippi

I. INTRODUCTION 152
II. FRONTAL CHROMATOGRAPHY 153
    A. Factors Influencing the Shape of a Frontal
       Chromatogram (Frontalgram) 153
    B. Modes of Operation 154
III. THEORY OF CHROMATOGRAPHY AT
     FINITE CONCENTRATIONS 155
IV. PREVIOUS CHROMATOGRAPHIC INVESTIGATIONS
     INVOLVING SIMPLIFYING ASSUMPTIONS 160
    A. Theories Utilizing Assumption I 161
    B. Theories Utilizing Assumption II 162
    C. Theories Utilizing Assumption III 162
    D. Numerical Solution of the Unrestricted Equations ... 163
V. DISCUSSION 164
VI. PRACTICAL APPLICATIONS OF
     FRONTAL CHROMATOGRAPHY 165
    A. Adsorption Isotherms 165
    B. Partition Isotherms 165
    C. Elemental Analysis 165
    D. Study of Ternary Systems 166
VII. SUMMARY 170
     ACKNOWLEDGMENTS 171
     REFERENCES 171

I. INTRODUCTION

Frontal chromatography is a mode of chromatographic operation in which the sample is fed into the column continuously at relatively high mole fraction. The feed of sample can be interrupted and renewed at intervals to produce "breakthrough" curves or frontalgrams. These profiles can be very abrupt or very diffuse and every range between, depending on the relation between the concentration of solute, i, in the stationary phase, Q_i, to the concentration in the mobile phase, C_i. This relation is usually described in terms of an isotherm equation $Q_i = f(C_i)$ or graphically as an equilibrium isotherm in the form of a plot of Q_i versus C_i at a given temperature. Because the exact shape of the frontalgram is at least partially determined by the functional form of the equilibrium isotherm, the analysis of experimental breakthrough curves can be used to measure adsorption or partition isotherms. At present this is the primary practical application of the technique of frontal chromatography. The purpose of this chapter is to review briefly the theory and experimental techniques of frontal chromatography and to summarize the most recent developments in this growing field.

In 1970, Dr. G. Schay, who is certainly one of the pioneers in the application of frontal chromatography, described the technique as a "stepchild of chromatography" and an "instructive curiosity" when compared to elution chromatography [1]. This comparison was valid in 1970 and may still be valid today when the comparison is made on the basis of analytical utility. However, in recent years the nonanalytical utility of frontal chromatography has been investigated and been found useful in many diverse fields. The three main areas of practical applications have been (a) the study of catalysts and solid adsorbents, (b) petrochemistry and petroleum refining, and (c) nonelectrolyte solution properties.

In 1967, both Habgood [2] and Kobayashi et al. [3] reviewed the applications of frontal chromatography to the measurement of adsorption isotherms of solid adsorbents. More recently, Saha and Mathur [4] have given a complete review of the applications of gas chromatography (including frontal) to the study of catalysis and catalysts. The primary application of chromatography in this field is the measurement of adsorption isotherms to investigate the relationship between the activity of the catalyst and its structural and chemical characteristics.

Another area in which frontal chromatography has been used extensively is in the petroleum industry. Berezkin [5] reviewed the articles published in this field up to 1974. The primary interest in this area is the measurement of the solubility of vapors and gases in high-molecular-weight coal and petroleum liquids.

By far the most active investigations have been in the study of the properties of lower-molecular-weight hydrocarbon solutions by physical chemists.

Giddings and Mallik [6], Conder [7], and Young [8] all wrote extensive reviews of this field between 1967 and 1968. More recently, Locke [9, 10] has summarized the work carried out in the area of physicochemical measurements by the various forms of chromatography. Prior to this, Dr. Conder published another review [11] concerning only thermodynamic measurements by frontal chromatography. The main applications in the field of physical chemistry have been the measurement of the vapor-liquid equilibria of nonelectrolyte solutions and the attempt to correlate these data with one of the many solution theories.

II. FRONTAL CHROMATOGRAPHY

A. Factors Influencing the Shape of a Frontal Chromatogram (Frontalgram)

There are three principal experimental forms of frontal chromatography: (a) frontal analysis with a sharp front, (b) frontal analysis with a diffuse boundary, and (c) elution on a plateau or vacancy chromatography. These various modes have been discussed extensively in the literature [12-15] and several authors have compared the various methods for accuracy [16-18]. The conclusions are somewhat vague but one general observation is that the three methods give comparably accurate data for low concentrations and/or systems with relatively linear isotherms.

Another important factor which has been investigated is the influence of the shape of the isotherm on the observed chromatogram. The conclusions here are sometimes contradictory although the consensus of opinion is the following:

1. A Langmuir-type isotherm will give a sharp sorption chromatogram and a diffuse desorption curve.

2. An anti-Langmuir isotherm will usually cause the sorption chromatogram to be diffuse and the desorption sharp. This effect may be offset by the sorption effect to produce the opposite situation for near-linear isotherms or high solute concentrations.

3. A linear isotherm will yield a sharp sorption frontalgram at high concentrations and a diffuse desorption profile.

Many authors [15, 16] have discussed the observed concentration profiles in terms of the variation of a "concentration velocity" with the concentration of the solute. However, this type of explanation is of limited utility because the velocity of a point of fixed solute concentration is determined by a number of factors. One of these factors is the shape of the isotherm, but other important factors are the variation of the gas phase viscosity with composition and the "sorption effect" [15, 19-21] caused by the tremendous difference in the molar volume of a component in a liquid state and the same component in the gaseous state. Thus an attempt to discuss the form of a frontalgram in terms of only the shape of the isotherm is doomed to failure when the other two effects are significant.

There are several other phenomena which may also influence the shape of a frontalgram, such as diffusion and mass-transfer kinetics. Valentin and Guiochon [22] have presented a thorough discussion of all these effects and demonstrated the second-order nature of the diffusion and kinetic effects. The primary factors controlling the shape of the frontalgram are those caused by the large concentration of solute in the mobile and stationary phases.

B. Modes of Operation

There are two distinct classifications of experimental chromatographic techniques for measuring solubility data. The first is the true frontal technique in which the breakthrough curves are recorded, and the second is an elution procedure in which a binary carrier gas is employed with a fixed concentration of a condensable component. These two techniques will be designated by the titles "frontal analysis" and "elution on a plateau," respectively, following the nomenclature of Conder and Purnell [12].

1. Frontal Analysis Techniques

This type of procedure can also be subdivided into two distinct types. These are the "single-point" and "locus" methods. The single-point methods give only one point on the isotherm per experiment and do not require a detailed analysis of the shape of the chromatogram but rather a measure of an area on the chromatogram. This type of measurement can be applied to sharp or diffuse frontalgrams although the mathematics is considerably simplified for the case of sharp breakthrough curves. Schay et al. [23] have proposed an alternative technique that does not depend on the shape of the frontalgram but requires an integral flowmeter.

The locus methods involve a detailed analysis of the detector response as a function of time but have the advantage of giving the isotherm over the entire range of concentration from zero to the concentration used in the experiment.

2. Elution Methods

An alternative method of frontal chromatography involves the measurement of the retention volume of a solute on a column that is partially saturated with a "moderator." A binary liquid phase or a single condensed liquid phase is obtained by using a binary carrier gas that contains a component which will condense on the stationary phase (liquid or solid) in the column. When the column has attained equilibrium, samples can be introduced as in normal elution chromatography.

Reilley et al. [13] studied the elution of a concentration pulse of the moderator and called the method "vacancy" chromatography. The application of this experimental procedure for the measurement of adsorption and partition isotherms has been investigated by several authors [12, 24, 25] and the technique has been called variously "elution on a plateau" and the "step and pulse" method.

In 1963, Helfferich and Peterson [26] proposed a similar method involving the elution of an infinitely dilute sample of the moderator which had been "tagged" with a radioactive atom. This method was very simple and accurate but required an elaborate detection system, and for this reason has not been widely accepted.

Recently, a third method has been investigated [27, 28]. In this technique the retention of an infintely dilute sample of a solute other than the moderator is monitored and related to the amount of moderator in the stationary phase. This technique requires a knowledge of the variation of a "foreign" solute's retention volume or partition coefficient with the composition of the stationary phase. Recent developments in the area of binary liquid phases [28-34] have made it possible to predict this variation for certain solutes.

III. THEORY OF CHROMATOGRAPHY AT FINITE CONCENTRATIONS

All the experimental techniques discussed in the preceding section can be interpreted in terms of the same basic theory with different conditions or simplifying assumptions.

The theory of mass transport in fixed bed reactors has been discussed in the engineering literature. The general case would involve a finite concentration of a single solute at large carrier gas flow rates. Ideally the flow rate should correspond to the optimum velocity on a van Deemter plot in order to minimize the contributions of axial diffusion and nonequilibrium or slow mass transfer to the shape of the breakthrough curves. Operating at optimum flow rates requires a finite pressure drop across the column and the viscosity of the carrier gas mixture may be composition dependent.

If the column operates isothermally, the concentrations, velocity, and quantities dependent on concentration will be functions of time, t, and position, z, in the column. Under these conditions the mass balance equations describing the system are [19, 35, 36]

Solute:

$$\frac{\partial}{\partial z}[C_A v] + \theta_g \left[\frac{\partial C_A}{\partial t}\right] + \theta_L \left[\frac{\partial Q_A}{\partial t}\right] = 0 \tag{1}$$

Total:

$$\frac{\partial}{\partial z}[C_t v] + \theta_g \left[\frac{\partial C_t}{\partial t}\right] + \theta_L \left[\frac{\partial Q_A}{\partial t}\right] = 0 \tag{2}$$

where C_A is the molar concentration of solute A in the gas phase, C_t the total molar concentration of carrier plus solute in the gas phase, Q_A the concentration of solute in the liquid phase in moles per liter of solvent, θ_L and θ_g the ratios of the volumes of the nonvolatile stationary phase and the mobile phase to the volume of the empty column, respectively, v the superficial velocity of the gas phase (cm/sec), and z and t the distance and time variables. These equations are completely general and have been presented by several authors. The equations are presented in terms of molar concentrations rather than mole fractions to avoid the unnecessary assumption of constant total concentration, i.e., zero pressure drop across the column. Q_A is defined as the concentration of solute per liter of solvent rather than per liter of solution.

The momentum balance for the flow of an ideal gas in a packed bed is given by [37]

$$v = \frac{\psi}{\eta_{mix}} \left(\frac{\partial C_t}{\partial z}\right) \tag{3}$$

$$\psi = 8.314 \times 10^6 K_p T \tag{4}$$

where η_{mix} is the viscosity of the gas phase mixture and K_p is the permeability of the packed column (cm^2) and is assumed constant.

None of the previous general theories of finite concentration chromatography have attempted to treat the variation of the viscosity of the gas phase with composition. As early as 1965, Dr. Pretorius and other workers in South Africa studied the effect of variations in the gas phase viscosity on the migration rate of sharp fronts [38-41]. Dyson and Littlewood measured the composition dependence of gaseous mixtures at low concentrations [42] and the effect of the perturbation of the flow rate caused by viscosity changes on the shape and area of eluted peaks [43]. They showed that an error of up to 10% can be caused by neglecting the viscosity effect. However, in spite of this early work, many of the previous theories and techniques for measuring equilibrium isotherms by frontal chromatography invoked the assumption of constant gas phase viscosity.

The common carrier gases have a wide range of viscosities, from helium with a viscosity of 1.94×10^{-2} cP at 20°C to hydrogen with a viscosity of 0.88×10^{-2} cP. Also, the viscosities of binary gas mixtures are seldom linear in composition because of the greater effect of the heavier molecule in a collision; a maximum is often observed in the viscosity composition diagram. This is most often observed with mixtures of polar and nonpolar gases of similar viscosities but different molecular weight [44]. This is exactly the situation encountered in frontal chromatography with hydrogen carrier gas. Figure 1 illustrates the composition dependence of several gas mixtures. It can be seen that there is a wide variety of types of behavior and in particular that the viscosity of a mixture of pentane and hydrogen is very composition dependent, especially at low concentrations of pentane, even though the viscosities of the pure components are very similar. Thus, it is not valid to assume constant gas phase viscosity for binary mixtures even when the pure component viscosities are equal if the components differ in molecular weight.

Wilke [45] has presented a general equation for the viscosity of binary gas mixtures as a function of molecular weights, composition, and pure component viscosities:

$$\eta_{mix} = \frac{\eta_1}{1 + Y_2 \phi_{12}/Y_1} + \frac{\eta_2}{1 + Y_1 \phi_{21}/Y_2} \tag{5}$$

$$\phi_{ij} = \frac{\left\{1 + (\eta_i/\eta_j)^{1/2} (M_j/M_i)^{1/4}\right\}^2}{\left\{8[1 + (M_i/M_j)]\right\}^{1/2}} \tag{6}$$

where η_{mix} and η_i are the viscosities of the mixture and pure component i, and Y_i and M_i are the mole fraction and molecular weight of component i,

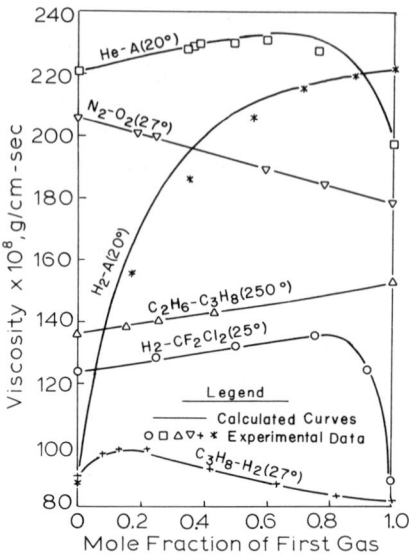

FIG. 1. Variation of the gas phase viscosity with composition for some binary mixtures. (Data from Wilke [45]. Courtesy of the Journal of Physical Chemistry.)

respectively. This correlation has been tested extensively [44] and found to be generally accurate to within 1 to 4%, even for systems with a maximum. When a theoretical model of this accuracy is available there is little need to assume constant gas phase viscosity.

The only other equation required to complete the formulation of the problem is an equation for the equilibrium isotherm. However, this can be a limiting factor because the accuracy of the mathematical solution is limited by the accuracy of the isotherm model. This is one of the serious disadvantages of the indirect chromatographic methods such as elution on a plateau and the locus methods. The exact mathematical form of the isotherm model will depend on the chromatographic system. In the case of partition isotherms, the best models are those derived from the Wilson equation [46, 47] or the Flory-Huggins equation [48-50], and numerous equations are available for adsorption isotherms of solids.

The set of equations given in Table 1 completely and rigorously describes the chromatographic system if certain additional assumptions or approximations are valid. Some of these assumptions are introduced simply for convenience and others are necessary to avoid the introduction of separate contributions to the shape of the isotherm. We have assumed that the gas phase is ideal and that the isotherm equation is independent of the total

TABLE 1

Basic Equations for Finite Concentration Chromatography of a Single Component (Solute or Moderator)

Mass balance equations

Solute: $\dfrac{\partial}{\partial z}\left[C_A v\right] + \theta_g\left[\dfrac{\partial C_A}{\partial t}\right] + \theta_L\left[\dfrac{\partial Q_A}{\partial t}\right] = 0$ (1)*

Total: $\dfrac{\partial}{\partial z}\left[C_t v\right] + \theta_g\left[\dfrac{\partial C_t}{\partial t}\right] + \theta_L\left[\dfrac{\partial Q_A}{\partial t}\right]$ (2)

Momentum balance equation: $v = \dfrac{\psi}{\eta}\left(\dfrac{\partial C_t}{\partial z}\right)$ (3)

Composition dependence of the gas phase velocity:

$$\eta_{mix} = \dfrac{\eta_1}{1 + Y_2 \phi_{12}/Y_1} + \dfrac{\eta_2}{1 + Y_1 \phi_{21}/Y_2} \quad (5)$$

Equilibrium isotherm equation: $Q_A = Q_A(C_A)$

---

*Equation numbers used in text.

pressure in the system. These assumptions could be eliminated by using the virial equation of state and an empirical expression for $Q_A = f(C_A, P)$ as proposed by Conder and Purnell [51]. The column is assumed to operate isothermally and the gas phase volume and column permeability are assumed to be independent of concentration. These assumptions are not critical and involve only minor perturbations on the system. However, there are two additional effects which must be treated: diffusion and nonequilibrium effects.

At very slow flow rates, molecular diffusion in the gas phase can influence the shape of the breakthrough curve. If the velocity is very high or the liquid layer is very thick, or for any reason the system departs from equilibrium, the simple isotherm model of $Q_A = f(C_A)$ will no longer be valid. Previous investigators have proposed certain experimental procedures to ensure the validity of these assumptions. The use of coarse packing has been advocated to ensure that the swelling of the liquid phase does not change the column permeability and to give a very low pressure drop across the column [52]. The use of long columns to reduce kinetic effects influencing the boundary shape and low flow rates to minimize the pressure drop are also common precautions.

Ideally, the frontal experiment should be run at a flow rate corresponding to the optimum velocity on a van Deemter plot. This choice of flow rate would minimize the influence of kinetic and diffusion effects on the shape of the frontalgram. This is not common practice because of the requirement for a minimum pressure gradient imposed by the theory used by previous investigators. However, if both the total material balance and momentum balance equations are utilized, then there is no need to restrict the pressure drop and the optimum flow rate can be employed.

IV. PREVIOUS CHROMATOGRAPHIC INVESTIGATIONS INVOLVING SIMPLIFYING ASSUMPTIONS

As stated earlier, a commonly employed assumption is constancy of the gas phase viscosity. This condition can be partially fulfilled in practice by limiting the experiment to low concentrations with a low-viscosity carrier gas. These are severe restrictions and limit the utility of the technique for isotherm measurement. However, the arithmetic is considerably simplified by the elimination of Eq. (5) and the resultant linear relation between velocity and the gradient in total concentration, $\partial C_t/\partial z$. All the theories discussed in this section take advantage of this assumption.

All the simplifying conditions previously employed have involved some assumed gradient in the total concentration. Three of the possible assumptions are as follows:

I. The total gas phase concentration is constant, i.e., $\partial C_t/\partial z = \partial C_t/\partial t = 0$.

II. The total gas phase concentration is a linear function of z, i.e., $\partial C_t/\partial z = $ const.

III. The square of the total gas phase concentration is a linear function of z, i.e., $\partial C_t^2/\partial z = $ const.

The third assumption is not common, but is physically realistic. Several authors [37, 53] have shown that for a compressible gas flowing at a constant molar flow rate, the pressure, P_z, at any point in the column is given by

$$\frac{z}{L} = \frac{(P_I/P_O)^2 - (P_z/P_O)^2}{(P_I/P_O)^2 - 1} \tag{7}$$

where P_I and P_O are the inlet and outlet pressures, respectively, and L is the column length.

For an ideal gas, the pressure is directly related to the total concentration and Eq. (7) can be rearranged to show that C_t^2 is a linear function of z:

$$C_t^2 = C_{t,0}^2 - \frac{z}{L}[C_{t,0}^2 - C_{t,L}^2] \tag{8}$$

where $C_{t,0}$ and $C_{t,L}$ are the total gas phase concentrations at the inlet and outlet of the column. Equation (8) is only valid at constant molar flow rate and this condition is not fulfilled in a column with a significant concentration gradient due to the sorption effect. However, the condition is valid for certain types of chromatography such as elution of small samples or elution on a plateau.

A. Theories Utilizing Assumption I

If the momentum balance equation is replaced by assumption I and the gas phase viscosity is assumed invariant, then Eqs. (1) and (2) can be simplified:

$$v\left(\frac{\partial C_A}{\partial z}\right) + C_A\left(\frac{\partial v}{\partial z}\right) + \left[\theta_g + \theta_L\left(\frac{\partial Q_A}{\partial C_A}\right)\right]\frac{\partial C_A}{\partial t} = 0 \tag{9}$$

$$C_t\left(\frac{\partial v}{\partial z}\right) + \theta_L\left(\frac{\partial Q_A}{\partial C_A}\right)\left(\frac{\partial C_A}{\partial t}\right) = 0 \tag{10}$$

Since it is embarrassing to talk about the variation of the velocity in a column with no pressure drop, this troublesome derivative can be eliminated to yield

$$\frac{v}{(\partial z/\partial t)_{C_A}} = \theta_g + \left\{1 - \left(\frac{C_A}{C_t}\right)\right\}\theta_L\left(\frac{\partial Q_A}{\partial C_A}\right) \tag{11}$$

The velocity v is the superficial velocity of the gas phase and is equal to the flow rate divided by the cross-sectional area of the <u>empty</u> column. The derivative $(\partial z/\partial t)_{C_A}$ is the linear velocity of a fixed concentration, C_A. If these two velocities were constant, the ratio given by the left-hand side of Eq. (11) would be equal to the ratio of the retention volume of the solute to the retention volume of an inert sample, V_R/V_m, or if we could integrate the velocities from 0 to L, the same ratio V_R/V_m could be obtained. Several workers have utilized this concept [16, 19] and it is the most commonly used retention volume equation for finite concentration chromatography:

$$V_R = V_M + (1 - Y) \frac{\partial Q_A}{\partial C_A} V_L \tag{12}$$

It is difficult to evaluate V_R experimentally because the velocity v is a function of z and t. Guiochon and Jacob [54] avoided the use of a retention volume by solving Eqs. (9) and (10) using the method of characteristics; however, this method still requires the use of assumption I.

B. Theories Utilizing Assumption II

The overall material balance equation may be replaced by the assumption of a linear pressure gradient. Darcy's law requires that the velocity is constant in this case and the solute mass balance equation can be used to give another form of the retention volume equation:

$$V_R = V_M + \frac{\partial Q_A}{\partial C_A} V_L \tag{13}$$

If the isotherm is linear at the concentrations of interest, then Eq. (13) reduces to the popular form of the retention volume equation $V_R = V_M + KV_L$, where K is the "chromatographic" partition coefficient, Q_A/C_A.

C. Theories Utilizing Assumption III

The assumption that C_t^2 is a linear function of z [Eq. (8)] is the most realistic simplifying assumption, and Guiochon [22] has derived a complete set of equations for finite concentration chromatography based on this assumption.

Another set of equations can be obtained by replacing the overall material balance equation with Eq. (8) to give a single differential equation for the solute:

$$\alpha' \left(\frac{\partial C_A}{\partial z} \right) + \beta' C_A = \frac{\partial C_A}{\partial t} \tag{14}$$

FINITE CONCENTRATION FRONTAL CHROMATOGRAPHY

$$\alpha' = \frac{-\psi}{4\eta C_t^3}\left(\frac{\partial C_t^2}{\partial z}\right) \Big/ \left\{\theta_g + \theta_L\left(\frac{\partial Q_A}{\partial C_A}\right)\right\}$$

$$\beta' = \frac{-2C_t^2 \alpha'}{\partial C_t^2/\partial z}$$

$$\frac{\partial C_t^2}{\partial z} = \frac{-(C_{t,0} - C_{t,L})}{L}$$

Although assumption III is the most realistic assumption, it does not reduce the mathematical complexity of the system to the point at which an analytical solution can be obtained. Neither Eq. (14) nor Guiochon's equations have been used as a practical approach for the measurement of equilibrium isotherms.

D. Numerical Solution of the Unrestricted Equations

In many cases, none of the simplifying assumptions is realistic and a numerical solution of the full set of equations must be utilized. Also, all the equations discussed in Sec. IV have assumed a constant gas phase viscosity, and this is very often an unrealistic assumption.

The full set of equations can be cast in a workable form by eliminating v between Eqs. (1), (2), and (3) and writing the total mass balance in terms of C_t^2.

$$\alpha\left(\frac{\partial C_A}{\partial z}\right) + \beta C_A = \frac{\partial C_A}{\partial t} \tag{15}$$

$$\gamma\left(\frac{\partial^2 C_t^2}{\partial z^2}\right) + \delta\left(\frac{\partial C_t^2}{\partial z}\right) + \epsilon = \frac{\partial C_t^2}{\partial t} \tag{16}$$

$$\alpha = \frac{\psi(\partial C_t^2/\partial z)}{2\eta C_t\{\theta_g + \theta_L(\partial Q_A/\partial C_A)\}}$$

$$\beta = \frac{\alpha}{(\partial C_t^2/\partial z)}\left[\frac{\partial^2 C_t^2}{\partial z^2} - \frac{1}{2C_t^2}\left(\frac{\partial C_t^2}{\partial z}\right)^2 - \frac{1}{\eta}\left(\frac{\partial C_t^2}{\partial z}\right)\left(\frac{\partial \eta}{\partial z}\right)\right]$$

$$\gamma = \frac{\psi C_t}{\eta \theta_g}, \qquad \delta = \frac{-\gamma}{\eta}\left(\frac{\partial \eta}{\partial z}\right), \qquad \epsilon = -\frac{2\theta_L C_t}{\theta_g}\left(\frac{\partial Q_A}{\partial t}\right)$$

Application of the proper boundary conditions and equations for the gas phase viscosity and the pártition isotherm completes the problem formulation. The system of equations is very complex but perfectly rigorous aside from the usual assumptions concerning the partition or adsorption kinetics and the contribution of diffusion and mass transfer to the shape of the elution peak or breakthrough curve.

A numerical solution of Eqs. (15) and (16) has been developed and the isotherms of several reference systems have been measured [55]. The general approach is to make an initial estimate of an isotherm equation, solve Eqs. (15) and (16), for a profile of C_A as a function of time at the detector, and to adjust the isotherm parameters to obtain coincidence of the calculated and observed chromatograms, using a nonlinear least-squares routine. Because there are no restrictions on the pressure drop or the gas phase viscosity, any suitable carrier gas can be used and the column can be operated at the optimum flow rate.

V. DISCUSSION

The rigorous numerical solution of Eqs. (15) and (16) can be used to calculate an equilibrium isotherm under any experimental conditions, such as a large pressure drop, high solute concentration, high flow rate. However, this method may require an exorbitant amount of computer time. Thus it is often advantageous, from a practical standpoint, to arrange the experimental conditions to ensure validity of one or more of the simplifying assumptions. The investigator can thus obtain a compromise between the rigor and accuracy of the results and the complexity and difficulty of the mathematical solution.

Another parameter which can influence the operator's choice of experimental techniques is the type of isotherm data required. The single-point method discussed in Sec. II.B.1 (tracer pulse, frontal analysis with a sharp front, and the "foreign" solute methods) all give a Q_A value directly and the arithmetic is relatively simple. However, the experimental technique is complicated and several experiments must be carried out to obtain an isotherm over a useful range of concentrations.

The "locus" methods (frontal analysis with a diffuse front, elution on a plateau, and the numerical solution of the rigorous equations) give the derivative of the isotherm, Q_A / C_A, and the data must be fit to an isotherm equation. Thus, the mathematics is far more difficult but an entire isotherm may be obtained in one or two experiments.

VI. PRACTICAL APPLICATIONS OF FRONTAL CHROMATOGRAPHY

A. Adsorption Isotherms

The use of frontal chromatography to measure the adsorption isotherms of solutes on solid surfaces is the oldest and most common application of this technique. Many authors have reviewed this field [4, 16, 18] and the interested reader is referred to the original literature for the myriad examples of successful application of this technique.

B. Partition Isotherms

Determination of the solubility of a solute in a stationary liquid phase by frontal chromatography is more difficult than the determination of adsorption isotherms for a variety of reasons. The primary difference is the greater amount of solute dissolved in a liquid as opposed to that adsorbed on a solid. However, several investigators have measured partition isotherms [16, 22, 52] and shown the results to be accurate compared with static measurements.

An interesting application of this technique has been an investigation of liquid surface adsorption of alcohols on hydrocarbon solutes. Elution studies [56] and static measurements [57] indicated that there was significant adsorption of alcohols at the gas-liquid interface. Frontal chromatographic measurements of the partition isotherms, however, have shown that there was no indication of such interfacial adsorption for alcohol-hydrocarbon systems [58]. Conder [59] has recently proposed one plausible explanation for this apparent contradiction in terms of the different concentration ranges covered in elution and frontal experiments.

C. Elemental Analysis

Although the nonanalytical applications of frontal chromatography are predominant, Rezl and Janak have developed a technique and the instrumentation for elemental (CHN) analysis by frontal chromatography [60-62]. The sample is pyrolized with a catalyst and the gaseous products are swept into a chamber and diluted with an inert gas. The diluted sample is then pumped into a chromatographic column and the composition of the mixture is determined from the heights of the plateaus on the frontalgram. The authors have shown that the method is fast, accurate, and simple compared to an elution technique. This approach has also been used to measure diffusible hydrogen and oxygen in metals [63].

D. Study of Ternary Systems

A binary carrier gas with one condensable component can be used in a chromatographic system with either a solid adsorbent or a nonvolatile liquid as the stationary phase. If the condensable component of the carrier gas adsorbs on a solid adsorbent, elution of another solute from the equilibrated column would involve three chromatographically active phases (binary carrier gas, adsorbed liquid, and solid adsorbent). If the condensable component of the carrier gas forms a solution with a nonvolatile liquid phase, there will only be two phases which affect the retention of another solute (binary carrier gas and binary liquid phase). Elution of a "foreign" solute would mean the addition of one component to each of the active phases even though the solute is "infintely dilute."

1. Three-Phase Systems (Gas, Liquid, and Solid)

The most common application of this mode of operation is the deactivation of solid supports by water, steam, volatile acids, or amines in GLC. Nonaka [64] has written an excellent review of this area with particular reference to the use of steam. Little work has been done in the study of the effect of adsorbed liquids on the adsorptive or catalytic properties of solids, and frontal chromatography is an excellent experimental technique for such investigations.

2. Two-Phase Systems (Gas and Liquid)

Drs. Kobayashi, Stalkup, and Deans have studied a large number of systems involving a condensable component in the carrier gas. They studied the retention characteristics of the C_2-C_4 hydrocarbons at infinite dilution in n-decane and n-hexadecane using methane as the carrier gas [65, 66]. These authors measured the partition coefficient $K_i = y_i/x_i$ of the infinite dilution solutes as a function of the concentration of methane in the liquid phase by the relation

$$K_i = \frac{n_{NV}^L}{C_t V_{N_i}(1 - 1/K_{CH_4})} \tag{17}$$

where C_t is the total molar concentration of the gas phase (assumed constant), V_{N_i} the net retention volume of solute i at infinite dilution, n_{NV}^L the number of moles of the nonvolatile liquid phase in the column, and x_i and y_i the mole fraction of component i in the liquid and gaseous phases, respectively. This method requires an independent measurement of K_{CH_4} as a function of the pressure of methane in the carrier gas. This method can be used to measure the partition coefficient of a "foreign" solute in a binary liquid phase as a function of composition if the composition of the liquid phase can be determined by some independent means.

Other investigators have used the same theory to analyze data for CO_2 in hydrocarbon solvents with methane as the carrier gas [67, 68], and the n-decane system has been studied at finite concentrations of CO_2 [69] using the relation derived by Stalkup and Deans [65]:

$$V_{N_i} = \frac{n_{NV}^L K_i K_{CH_4}}{C_t} \left\{ \frac{K_{CH_4} - 1 + Y_i[K_i - K_{CH_4}]}{[K_i K_{CH_4} - K_i + Y_i(K_i - K_{CH_4})]^2} \right\} \quad (18)$$

The use of Eqs. (17) and (18) is limited by the requirement for an a priori knowledge of K_{CH_4} as a function of pressure.

It is possible to determine K for the condensable component of the carrier gas, i.e., the moderator, by the usual frontal analysis methods. This has recently been done [27, 28] for four chromatographic systems.

One of the most interesting areas of study for this type of investigation is the controversial area of the measurement of stability constants of weak complexes by chromatography. A great deal of work has been carried out by measuring the increase in the retention volume or partition coefficient of a solute caused by the addition of a complexing agent to an inert liquid phase. Equation (19) is the usual expression for the evaluation of the complex formation constant K_1:

$$(K_{R_i}^°)_M = (K_{R_i}^°)_I [1 + K_1 Q_A] \quad (19)$$

where $K_{R_i}^°$ is the "chromatographic" partition coefficient, Q_i/C_i, of component i at infinite dilution. The subscripts M and I refer to the binary liquid phase and the pure nonvolatile stationary phase, respectively.

This equation was originally proposed by Schubert [70, 71] for complexing in liquid systems. Many authors [29, 30, 71-74] have used this equation to calculate K_1 by measuring $(K_{R_i}^°)_M$ as a function of Q_A over wide ranges of Q_A. However, this equation predicted that $(K_{R_i}^°)_M = (K_{R_i}^°)_I$ when there was no possibility of complexing in the system and that $(K_{R_i}^°)_M \geq (K_{R_i}^°)_I$ in all cases. It soon became obvious that neither of these conclusions is valid. For example, Castells [75] measured the partition coefficients of a series of alkanes in mixtures of 1,3,5-trinitrobenzene (TNB) and dinonyl phthalate (DNP) and observed that $(K^\circ_{R_i})_M$ decreased with increasing mole fraction of TNB. It was observed that the activity coefficients of the alkane solutes (cyclohexane, methycyclohexane, n-hexane, n-heptane, and 2,2,4-trimethylpentane) increased with the mole fraction of TNB even though there was no complexing in the system.

In complexing systems there are two types of solute-solvent interactions, a chemical interaction (complex formation) and a physical interaction (mixed solvent effect) [29, 30]. The main problem then became that of finding an independent method for evaluating the solubility of a complexing solute in a binary solvent if a complex was not formed. Martire proposed an equation of the following form for a 1:1 complex [29, 76]:

$$\frac{(K^\circ_{R_i})_M}{(K^\circ_{R_i})_I} = 1 + (K_1 + \alpha_1)Q_D + \left[K_1(\alpha_1 + \beta_1) + \frac{\alpha_1^2}{2} + \alpha_2\right]Q_D^2 \qquad (20)$$

$$\alpha_1 = \frac{v_A(v_I - v_D)}{v_1} + v_I + v_D x_I^i - v_D x_D^i + v_A x_I^D$$

$$\alpha_2 = -v_A v_D x_I^D, \qquad \beta_1 = \frac{d\gamma_D}{dQ_D}$$

where the subscripts D, A, and I refer to the electron donor additive or moderator, the solute, and the inert, nonvolatile liquid phase, v_j is the molar volume of component j, Q_D the molar concentration of component D in the solution, γ_D the activity coefficient of D in the solution, and x_j^k the interaction parameter of component k in component j. The α and β terms of the equation are determined by the physical interactions between the solute and solvent.

It is difficult to evaluate the higher order term of the equation with mixed liquid phases because of the conditions $v_I \approx v_D \gg v_A$ and $x_D \approx \phi_D$. Purnell and Vargas de Andrade [31, 32] have observed that $(K^\circ_{R_i})_M$ is almost always a linear function of Q_D for mixed nonvolatile liquid phases.

Frontal chromatography provides an excellent method for studying the nonspecific interactions because one component of the liquid phase is volatile; thus $v_A \approx v_D \ll v_I$ and often $\ln\,[(K^\circ_{R_i})_M/(K^\circ_{R_i})_S] > 1$. These conditions invalidate some of the assumptions used in the derivation of Eq. (20), and the logarithmic form must be used:

$$\ln\left\{\frac{(K_{R_i}^\circ)_M}{(K_{R_i}^\circ)_I}\right\} = \alpha_1 Q_D + (\alpha_2 + K_1\alpha_1)Q_D^2 \tag{21}$$

Equation (21) can be cast in a more convenient form by using volume fractions, ϕ_i. If the excess volume of mixing for the binary liquid phase is negligible, then $\phi_i = C_i v_i$ and

$$\ln\left\{\frac{(K_{R_i}^\circ)_M}{(K_{R_i}^\circ)_I}\right\} = \left(\frac{\alpha_1}{v_D}\right)\phi_D + \left(\frac{\alpha_2}{v_D^2} + \frac{K_1\alpha_1}{v_D^2}\right)\phi_D^2 \tag{22}$$

If the coefficient of the ϕ_D^2 term is negligible and $\alpha_1\phi_D/v_D \ll 1$, then Eq. (22) reduces to

$$\frac{(K_{R_i}^\circ)_M}{(K_{R_i}^\circ)_I} = 1 + \left(\frac{\alpha_1}{v_D}\right)\phi_D \tag{23}$$

$$(K_{R_i}^\circ)_M = (K_{R_i}^\circ)_I \phi_I + \left[(K_{R_i}^\circ)_I\left(1 + \frac{\alpha_1}{v_D}\right)\right]\phi_D \tag{24}$$

This is the equation proposed by Purnell and Vargas de Andrade [31, 32]. Frontal chromatography can be used to test the validity of Eqs. (22) and (24) by studying systems in which $\alpha_1\phi_D/v_D \geq 1$ and the ϕ_D^2 term is not negligible. Also, the value of χ_I^D can be evaluated directly from the infinite dilution activity coefficient of the volatile moderator D in the inert solvent I. Thus α_1 and α_2 can be measured explicitly for noncomplexing systems and the significance of the ϕ_D^2 term of Eq. (22) evaluated. This has been done by the author [28] and an example of the results obtained are given in Fig. 2. The ϕ_D^2 term is insignificant for acetone with n-heptane

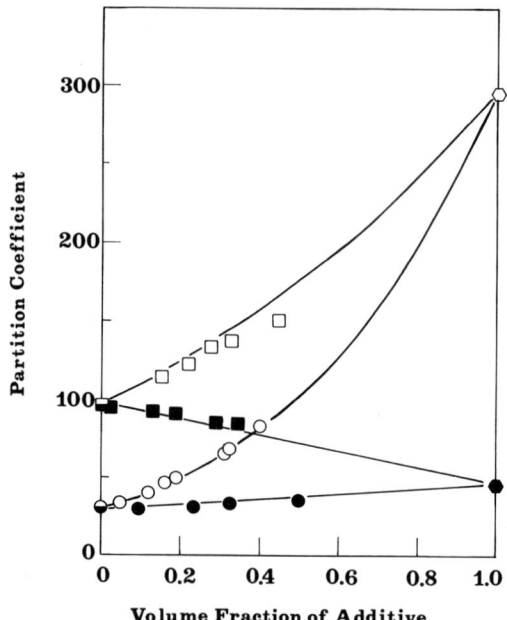

FIG. 2. Chromatographic partition coefficient of acetone as a function of liquid phase composition. □, Benzene in dinonyl phthalate; ○, benzene in squalane; ■, n-heptane in dinonyl phthalate; ●, n-heptane in squalane; ○, value calculated from the infinite dilution activity coefficient of acetone in benzene; ●, value calculated from the infinite dilution activity coefficient of acetone in n-heptane.

as the moderator for either squalane or dinonyl phthalate as the inert liquid phase. Thus this system obeys an equation of the form proposed by Purnell. The ϕ_D^2 term is obviously significant when benzene is used as the moderator. In this case, Eq. (24) is inadequate and the full expression, Eq. (22) must be used.

VII. SUMMARY

Frontal chromatography is no longer an "instructive curiosity" but is a supplementary technique for normal elution chromatography. In any investigation requiring data at finite solute concentrations or with multiple-component systems, frontal chromatography can be used to advantage. The sophistication of the instrumentation, experimental technique, and the mathematics depends on the particular application and can vary from very simple to very complex.

Recent developments in this field should allow future investigators to concentrate on the practical applications of the technique without becoming embroiled in the complex theory of finite concentration chromatography.

ACKNOWLEDGMENTS

The author would like to express his appreciation to Dr. H. W. Haynes of the Chemical Engineering Department of the University of Mississippi and Dr. D. E. Martire of Georgetown University for many helpful discussions, and the National Science Foundation (Grant No. CHE73-08458 A03) for financial support.

REFERENCES

1. G. Schay, Chromatographia, 3, 203 (1970).
2. H. W. Habgood, in The Gas-Solid Interface, Vol. 2 (E. A. Flood, ed.), Marcel Dekker, New York, 1967, p. 611.
3. R. Kobayashi, P. S. Chappelear, and H. A. Deans, Ind. Eng. Chem., 59(10), 63 (1967).
4. N. C. Saha and D. S. Mathur, J. Chromatogr., 81, 207 (1973).
5. V. G. Berezkin, J. Chromatogr., 91, 559 (1974).
6. J. C. Giddings and K. L. Mallik, Ind. Eng. Chem., 59(4), 18 (1967).
7. J. R. Conder, in Progress in Gas Chromatography (J. H. Purnell, ed.), Wiley-Interscience, New York, 1968, p. 209.
8. C. L. Young, Chromatogr. Rev., 10, 129 (1968).
9. D. C. Locke, Am. Lab., 7 (5), 17 (1975).
10. D. C. Locke, in Advances in Chromatography, Vol. 14 (J. C. Giddings, ed.), Marcel Dekker, New York, 1976, p. 87.
11. J. R. Conder, Chromatographia, 7, 387 (1974).
12. J. R. Conder and J. H. Purnell, Trans. Faraday Soc., 65, 824 (1969).
13. C. N. Reilley, G. P. Hildebrand, and J. W. Ashley, Anal. Chem., 34, 1198 (1962).

14. F. Helfferich, J. Chem. Educ., **41**, 410 (1964).
15. D. L. Peterson and F. Helfferich, J. Phys. Chem., **69**, 1283 (1965).
16. C. J. Chen and J. F. Parcher, Anal. Chem., **43**, 1738 (1971).
17. P. A. Sewell and R. Stock, J. Chromatogr., **50**, 10 (1970).
18. J. F. K. Huber and R. G. Gerritse, J. Chromatogr., **58**, 137 (1971).
19. J. R. Conder and J. H. Purnell, Trans. Faraday Soc., **64**, 3100 (1968).
20. C. H. Bosanquet and G. O. Morgan, in Vapor Phase Chromatography (D. H. Desty, ed.), Butterworths, London, 1957, p. 35.
21. J. F. Parcher, J. Chem. Educ., **49**, 472 (1972).
22. P. Valentin and G. Guiochon, Separation Sci., **10**, 245 (1975).
23. G. Schay, L. G. Nagy, and G. Racz, Acta Chim. Hung., **71**, 23 (1972).
24. P. Valentin and G. Guiochon, J. Chromatogr. Sci., **14**, 56 (1976).
25. P. Valentin and G. Guiochon, J. Chromatogr. Sci., **14**, 132 (1976).
26. F. Helfferich and D. L. Peterson, Science, **142**, 661 (1963).
27. J. F. Parcher and T. N. Westlake, J. Chromatogr. Sci., **14**, 343 (1976).
28. J. F. Parcher and T. N. Westlake, J. Phys. Chem., **81**, 307 (1977).
29. D. E. Martire, Anal. Chem., **46**, 1712 (1974).
30. D. E. Martire, Anal. Chem., **48**, 398 (1976).
31. J. H. Purnell and J. M. Vargas de Andrade, J. Am. Chem. Soc., **97**, 3585 (1975).
32. J. H. Purnell and J. M. Vargas de Andrade, J. Am. Chem. Soc., **97**, 3590 (1975).
33. R. J. Laub and J. H. Purnell, J. Am. Chem. Soc., **98**, 30 (1976).
34. R. J. Laub and J. H. Purnell, J. Am. Chem. Soc., **98**, 35 (1976).
35. J. J. Haydel and R. Kobayashi, Ind. Eng. Chem., Fundam., **6**, 546 (1967).
36. P. Fejes and G. Schay, Acta Chim. Hung., **17**, 377 (1958).
37. G. Guiochon, in Chromatographic Reviews, Vol. 8 (M. Lederer, ed.), Elsevier, Amsterdam, 1966, p. 1.
38. G. J. Krige and V. Pretorius, Anal. Chem., **37**, 1186 (1965).
39. G. J. Krige and V. Pretorius, Anal. Chem., **37**, 1191 (1965).

40. G. J. Krige and V. Pretorius, Anal. Chem., 37, 1195 (1965).
41. G. J. Krige and V. Pretorius, Anal. Chem., 37, 1202 (1965).
42. N. Dyson and A. B. Littlewood, Trans. Faraday Soc., 63, 1895 (1967).
43. N. Dyson and A. B. Littlewood, Anal. Chem., 39, 638 (1967).
44. R. C. Reid and T. K. Sherwood, The Properties of Gases and Liquids, 2nd ed., McGraw-Hill, New York, 1966, p. 420.
45. C. R. Wilke, J. Chem. Phys., 18, 517 (1950).
46. G. M. Wilson, J. Am. Chem. Soc., 86, 127 (1964).
47. R. V. Orye and J. M. Prausnitz, Ind. Eng. Chem., 57, 18 (1965).
48. P. J. Flory, J. Chem. Phys., 10, 51 (1942).
49. M. L. Huggins, Ann. N.Y. Acad. Sci., 43, 1 (1942).
50. G. M. Janini and D. E. Martire, J. Chem. Soc., Faraday Trans. II, 70, 837 (1974).
51. J. R. Conder and J. H. Purnell, Trans. Faraday Soc., 64, 1505 (1968).
52. J. R. Conder and J. H. Purnell, Trans. Faraday Soc., 65, 839 (1969).
53. A. I. M. Keulemans, Gas Chromatography, 2nd ed., Reinhold, New York, N.Y., 1959, p. 142.
54. G. Guiochon and L. Jacob, Chromatogr. Rev., 14, 77 (1971).
55. J. F. Parcher, T. H. Ho, and H. W. Haynes, J. Phys. Chem., 80, 2656 (1976).
56. H. L. Liao and D. E. Martire, Anal. Chem., 44, 498 (1972).
57. R. L. Pecsok and B. H. Gump, J. Phys. Chem., 71, 2202 (1967).
58. J. F. Parcher and C. L. Hussey, Anal. Chem., 45, 188 (1973).
59. J. R. Conder, Anal. Chem., 48, 917 (1976).
60. V. Rezl and J. Uhdeova, Am. Lab., 8(1), 13 (1976).
61. V. Rezl and J. Janak, J. Chromatogr., 81, 233 (1973).
62. V. Rezl, J. Chromatogr. Sci., 10, 419 (1972).
63. V. Rezl, B. Kaplanova, and J. Janak, Anal. Chem., 47, 159 (1975).
64. A. Nonaka, in Advances in Chromatography, Vol. 12 (J. C. Giddings, ed.), Marcel Dekker, New York, 1975, p. 223.
65. F. I. Stalkup and H. A. Deans, A.I.Ch.E. J., 9, 106 (1963).
66. F. I. Stalkup and R. Kobayashi, A.I.Ch.E. J., 9, 121 (1963).

67. K. Asano, T. Nakahara, and R. Kobayashi, J. Chem. Eng. Data, 16, 16 (1971).
68. A. Yudovich, R. L. Robinson, and K. C. Chao, A.I.Ch.E. J., 17, 1152 (1971).
69. F. Khoury and D. B. Robinson, J. Chromatogr. Sci., 10, 683 (1972).
70. J. Schubert, J. Phys. Chem., 56, 113 (1952).
71. J. H. Purnell, in Gas Chromatography 1966 (A. B. Littlewood, ed.), Institute of Petroleum, London, 1967, p. 3.
72. D. F. Cadogan and J. H. Purnell, J. Phys. Chem., 73, 3849 (1969).
73. C. Eon, C. Pommier, and G. Guiochon, J. Phys. Chem., 75, 2632 (1971).
74. R. Vivilecchia and B. L. Karger, J. Am. Chem. Soc., 93, 6598 (1971).
75. R. C. Castells, Chromatographia, 6, 57 (1973).
76. J. M. Janini, J. W. King, and D. E. Martire, J. Am. Chem. Soc., 96, 5368 (1974).

Chapter 6

THE LIQUID-CHROMATOGRAPHIC RESOLUTION
OF ENANTIOMERS

Ira S. Krull*

Environmental Biology Department
Boyce Thompson Institute
Yonkers, New York

I. INTRODUCTION AND BACKGROUND 176
II. LIQUID-CHROMATOGRAPHIC SEPARATION
OF DIASTEREOMERS............................. 177
III. A POTENTIAL OPTICALLY ACTIVE RESOLVING
COLUMN FOR LIQUID CHROMATOGRAPHY 183
IV. DIRECT RESOLUTIONS BY LIQUID CHROMATOGRAPHY
USING OPTICALLY ACTIVE ELUENTS 184
V. DIRECT RESOLUTIONS BY LIQUID CHROMATOGRAPHY
USING AN OPTICALLY ACTIVE SUBSTRATE 186

   A. Optically Active Packing Materials in General 186
   B. Optically Active Crown Ether-Bonded Supports 188
   C. Specifically Adsorbing Silica Gels 192
   D. Optically Active Polymeric Supports 193
   E. Ligand-Exchange Resolutions 196
   F. Sephadex-Type Supports......................... 200
   G. Enzyme-Based Supports......................... 201

VI. DIRECT RESOLUTIONS USING ION-PAIR
PARTITION CHROMATOGRAPHY...................... 202
VII. CONCLUSIONS 203
   ACKNOWLEDGMENTS 204
   REFERENCES 204

*Current affiliation: Cancer Research Center, Thermo Electron
Corporation, Waltham, Massachusetts

I. INTRODUCTION AND BACKGROUND

There has been a recent upsurge in the development of chromatographic methods for the resolution of enantiomers, including vapor phase chromatography (VPC), thin-layer chromatography (TLC), and liquid chromatography (LC). As a result of the increase in the number of publications dealing with this subject, there have appeared a rather large number of excellent review papers [1-6]. The purpose of this chapter is to indicate those areas of LC resolutions that have been of major interest, and which should offer improved areas of study for the future. It is intended to present certain untested methods of resolution using LC, in the hope of contributing to the future development of this subject.

At the same time, it is not my intention to describe the latest developments in instrumentation and general applications of high-pressure liquid chromatography (HPLC), since a number of very recent books and articles dealing with this subject are available [7-13]. Rather, it is hoped to describe recent approaches involving various aspects of liquid chromatography that have resulted in either partial or complete resolutions of enantiomers. Detailed discussions of the various terms used by the organic chemist in dealing with optical isomers can be found in any of a number of books and survey articles [14-19]. Of particular interest are the references of Mislow and Raban [15, 17, 18].

In dealing with optical isomers, i.e., nonsuperimposable mirror images, such as a "right and left hand," all physical and spectral properties are identical. This is true only in the absence of other optically active substances or reagents. The single physical property that differs for any two enantiomers (mirror images) is their interaction with the plane of polarized light (plain). Thus, two enantiomers will rotate this plane of light to the same extent if each is 100% pure, but they will do so in entirely different directions. One of the two enantiomers will rotate the light in a right-handed direction, and is termed dextrorotatory (d-), whereas the other enantiomer will rotate the same light in a left-handed direction, and is termed levorotatory (l-). All enantiomers will react with an optically inactive reagent to the same extent, i.e., they will have the same reaction rate constants and equilibrium constants. For the case of an optically active or chiral reagent (or light), two enantiomers may have quite different rate constants, and the reaction with such a reagent will usually result in unequal amounts of the two diastereomeric products.

The separation of two enantiomers (racemates) is termed resolution, and the conversion of one enantiomer into a mixture of both enantiomers is termed racemization. Since all the physical properties of two given enantiomers will be the same in the absence of an optically active medium, the separation of these enantiomers becomes a much more difficult problem in chromatography than the relatively simple separation of geometrical isomers, stereoisomers, positional isomers, or even the separation of compounds differing solely in their isomeric distributions. In most of the examples to be discussed regarding the resolution of enantiomers, the optical activity of the materials is usually due to the presence of one type of asymmetric carbon atom. However, optical activity is often due to structural parameters other than an asymmetric center; some of these parameters are (a) an allene or spirane grouping; (b) a hindered biphenyl arrangement; (c) a rigid, cyclic, trans-olefin; or (d) another unusual atomic arrangement (e.g., sulfoxides). Most of the compounds of interest to organic or biological chemists are optically active because of the presence of an optically active carbon atom, and most of the liquid-chromatographic resolutions described deal with these types of compounds. Resolutions can be divided rather arbitrarily into two major types, direct and indirect [1]. A direct resolution of enantiomers usually means the physical separation of such isomers without the formation of separate, distinct diastereomers by the use of an optically active resolving agent. Indirect resolutions involve the use of such diastereomers, and some examples of this method, the use of which has been traditional for resolving enantiomers, are briefly described. However, direct resolutions are emphasized herein, mainly because these methods are usually simpler, faster, less tedious, and often less expensive to perform. These also appear to be the more elegant method of resolution, on which the emphasis for future work will undoubtedly be placed.

Mention should be made at the outset of the early attempts at resolutions, most notably those of Pasteur [20]. Pasteur accidentally chose to work with certain salts of tartaric acid, and because of certain physical properties of these salts in solution below 27°C, he was able to obtain two different crystalline modifications. It was then a rather simple matter to separate the two forms physically, and thereby obtain the pure acids by acidification of the salts [20]. Other more recent applications of the preferential crystallization procedure used by Pasteur have appeared sporadically throughout the literature [21, 22]. One other popular method utilized for the resolution of various olefins was complexation with a metal complex that already contained an optically active organic ligand [23-27]. Much of this particular area has been reviewed [23]. Resolutions by VPC have also been dealt with rather extensively in the survey article by Gil-Av and Nurok [2], and shall be reviewed again shortly [28].

II. LIQUID-CHROMATOGRAPHIC SEPARATION OF DIASTEREOMERS

The most general procedure for the resolution of enantiomers has involved the formation of appropriate diastereomeric mixtures using an optically active resolving agent [16], many of which are commercially available.

The formation of diastereomers for the case of a racemic organic acid and an optically active organic base is depicted in Fig. 1. Under the appropriate reaction conditions, all of the acid to be resolved will be converted to two distinct diastereomers, represented in Fig. 1 by (D-acid + D-base) and (L-acid + L-base). Diastereomers are actually two different compounds, and can be separated by many different physical methods, e.g., fractional crystallization, VPC, distillation, TLC, or LC [1, 2, 18, 43, 44, 137, 138]. The choice of a resolving agent to be used for a particular set of enantiomers must be based largely on the nature of the functional group(s) present, the availability of the desired resolving agent, and a knowledge of suitable reaction conditions necessary for the reaction(s). Several typical resolving agents in common use are indicated in Fig. 2; several of these may be used with alcohols, amines, carboxylic acids, or anhydrides. The method to be chosen for the separation of the diastereomers depends on several factors; often, the formation of a mixture of diastereomers using a certain resolving agent does not guarantee an eventual separation.

Many separations of diastereomers have been routinely done by VPC [2]. This is acceptable if the materials are volatile enough, and if they are stable at the separation conditions required. It is also usually necessary to obtain large-scale amounts of the resolved enantiomers for any biological or physical testing; thus the diastereomers must be subjected to preparative VPC with the need for baseline or near-baseline separations. Much recent work has used HPLC for the separation of certain diastereomers, and this method appears to be gaining in popularity. One of the advantages of HPLC over VPC is the ability to recover quantitatively all of the diastereomers placed on the column. Other advantages involve the ability to easily scale-up, to recycle, to change separation conditions rapidly, and to use low temperatures. With the advent of microparticle packings, and the technical perfection of recycling and preparative LC equipment by various manufacturers, HPLC is fast becoming the method of choice for the separation of large amounts of diastereomers. However, it is still necessary for the chemist to quickly work out the best separation conditions within HPLC, the best packing material, solvent conditions, flow rate, column size, temperature, and other factors. All of this requires a good deal of time, equipment, solvents, effort, and money. In addition to the work required with the HPLC method itself, one must still prepare the needed diastereomers, ensure the 100% optical purity of the resolving agents used, and, after the actual

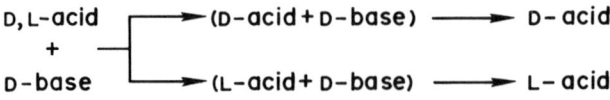

FIG. 1. Formation of diastereomers.

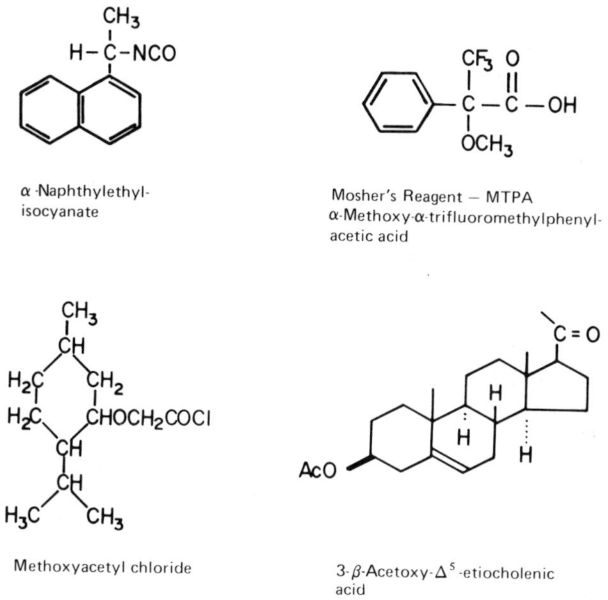

FIG. 2. Typical optically active resolving agents.

separation of the diastereomers, one must then recover the resolved enantiomers by a reaction on the individual diastereomers. In some cases, the recovered optically active enantiomers must be purified further before any physical or biological studies can be done. Success with one particular class of organic compounds using a particular resolving agent and set of HPLC conditions does not guarantee success for another member of the same series. Some recent examples involving the HPLC separation of diastereomers are illustrated herein; however, a complete summation of all that has been done in this area requires a separate review paper [28].

Valentine et al. have recently described the successful resolution, on an analytical scale, of citronellic acid and various related compounds, using the diastereomeric amides formed from (R)-(+)-α-methyl-p-nitrobenzylamine and the acid chlorides [29]. Valentine's direct determination of the optical purity for his compounds made use of HPLC on Partisil 10 (Reeve Angel, Inc.), a 10-μm silica gel packing material, with a mobile phase of 20% (v/v) THF in n-heptane, with a flow rate of 1.5 ml/min. The exact composition of enantiomers in this study could be easily determined by the relative peak areas of the diastereomers in the HPLC. These results agreed extremely well with a determination of enantiomeric composition done using an optically active NMR shift reagent [29-32, 64]. In using optically active resolving agents of commercial origin, it is essential to demonstrate the complete optical purity for each reagent used, as well as the chemical purity of the materials. Thus, each sample should be analyzed first by VPC or HPLC for chemical purity, and then by its optical rotation, ORD spectrum, or an optically active shift reagent in the NMR analysis [14, 15, 29-32].

Nakanishi et al., at Columbia University, have been able to separate a number of diastereomeric mixtures of natural products, such as (+)-abscisic acid and demosterol [33, 34]. Much of the recent work regarding the separation of diastereomers has involved the commercially available resolving agent (+)-MTPA, α-methoxy-α-trifluoromethylphenylacetic acid (Aldrich Chemical Company; see Fig. 3). This same resolving agent has been applied in a number of other publications, especially in the indirect resolution of juvenile hormones and insect pheromones [35-37]. The acid chloride, MTPA-Cl, will form diastereomers with virtually any alcohol, amine, or carboxylic acid, as well as with other suitable functional groups. The resulting mixtures of diastereomers are usually oils, which can be successfully separated by either VPC, TLC, or HPLC. As depicted in Fig. 3, Nakanishi was able to separate, on a preparative scale, the MTPA derivative of a certain cis-diol using a series of recycling steps with an LC column of Porasil T (3 × 3 ft) with a 1% IPA/hexane solvent, coupled with uv detection at 254 nm. Most of the currently used resolving agents absorb strongly in the uv, and for this reason, they find wide usage with HPLC equipment [33-37]. Other strongly uv-absorbing resolving agents have been used recently in various resolutions, and some of these have yet to be described [38].

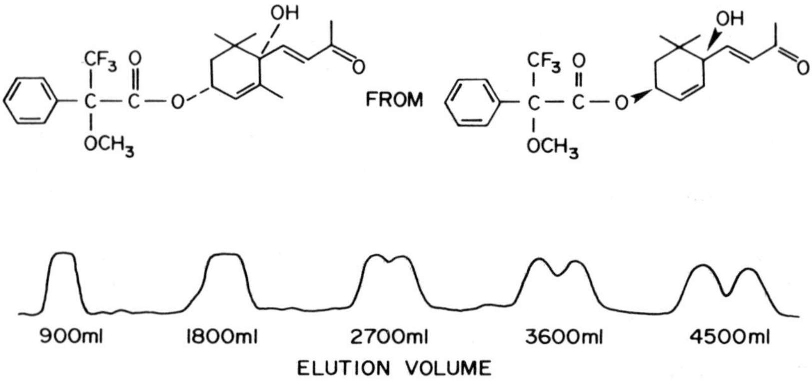

FIG. 3. Separation of MTPA diastereomers by HPLC. (From Refs. 34 and 34a. Reprinted with permission of the authors.)

With the introduction of newer preparative LC equipment by various manufacturers, as well as the introduction of new column packings, there is every reason to expect that very large amounts of enantiomers can be obtained by use of diastereomeric separations with HPLC [39]. The final optically and chemically pure enantiomers must eventually be obtained by suitable chemical reactions on the fully separated diastereomers, completing the indirect resolution.

In recent years, Professor W. H. Pirkle and co-workers of the University of Illinois at Urbana have been involved in the applications of optically active solvents in the NMR determination of enantiomeric composition [40-42]. The success of this work with respect to certain fluorinated NMR solvents led to their widespread use in NMR analyses of enantiomeric mixtures and structure determinations. Much of this work and an understanding of the physical interactions involved in the NMR studies led Pirkle et al. to develop both direct and indirect LC methods for the resolution of enantiomers. One of the newer resolving agents applied by Pirkle was (-)1-(1-naphthyl)-ethyl isocyanate (Fig. 4), prepared by a simple reaction on commercially available (R)-(+)-1-(1-naphthyl)ethylamine (Norse Chemical Company) [43].

As depicted in Fig. 4 [44], this particular resolving agent reacts readily with most alcohols in a short period of time to give a mixture of diastereomeric carbamates. Pirkle has been able to separate these mixtures of carbamates by LC in a preparative mode using a ratio of 2500:1

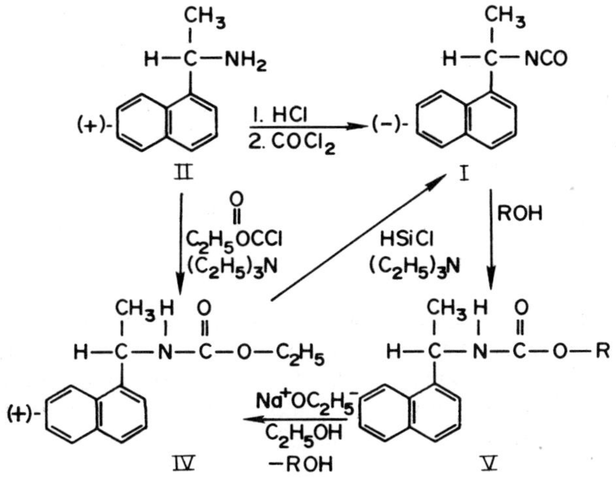

FIG. 4. Use of the resolving agent (-)1-(1-naphthyl)ethyl isocyanate (I). (From Ref. 43. Reprinted with permission from the Journal of Organic Chemistry. Copyright by the American Chemical Society.)

acidic (neutral, basic) alumina:compounds, with benzene as the eluent in most cases. Other mixtures of solvents can also be used, as can other types of adsorbents, naturally resulting in different α values (resolution factors). Detection of the two separated diastereomers was by uv at 254 nm [43, 44]. Using a newly developed recycling apparatus for the solvents alone, it was possible to separate upwards of 1 g of carbamates in about 3 hr. As indicated in Fig. 4, the resolving agent is totally recoverable, and can be used repeatedly without any loss of optical or chemical purity [43]. Typical LC chromatograms developed using this method are shown in Fig. 5. Pirkle's method probably represents the most efficient, fastest, and generally applicable method known for the routine resolution of those compounds that

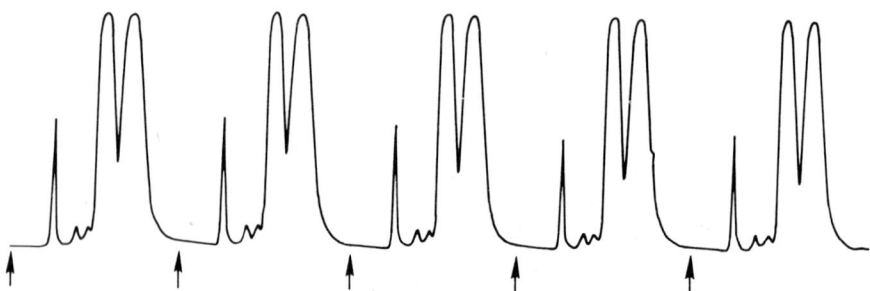

The automated repetitive chromatographic separation of diastereomic carbamates Va and Vb on acidic alumina with benzene. The separability factor, α, is 1.37. The R,R diastereomer Va is the first of the two major bands; minor absorptions are caused by impurities. Sample injections of 1 g (arrows) occur every 3 hr. Because of saturation of the 280-nm detector, the extent of peak overlap appears to be greater than is actually the case.

FIG. 5. Automated liquid chromatography of diastereomeric carbamates. (From Ref. 43. Reprinted with permission from the Journal of Organic Chemistry. Copyright by the American Chemical Society.)

can react irreversibly with the indicated resolving agent (isocyanate). However, since the observed α values vary for each set of diastereomers, improved conditions need to be developed, and there may be cases in which the α values obtained are too low for preparative applications. With the inclusion of solute recycling capabilities to those already indicated, these low α values may be improved upon. Thus far, Pirkle has used only a single pass of the diastereomers through the column in order to effect satisfactory separations.

III. A POTENTIAL OPTICALLY ACTIVE RESOLVING COLUMN FOR LIQUID CHROMATOGRAPHY

For any successful indirect resolution, the resolving agent must be 100% pure (optically and chemically), the racemic mixture must react completely with the resolving agent, a complete LC separation of the diastereomers must be devised, each diastereomer must be collected and then converted to the desired enantiomer, and finally the resultant enantiomers must usually be purified once more for subsequent studies. There is a potential method of employing an optically active resolving agent attached to an LC support, thereby simplifying the overall indirect procedure. Many publications have appeared since the mid-1960s regarding the use of polymeric reagents in organic synthesis, i.e., organic reagents that are attached (leashed) to a polymeric backbone, which is then used to perform a number of different types of reactions [45-49, 117-119]. Most of the polymeric matrices used have been various mixtures of polystyrene-divinylbenzene copolymers (gels) with varying amounts of cross-linking present to allow for a slight amount of swelling in going from one solvent to another. A chloromethyl (bromomethyl) group is often attached to a fraction of the aromatic groups on the polystyrene backbone of the polymer. Such gels can now be obtained commercially (Bio-Rad Labs, Aldrich Chemical, Realco Chemical Company). Various chemical reagents are then attached to the polymer by reactions with the chloromethyl groups, finally yielding materials that can be used repeatedly with minimal reactivation or cleaning. By analogy with what has been done already, it should be possible to incorporate, in a number of ways, Mosher's reagent, MTPA, or an analogous resolving agent into such a polymeric gel, whereby the asymmetric center is preserved, and the reacting functionality is free enough to react further with added racemates, exactly as is the case in solution reactions.

With the enantiomers to be resolved firmly anchored to the polymeric support, an LC column can then be packed with this material and washed with a suitable solvent to remove any impurities or unreacted materials. In essence, one is then left with a collection of diastereomeric materials (esters, amides, etc.). The two diastereomers present on the polymer should theoretically have two different rates of formation and hydrolysis (esters, amides). In fact, there should be a different set of conditions (e.g., pH) necessary to saponify each of the two diastereomers on the polymer. Using a sufficiently slow gradient elution procedure, it could be possible to alter the hydrolysis conditions within the eluting mixture slowly enough so that one, and only one, of the two diastereomers would undergo saponification (esters) initially. The freed alcohol will then elute from the column, leaving behind the remaining diastereomer still leashed to the polymeric support. Further changes in the hydrolysis conditions of the eluting solvent would then release the other ester, and the second desired enantiomer would elute. Using a suitable detector, it will be possible to determine exactly when each of the two enantiomers has eluted from the column, and the necessary hydrolysis conditions for each diastereomer. Subsequent preparative work would follow immediately, since the necessary conditions for the resolution would be known exactly. The optically active polymeric reagent would be regenerated in a subsequent step by passing through the column a solution of the appropriate reagent needed to regenerate the acid chloride (MTPA-Cl). In principle, the same polymeric material can be used a number of times, thus reducing the overall costs, time, and materials needed for each resolution. Such a method as that described here is entirely feasible, and the necessary knowledge and materials are already available.

IV. DIRECT RESOLUTIONS BY LIQUID CHROMATOGRAPHY USING OPTICALLY ACTIVE ELUENTS

Direct resolutions have not been as evident in the chemical literature as those involving the physical separation of diastereomers. There are a number of different approaches to the direct LC resolution of enantiomers, but from the chromatographer's point of view, there are really only two variables to consider. One might use an optically active packing material, or one might employ an optically active solvent. The majority of papers in recent years have discussed the use of optically active packing materials, but there are some references wherein optically active solvents have been employed as the resolving agent. Let us consider this approach first.

Pirkle's results in the use of optically active carbinols as NMR solvents for the direct determination of enantiomeric purity of various sulfoxides suggested the possibility of applying this knowledge to LC resolutions of these materials [50]. It has now been possible to plan a rational chromatographic resolution which can afford preparatively useful amounts of two sulfoxides in enantiomeric purities unobtainable by any other known method [51]. In addition to the sulfoxides indicated in Fig. 6, it has also been possible to resolve fully certain five- and six-membered lactones containing a

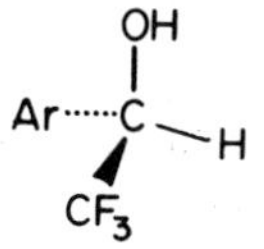

FIG. 6. Resolution of sulfoxides using achiral solvents. (From Ref. 51. Reprinted with permission of the authors and the Journal of Chromatography. Copyright by Elsevier Scientific Publishing Company.)

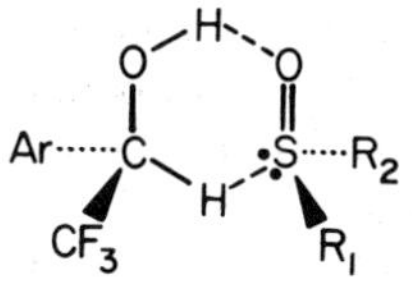

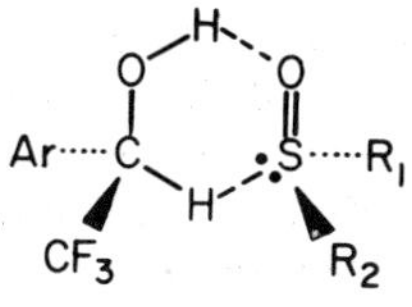

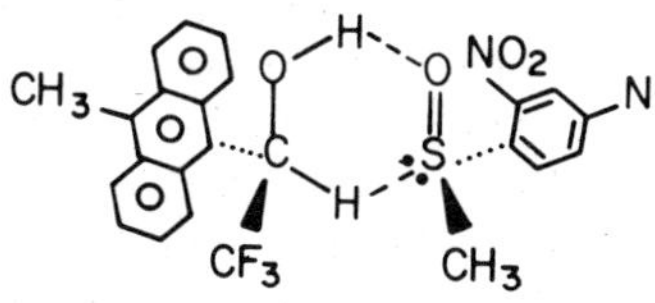

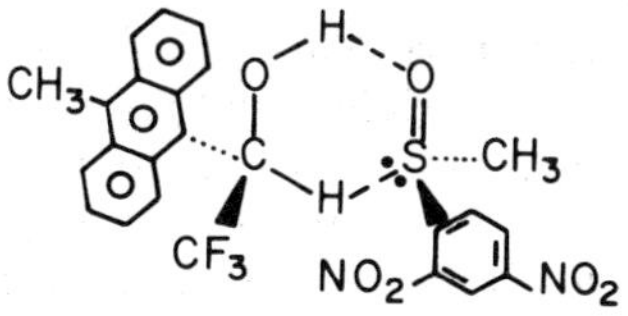

2,4-dinitrophenyl moiety within the molecule [51]. In this case, a complete resolution, α value = 1.50, was realized for a 50-mg sample of enantiomers in a single pass through the column. In the original LC resolutions, using an optically active solution as the eluting solvent, it was possible to resolve partially the sulfoxide enantiomers (Fig. 6) [51].

With the anthryl carbinol indicated, diastereomeric solvates can be formed with a wide variety of solute molecules. Owing to different two-point (VIIIb) versus three-point (VIIIa) interactions between certain conformations of the solute and solvent, one diastereomeric solvate will be formed preferentially over the alternative solvate. In an actual case (Fig. 6), the solvates formed were VIIIa and VIIIb, using the 9-anthryl carbinol and the methyl-2,4-dinitrophenyl sulfoxide. The support used for these direct resolutions was silica gel packed in a standard glass column, with the eluent being a 50-ml solution of S-(+)-2,2,2-trifluoro-(9-anthryl)ethanol (0.1 M).

The eluent was recycled through the column for 2 hr at a rate of 500 ml/hr, and the monitor for the resolutions was a uv detector. When the column had become equilibrated with the solution being recycled, a sample of the racemic sulfoxide in carbon tetrachloride, together with some of the optically active anthryl carbinol solution, was injected onto the column. The recycling of the solution was then continued for about 1 day, at which point the excess carbinol was eluted from the column with pure carbon tetrachloride. Continued elution of the entire column with carbon tetrachloride or with a stronger solvent, or removal of the packing material from the column, followed by elution of the colored bands with diethyl ether, yielded various fractions enriched in one enantiomer of the sulfoxide or its mirror image. The carbinol remaining in each fraction could then be separated by a subsequent column chromatography using regular silica gel with methylene chloride as the eluent [51].

The moving diastereomeric solvates are colored differently, depending on the particular solute-solvent π-π interactions involved, and these materials can be seen visually or with the use of a typical uv detector in the LC setup. Fractions can also be taken automatically to improve the optical purity of the various cuts. An alternative approach would be to anchor the optically active carbinol to the silica gel or another more inert matrix, and then pass the solution of the sulfoxide repeatedly through this column. Such a process has not yet been studied [51]. This might also help to indicate the exact nature of the molecular interactions responsible for the observed resolutions.

V. DIRECT RESOLUTIONS BY LIQUID CHROMATOGRAPHY USING AN OPTICALLY ACTIVE SUBSTRATE

A. Optically Active Packing Materials in General

Gil-Av has been extensively involved in the gas-chromatographic resolution of enantiomers for many years [2], and has realized some degree of qualitative success in this area. In more recent years, however, he has applied his efforts more toward liquid-chromatographic resolutions, and the initial results of this work have recently appeared [130, 131]. The most difficult resolutions to perform are those involving racemates that do not possess any handles, i.e., compounds such as saturated and unsaturated hydrocarbons, helicenes, and biphenyls. Such materials are customarily obtained optically pure by fractional crystallization, or more commonly via the resolution of a suitable precursor. Gil-Av et al. [130] have been able to resolve a number of helicenes using as the coating on silica gel either 2-(2,4,5,7-tetranitro-9-fluoroenylideneaminoxy)propionic acid (TAPA), or the 2-butyric acid analog (TABA), or other carboxylic acid analogs [131]. These optically active resolving agents form charge transfer complexes with both enantiomers of the helicenes. Since these charge transfer complexes are really diastereomeric complexes, they will have different equilibrium constants for their formation and dissociation on the silica gel support. The actual loading of the optically active resolving agent was in the range of 10 to 25% of either TABA or TAPA, and the eluting solvent was a mixture of cyclohexane and methylene chloride. It is necessary to use a solvent mixture wherein the resolving agent is completely insoluble. The actual resolutions were done at room temperature for most of the helicenes, but for the 5-helicene, it was necessary to go to 0°C and to use recycling. The resolution coefficients (α values) as determined by Gil-Av ranged from 1.112 to 1.268 for the work done at room temperature (6- to 14-helicenes) [130]. Additional work in this area by Gil-Av resulted in the direct attachment of a number of chiral, charge transfer complexing agents to the silica gel supports, and the subsequent direct resolution of a number of helicenes on these columns [131].

The direct attachment of a large number of organic, inorganic, and organometallic materials to silica gel has been described in recent years [132-135]. The variety of such chiral ligand to silica gel supports is almost endless, and many such materials may become commercially available in the near future [136]. Gil-Av has bonded his optically active charge transfer materials to the silica gel via a linkage molecule of 3-aminopropyltriethoxysilane. This step was followed by the attachment of the chiral carboxylic acid [e.g., R(-)-TAPA] to the aminated silica gel using an appropriate coupling reagent [131]. The final coupling step could be done before the silica gel was loaded into the column, or in situ, after the column was actually packed. The observed resolution coefficients for the helicenes were somewhat lower on the covalently bonded silica gel supports as compared with the coated columns [130]. As Gil-Av et al. note: "Modified supports never pack as efficiently as untreated ones, and they present higher resistance to mass transfer" [131]. This is not to say that in the future improved methods of preparation for bonded supports will be able to yield as good or better resolutions than simple coated ones. The nature of the linking molecule, the amount of material finally bonded to the silica gel, and the very nature of the supporting material itself (silica gel or polymeric gel) can all affect the final results obtained.

The use of polymeric materials as LC supports has been studied for a number of years, and the literature contains a number of references to this approach. Among the earliest materials tried were the cyclodextrins, but they were not widely accepted [52, 53]. Materials similar to the cyclodextrins are discussed further in Sec. V.F.

Jonas and Norden have obtained partial resolutions for enantiomeric mixtures of organometallic complexes of Al(III) and Fe(III). Using a support of D-lactose and aluminum oxide with isopentane/ethyl ether as the eluent, and temperatures between -30 and -100°C for the chromatography, it was possible to resolve these complexes partially [54]. The effect of a lowered temperature during the LC process may be applicable in other instances as well (e.g., the work by Gil-Av described earlier).

B. Optically Active Crown Ether-Bonded Supports

During the past decade a new area of organic synthetic reactions has made its appearance, that of phase transfer catalysis [120, 121]. In its early years, this new technique made extensive use of quaternary ammonium or phosphonium cations to transfer usually insoluble anions into nonpolar solvents. In the course of developing various effective phase transfer agents, a number of macrocyclic crown ethers were prepared [122-124]. A very large number of different crown ethers, having either oxygen, sulfur, or nitrogen as the usual heteroatoms, have now been reported with regard to their catalytic and complexing properties [122]. The synthetic usefulness of the various crown ethers has been incorporated into many new and significant reaction sequences, and they will remain an important tool in organic synthesis. Crown ethers function by complexing a cation so effectively, via use of their nonbonded electrons on the heteroatoms, that the remaining anion is effectively "naked," and thus becomes extremely reactive. By incorporating optically active ligands on the crown ether backbone, it was then possible to prepare optically active (chiral) 18-crown-6 molecules [124, 125]. Stoddart and co-workers then demonstrated a selective extraction-complexation property of this crown ether toward various racemic mixtures of ammonium salts [125]. In certain cases, the optically active ligands were able to complex one enantiomer of the amine salts more effectively than the other, thus leading the way for the application of suitable crown ethers in both solution and liquid-chromatographic resolutions of enantiomers [122]. In all such cases studied, the degree of differentiation exhibited for a pair of enantiomers depends on the formation of temporary diastereomeric complexes usually held together by numerous hydrogen bonds. The phenomenon depends mainly on differences in steric relationships between the optically active crown ether and the two enantiomers being resolved [122].

Cram et al. have investigated the use of optically active eluents for the direct resolution of enantiomers, as well as the use of similar optically active polymeric supports for the same purpose [55-57]. As discussed in the work of Stoddart, it was possible for Cram, using suitable models, to study the relationship between host-guest molecules in structured molecular complexes. A large variety of host-guest complexes have now been studied, with a view toward understanding the observed association constants and their responses to changes in temperature. In all host-guest complexes an equilibrium exists between the free host and the free guest molecules with that of the complexed pair. If those factors that improve K_a (association constants) are fully understood, then host molecules could be designed which will have the best fit for one particular enantiomeric guest over the other [55-57]. This is really the basis for nature's enzymatic, regulatory, and transport systems, and has been realized by chemists for many years. Cram et al. have attempted to judiciously design optically active host molecules that will selectively prefer one enantiomer of a pair of racemates over its mirror image. For one particular host molecule, each possible guest pair will have different K_a values, and different degrees of resolution will be obtained. Having such models prepared, it was then possible to use such optically active host molecules in a solution as the eluting solvent in LC, similar to what Pirkle had done (see earlier), and thereby attempt direct resolutions of various alkyl ammonium or amino ester salts [56]. The equilibrium existing between any host and guest molecules with the complexed pair is illustrated by [56]

$$(\text{host}) + [(CH_3)_3\overset{+}{N}H_3 \cdot SCN^-] \underset{CHCl_3}{\overset{K_a}{\rightleftharpoons}} [(CH_3)_3\overset{+}{N}H_3 \cdot SCN^-] \cdot \text{host}$$

Figure 7 illustrates several resultant LC traces obtained for various alkyl ammonium salts, e.g., α-phenylethylammonium hexafluorophosphate, methylphenylglycinate hexafluorophosphate salts. The resolving agent used in these studies, i.e., the host compound, was the macrocyclic tetranaphthyl ether, (RR)-1, in a solution of chloroform [56]. The LC was run with water-NaPF$_6$ or water-LiPF$_6$ on Celite or silica gel as the stationary phase, making use of a conductivity cell for detection of the eluted materials. Typical α values ranged anywhere from 1.52 to 3.60, and the temperatures were usually kept below room temperature (-15 to +25°C). A relatively large number of racemates have been completely resolved by Cram's procedure, but thus far the approach has been found to be most effective for amine salts or amino ester salts when silica gel is used.

Cram's next approach to direct LC resolutions was to attach the host molecule directly to the silica gel (Fig. 8). This is actually an example of solid-liquid chromatography, with the macrocyclic crown ether attached to the silica gel through only one Si-O-Si bond. The other three sites on the crown ether molecule were capped with Si(CH$_3$)$_2$OCH$_3$ groups to decrease or prevent entirely any tailing of the solutes during the elution process. Figure 8 also indicates one of the best resolutions of enantiomers obtained using this approach, with an α value of 4.4 and complete baseline separation [57]. For the majority of the compounds studied with this particular LC support, the overall resolutions were not this good. The tailing observed with silica gel as the support may be reduced by use of macroreticular polystyrene resins (gels) that could yield higher amounts of host per unit weight of support, as well as a support free of unprotected binding sites.

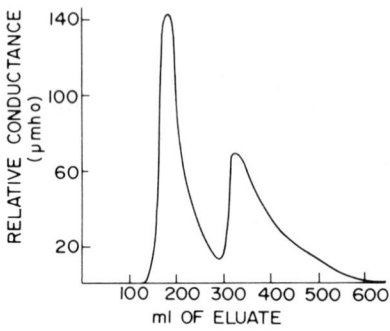
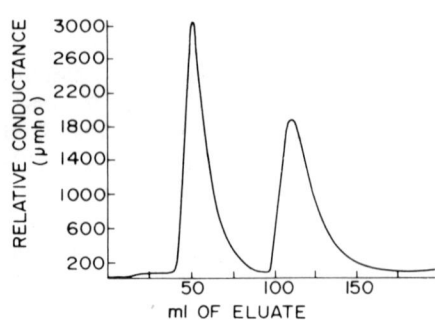

Chromatographic optical resolution by (RR)-1 of α-phenylethylammonium hexafluorophosphate.

Chromatographic optical resolution by (RR)-1 of methylphenylglycinate hexafluorophosphate salt.

Complexes of (RR)-1 with generalized salts

More stable complexes of (RR)-1
(S)-2, L = C_6H_5, M = CH_3, S = H
(R)-3, L = C_6H_5, M = CO_2CH_3, S = H
(R)-4, L = p-HOC_6H_4, M = CO_2CH_3, S = H

FIG. 7. Direct resolutions of enantiomers using host-guest molecules. (From Ref. 56. Reprinted with permission from the Journal of the American Chemical Society. Copyright by the American Chemical Society.)

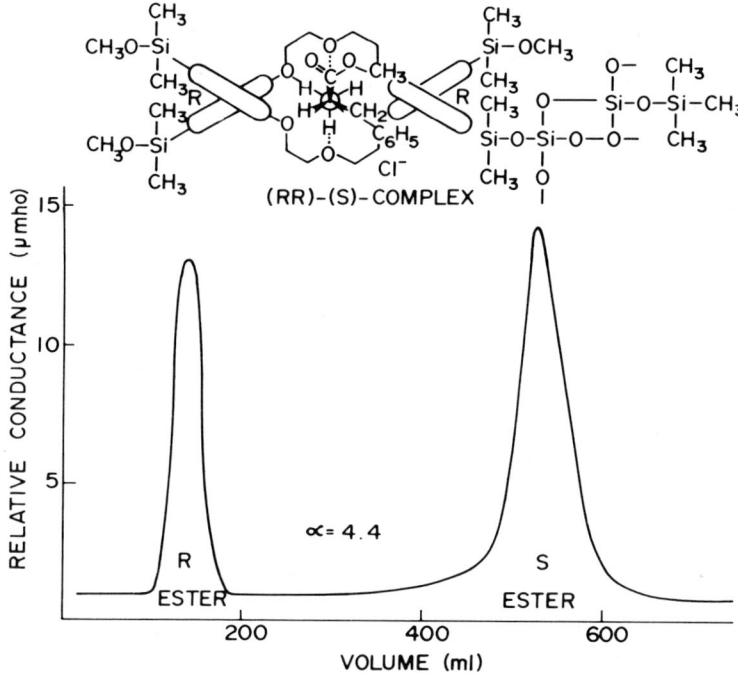

Chromatographic optical resolution by host-bound silica gel of methyl phenylalaninate hydrochloride salt.

FIG. 8. Direct resolution of an amino ester salt using a host molecule attached to silica gel. (From Ref. 57. Reprinted with permission from the Journal of the American Chemical Society. Copyright by the American Chemical Society.)

Indeed, recently Cram et al. have reported on some results using just such polymeric gels as the supporting matrix for the chiral crown ethers [126, 127]. Some of the best resolutions of amine salts and amino acids were realized by chiral recognition of the enantiomers by a suitable host covalently bonded to a polystyrene resin [127, 128]. Since the supporting material was a polymeric gel in this case, it was possible to resolve the α-amino acids directly, as their salts, rather than protecting the free carboxylic acid as the ester, as was done previously [127]. With the use of noninterfering, inert gels as supports, it should be possible to improve future resolutions beyond those realized thus far.

As in Pirkle's studies described earlier, Cram has rationalized the preferred complexation of certain amine ester salts as a result of fourpoint binding being the more stable disastereomeric complex. For those amino ester salts not fully resolved, e.g., phenylglycine esters or tryptophan esters, only a three-point binding model is possible [57]. Again, we are talking about differences in hydrogen bonding between the ammonium protons and the nonbonded electrons on the heteroatoms in the crown ether backbone.

C. Specifically Adsorbing Silica Gels

A number of other unusual supports for LC have been described in the literature, one of the more interesting being the specifically adsorbing silica gels. A detailed review of these materials and the methods used in their preparation has appeared [58]. The exact nature of the silica gel used in these studies is not fully understood, although certain diagrammatic representations of the silica surfaces have appeared [58]. Specifically adsorbing silica gels will adsorb certain compounds in preference to others when solutions of these mixtures are passed through the carefully prepared silica gel. The final material is prepared by coagulation of the silicic acid in the presence of a small amount of the substance to be adsorbed. The prepared silica gel is then eluted with various solvents in order to remove totally the organic substance used to prepare it. When all this is removed, the silica gel that remains will have a memory of its formation, i.e., it will have "footprints" of the organic molecules. If a solution of various compounds is eluted very slowly through the silica gel, those molecules that fit best into the holes left in the silica during its preparation will be held back longer than the other molecules present. When all the nonretained molecules have fully eluted from the gel column, a change in the solvent being used for the elution will usually remove the compound that was specifically adsorbed initially. Numerous examples have been reported whereby optically active materials can be used as the imprinting agent during the preparation of the silica gel, and the resulting materials will then have a memory for that optical isomer rather than for the enantiomer. It has been possible at times to partially resolve mixtures of enantiomers using this approach, although the success or failure seems to vary with the particular individual performing the resolution, conditions of the gel formation, and the nature of the particular optically active material used. Of course, one must have an optically and chemically pure sample present beforehand for the gel formation step. The particular material must also have a rather rigid structure; it must be basic in nature, soluble in the aqueous medium used for the gel formation, and must be unaffected chemically by the polymerization step. Although this process has been known for a number of years, the actual number of reasonable successes realized with it have been minimal, so that it has not been widely accepted.

D. Optically Active Polymeric Supports

Continuing along the lines of optically active LC supports brings us to the work of a German group under the able direction of G. Blaschke of the University of Bonn [59-61]. All the optically active polymers described by this group of workers have been prepared by the polymerization of optically active olefinic amides acting as the monomers. As depicted in Fig. 9 a very large number of different optically active amines were prepared, and reacted with various allylic acid chlorides, to yield unsaturated amides. These amides were then polymerized under a wide variety of conditions, to give various popcorn-type polymers (Fig. 9, XI and XIII) [59]. Specific examples of these polymers are also indicated in Fig. 10 (XIV, XV, and XVI). These were used successfully for the resolution of optically active mandelamide and mandelic acid, the latter acting as markers for the effectiveness of the resolutions possible on each polymer. It was also possible to resolve many other materials partially, such as α-substituted phenyl amides (XVII), N-substituted acetamides (XVIII), N-acylamino acid esters (XIX), and benzoin (XX) (Fig. 10) [60]. Typical eluting solvents have been benzene, dioxane, and benzene/cyclohexane. The optical purity of each fraction collected could be determined by conventional methods of weighing and measuring the optical rotation, as well as by other methods [62]. In certain instances, the amount of the resolution was very good; however, for the majority of compounds studied, the resolutions were only partial. Fig. 9 indicates the almost complete resolution of mandelamide on polymer XIIIe (5.6 g), with 0.300 mg of the racemate being chromatographed. Other typical resolutions are indicated in the table in Fig. 10 [60].

Blaschke's approach to chromatographic resolutions appears to offer some degree of success; however, many unresolved problems remain. The resolving ability of the polymeric material depends on many factors, not the least of which is the basic structure of the monomer used. Many factors in the polymerization process are variables, and these variables strongly affect the final nature of the optically active polymers formed. Some of these items are the nature of the copolymer used, the initiating agent, solvents, and temperature. It may be difficult for the novice to prepare several batches of a given polymer having the same resolving ability and physical properties. Also, the swelling properties of the polyamides just described vary with the nature of the solvent used as the eluent. The amount of cross-linking present must be varied in order to minimize the final swelling properties of the polymers, or the final resolutions will suffer.

There are several explanations for the resolutions observed by Blaschke and co-workers, one of which may be the presence of asymmetric microholes in stereoregular patterns within the polymeric matrix. Another suggestion

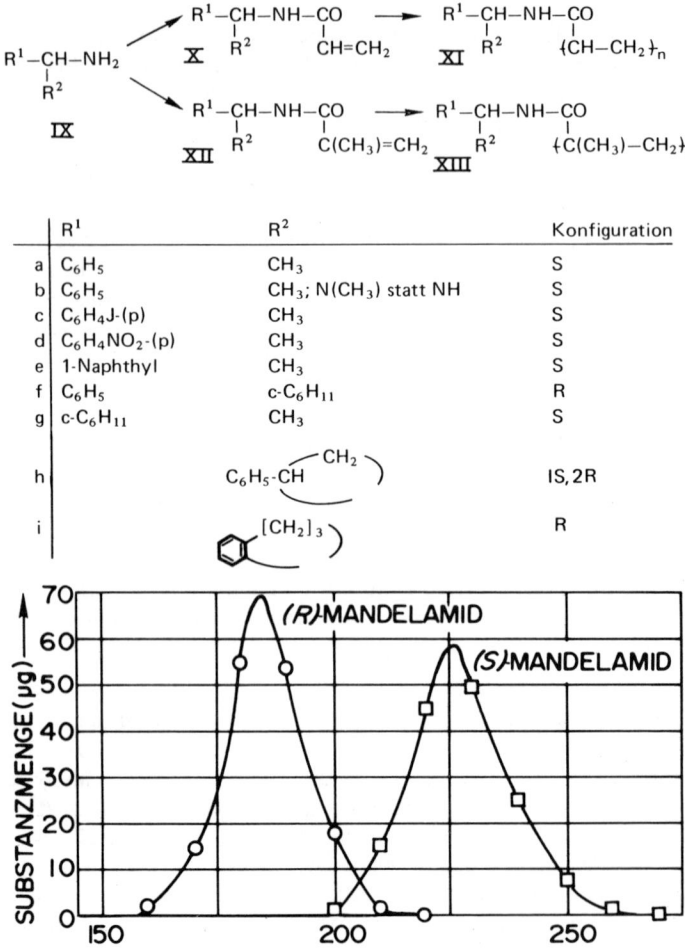

FIG. 9. Direct resolutions on optically active polymers. (From Ref. 59. Reprinted with permission from Verlag Chemie GMBH, and the author.)

Chromatographische Racemattrennungen an den S-konfigurierten Adsorbentien (xiv),(xv),und (xvi).

Racemat	R	Optische Ausbeute (Konfiguration des starker adsorbierten Enantiomers) am Adsorbent		
		(xiv)	(xv)	(xvi)
XVIIa	OH	60.3 (S)	38.9 (R)	28.7 (S)
XVIIb	CH_2OH	68.8 (S)	18.9 (S)	17.9 (S)
XVIIc	$OCOCH_3$	51.7 (R)	5.7 (S)	2.6 (R)
XVIId	OCH_3	12.2 (S)	48.2 (R)	7.3 (S)
XVIIe	C1	11.6 (R)	29.2 (R)	12.5 (R)
XVIIIa	CH_3	34.2 (S)	26.0 (S)	36.2 (S)
XVIIIc	CN	20.0 (R)	32.9 (R)	49.1 (R)
XIXa	H	0.0	16.9 (S)	30.2 (S)
XIXb	C_6H_5	33.6 (R)	18.0 (S)	32.4 (S)
XIXc	C_6H_4OH (p)	24.7 (R)	26.9 (S)	33.3 (S)
XX		7.2 (S)	0.6 (R)	21.6 (R)

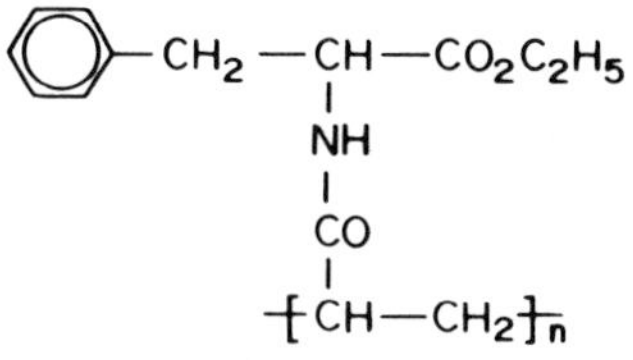

XIV

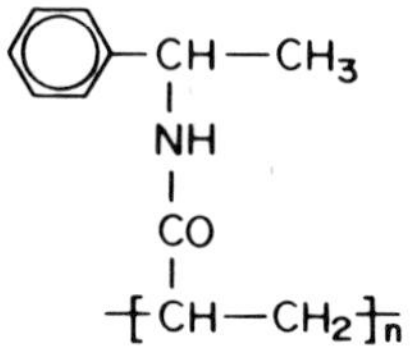

XV

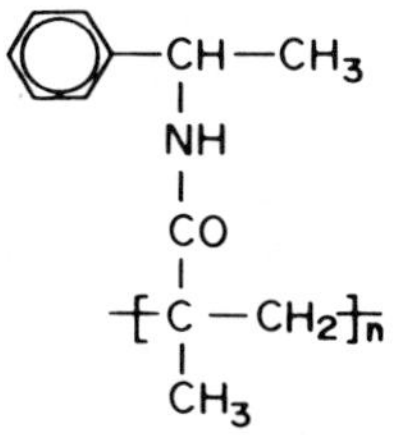

XVI

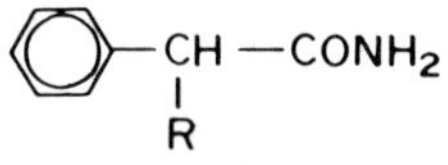

XVII

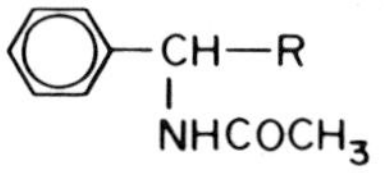

XVIII

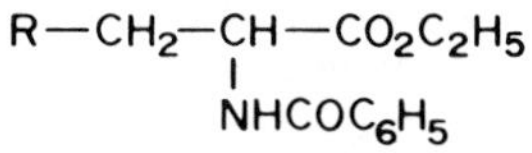

XIX

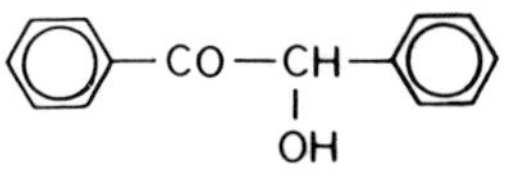

XX

FIG. 10. Optically active polymers and various resolutions. (From Ref. 60. Reprinted with permission from Verlag Chemie GMBH, and the author.)

is the formation of temporary diastereomers through hydrogen bonding with the amide groups of the polymer and the enantiomers. If the extent of hydrogen bonding differs for the two enantiomers, one may be eluted faster through the column than the other. All optically active adsorbents in LC function in very much the same way. Thus, if we have (-)Ads as the substrate, a racemate ($\pm$)X will form two separate adsorbates, (-)Ads(-)X and (-)Ads(+)X. Since these two adsorbates are really diastereomeric, they are not equally stable, and one enantiomer (that forming the less strong adsorbate) will pass through the column faster than the other. A partial or complete resolution may occur in this way. It is difficult to predict beforehand which polymers will yield the best resolving ability for a given mixture of racemates. The exact nature of the interactions in a polymeric matrix during the LC process is still largely unknown.

E. Ligand-Exchange Resolutions

Optically active LC supports have been used for direct resolutions in ligand-exchange chromatography [63]. Much of the work in this field has been done by Davankov and Angelici. In typical ligand-exchange chromatography, a polymeric backbone contains a cation-exchange material complexed with a metal ion capable of undergoing cation interchange [63]. The metal employed may be copper(II), nickel(II), or zinc(II), and may form part of the actual stationary phase. The polymeric backbone varies, and several different types have been used. The mobile phase used is usually ammonia, pyridine-water, or an amine, since these easily form complexes with the metal ion. A solution of various ligands to be separated, which usually contain basic nitrogen atoms, will displace ammonia from the stationary phase and form coordinates with the metal ion. The addition of more of the ammonia solution down the column will then displace the attached ligands, and the elution order will depend on the actual binding strengths of the ligand and the metal in the polymeric matrix. In ligand-exchange chromatography, the ligands directly attached to the counterions are exchanged, rather than the counterions themselves as in direct ion-exchange LC. Ligands in the solution will replace ligands on the metal ions or solvent molecules in the metal ion solvation shell [63]. Since the number of variables in ligand-exchange chromatography is greater than in conventional ion-exchange LC, e.g., metal ion, exchanger, mobile phase, solvents, a large number of variations in selectivity is possible. In general, there is also an increased exchange capacity, since this is dependent on the number of available coordination sites. The basicity of the amines or amino acids being chromatographed will largely determine the elution order observed.

Ligand-exchange chromatography has been used in the separation of many types of mixtures, including polyhydric alcohols, alkanolamines, aliphatic diamines, amino acids, peptides, aromatic amines, aziridines, drugs, hydrazines, nucleosides, nucleic acid bases, and sugars. The basic differences between ion-exchange and ligand-exchange chromatography are illustrated in Fig. 11 [63].

Perhaps the most extensive research into the use of ligand-exchange LC for the direct resolution of enantiomers has been that of the Russian group under Davankov[4]. A large number of optically active ligand-exchange resins have been described, often containing optically active amino acids as the ligands, coupled with transition metal ions such as Cu(II) or Ni(II) [65-69]. In many instances, the optically active supports provide for the quantitative resolution of racemates that are capable of forming complexes (temporary) with the support. Most of the materials thus far resolved by this approach have been amino acid derivatives or amino acids themselves. Usually ammonia, pyridine, or an amine solution is used as the eluting solvent, and the temperature can also be varied to improve the overall resolution. According to the Russian workers, "there is no other chromatographic process which could compete with ligand-exchange chromatography in the separation of enantiomers on a preparative scale" [66]. It is important to remember that the metal ion forming the complex may be combined with either the stationary or moving ligand. The complex being generated must be kinetically labile, and readily decomposed and re-formed a number of times during the chromatographic process.

As a typical example of resolution by this method, let us consider the case of DL-proline. Davankov prepared a highly asymmetric sorbent by

Ion exchange
$$Zn^{2+} + 2RNH_4 = R_2Zn + 2NH_4^+ \tag{1}$$
$$H_3^+NCH_2CH_2CH_3 + RNH_4 = RH_3NCH_2CH_2CH_3 + NH_4^+ \tag{2}$$
Ligand exchange
$$R_2Zn(NH_3)_4 + 4H_2NCH_2CH_2CH_3 = R_2Zn(H_2NCH_2CH_2CH_3)_4 + 4NH_3 \tag{3}$$
$$R_2Cu(NH_3)_4 + 2H_2NCH_2CH_2NH_2 = R_2Cu(H_2NCH_2CH_2NH_2)_2 + 4NH_3 \tag{4}$$

[a]The water of hydration is omitted in the equations. The ammonium ion is the counter ion and the exchanger is represented as R.

FIG. 11. Ion-exchange versus ligand-exchange reactions. (From Ref. 63.)

the treatment of a chloromethylated styrene-p-divinylbenzene (0.8%) copolymer with L-proline. The sorbent was then treated with 0.1 N $CuSO_4$ solution in 1 N ammonia, washed with water, and then loaded into a chromatographic column. A second column of the same sorbent in the absence of any Cu(II) ions was placed directly below the first column in order to trap any eluted copper ions. An aqueous solution of DL-proline was then passed through the columns, followed by washings of water alone, which resulted in the complete elution of L-proline. A second solution of 1 N ammonia passed through the columns next eluted all remaining pure D-proline [65, 66]. The recovery was quantitative for each enantiomer, and the optical purity was 100% for the D- and L-prolines eluted. Workup was very minimal in each case, since the desired materials were recovered simply by evaporation of the solvents used for the elution.

Davankov has explained the ligand-exchange resolution process as illustrated in part in Fig. 12 [65-67]. A schematic representation of the actual resin used is indicated, where the stationary ligands are L-proline (R). When Cu(II) ions are added to this resin, the stationary complexes formed contain two fixed ligands per metal ion (R-Cu-R). When the proline

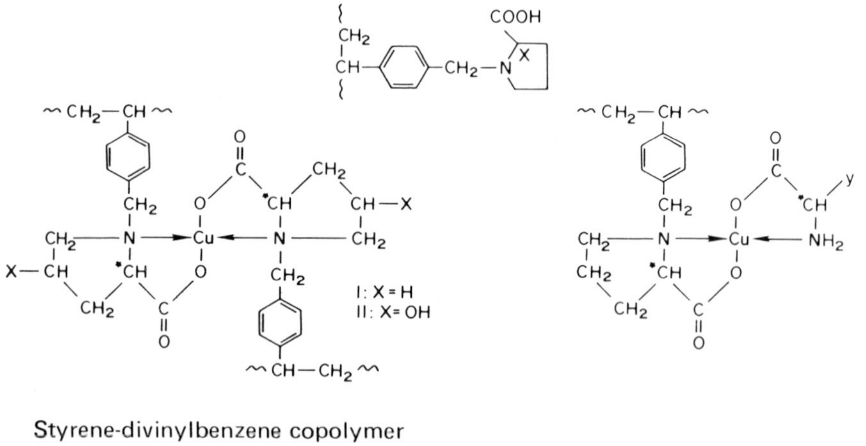

FIG. 12. Ligand-exchange chromatography supports. (From Ref. 67. Reprinted with permission from the Journal of Chromatography. Copyright by Elsevier Scientific Publishing Company.)

enantiomers are then added to the column resin during the resolution process, sorption takes place as indicated in Eqs. (1) and (2) of Fig. 13. The stationary complexes are then converted into mixed complexes (sorption), with the displaced R (L-proline) ligands still remaining attached to the column resin at all times. When the solvent is made more basic, by using pyridine as the eluent, for example, another displacement reaction takes place on the stationary complexes, as represented by Eqs. (3) and (4) of Fig. 13. The enantiomers are therefore released from the stationary complexes separately, depending on the concentration of the pyridine used in the aqueous eluent. Because of differences in the rates of formation and dissociation of the stationary complexes (diastereomeric complexes), the two enantiomers will be released from the column at different times. The selectivity for release of one enantiomer over the other is also a function of the temperature, and this will decrease with an increase in column temperature. In many instances, complete quantitative resolutions have been achieved for various amino acids and their derivatives having free amino groups.

It has also been possible to calculate selectivity factors ($\underline{a}$) as indicators of the selectivity of the support for a particular pair of enantiomers, as well as giving a good idea of the degree of resolution achieved. Thus $\underline{a}$ is really similar to the α values traditionally used in LC, $\underline{a} = V_D/V_L$, and has been found to vary from 1.67 to 3.50 for ligand-exchange LC. These values are much lower for simple ion-exchange LC or VPC resolutions. It has been possible to calculate that D-proline has a stronger adsorption to the resin phase than the L-proline by 300 to 400 cal/mol. When ammonia is used as the eluent for this resolution, 0.1 to 0.2 N solutions will remove the L-proline completely, but 1.0 to 2.0 N solutions are needed for elution

$$R-Cu-R+D\text{-}Pro \stackrel{K_D}{\rightleftharpoons} R-Cu-D\text{-}Pro+R \qquad (1)$$
$$R-Cu-R+L\text{-}Pro \stackrel{K_L}{\rightleftharpoons} R-Cu-L\text{-}Pro+R \qquad (2)$$
$$R-Cu-D\text{-}Pro+2\,Py \stackrel{K_{D'}}{\rightleftharpoons} R-Cu-(Py)_2+D\text{-}Pro \qquad (3)$$
$$R-Cu-L\text{-}Pro+2\,Py \stackrel{K_{L'}}{\rightleftharpoons} R-Cu-(Py)_2+L\text{-}Pro \qquad (4)$$

FIG. 13. Formation of diastereomeric complexes in ligand-exchange LC. (From Ref. 66. Reprinted with permission from the Journal of Chromatography. Copyright by Elsevier Scientific Publishing Company.)

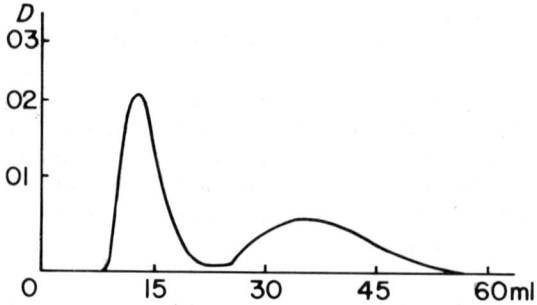

Resolution of 1.5 mg of DL-proline on the asymmetric resin II containing L-hydroxyproline-Cu(II) fixed complexes.

FIG. 14. A typical resolution by ligand-exchange chromatography. (From Ref. 67. Reprinted with permission from the Journal of Chromatography. Copyright by Elsevier Scientific Publishing Company.)

of the D isomer. However, the order of elution for a pair of enantiomeric amino acids depends apparently on the nature of the amino acid attached to the support, and the particular enantiomers being resolved. Thus, Angelici et al. [70] have found reverse order of elution, i.e., the D-amino acid was eluted first, for several amino acid racemates studied on a particular support. Figure 14 illustrates a typical ligand-exchange chromatogram for DL-proline according to Davankov et al. [67]. The same chromatographic sorbents used here, L-proline or L-hydroxyproline, can also be used successfully for resolving racemic diamines, hydroxyamines, and hydroxy acids, as well as their derivatives.

F. Sephadex-Type Supports

Mention should be made of the various reports in the literature involving the use of optically active ion exchangers, such as derivatives of DEAD-Sephadex, QAE-Sephadex, CM-Sephadex, Sephadex G-25. In the last case mentioned, an optically active support was prepared by chemically bonding L-arginine through a cyanuric chloride linkage to Sephadex G-25 [71]. With water as the eluent, it was possible to resolve DL-β-3,4-dihydroxyphenylalanine in one pass, $\alpha = 1.60$ [71]. Here again it was suggested that a three-point interaction between the sorbent and the sorbate was necessary for a successful partial resolution. Numerous other reports have appeared utilizing variants of these packing materials, some with excellent distribution coefficients for the enantiomers involved [72-77].

Gaal and Inczedy [72] were able to resolve certain aminodiols using an ion-exchange resin (Dowex 50-X8), with an eluent that contained Cu(II) ions plus the d-aminodiol. In an analogous study, the same authors [73] were able to resolve fully the racemates of d- and l-threo-1-(p-nitrophenyl)-2-aminopropanediol-1,3 by use of an anion-exchange resin (Dowex) containing d-tartrate ions. A similar result was obtained by the use of Dowex 1-X2 resin saturated with $(Cu-d-PDTA)^{2-}$ for the partial resolution of racemic 1-phenylethylamine, and other cations [Ni(II) and Zn(II)] were studied [74]. A partial chromatographic resolution of mandelic acid on various Sephadex gels was given by Rieman et al. [75], in whose report additional references can be found for the use of starch and cellulose as the stationary phases. The use of dextran-based ion exchangers, such as the Sephadex gels, for the resolution of racemic bases or acids was also studied by Popova et al. [76,77]. These investigators were able to show that the distance between the asymmetric center in the racemates and the chiral center of the ion exchanger probably affects the overall chromatographic resolution obtained [77].

G. Enzyme-Based Supports

Mention should be made here of attempts to utilize enzymes for the direct resolution of enantiomers, as well as for the commercialization of this process. Much work has been done in the area of affinity chromatography, especially with regard to the attachment of enzymes to various support materials [78-83]. A Japanese industrial firm involved in the synthesis of various amino acids has developed a large-scale industrial process for the separation of enantiomers using, in part, immobilized enzymes. The enzyme to be utilized is covalently attached to an inert support, such as glass beads, and in an automated, continuous process, mixtures of two racemic esters are passed through the column [82, 83]. The resultant solution contains one unprotected amino acid and the untouched amino acid ester. At this point separation of these two materials is a relatively simple process, and the amino acid ester is simply hydrolyzed by conventional means to the final optically pure amino acid in excellent overall yields. The immobilized enzyme systems are easily cleaned and prepared for subsequent runs in a continuous process [82, 83]. A similar method for direct resolutions of certain amino acids has been described by Stewart and Doherty [81] who have used bovine serum albumin-succinoylaminoethyl-Sepharose as the chromatographic support material to resolve DL-tryptophan. Other enzymatic systems may also work in particular instances [81]. Strictly speaking, enzymatic methods for resolving enantiomers are not direct resolutions, since the actual separation is taking place on the free amino acid and the still protected amino acid ester.

VI. DIRECT RESOLUTIONS USING ION-PAIR PARTITION CHROMATOGRAPHY

In recent years, the idea of using ion-pair partition chromatography as an alternative approach to ion-exchange chromatography [84] has attracted a number of investigators. Thus, Karger et al. have described the application of this type of chromatography for the separation of biogenic amines and their metabolites [85], as well as for the separation of thyroid hormones and sulfa drugs [86]. Caldwell has described the use of ion-pair chromatography for the separation of catecholamines and amino acids [87]. Although not directly related to HPLC separations or resolutions, Maestas and Morrow [88] have described the gas-chromatographic direct resolution of bicyclic alcohols and ketones. This approach to direct resolutions utilized an achiral support column and the coinjection of the racemate to be resolved along with a suitable, volatile, optically active resolving agent [88]. It is also possible that use of an optically active carrier gas might accomplish similar direct resolutions in VPC, although this has not been described in the literature.

Waters Associates describe ion-pair chromatography as paired-ion chromatography (or PIC for short), and their procedure calls for a large counterion in the mobile phase which enters into an equilibrium complex with the ionic sample being separated [89]. Such species, once formed, can be easily chromatographed by using reverse-phase LC methods, e.g., on Waters μ-Bondapak/C_{18} or an analogous column packing [89, 90, 93, 110]. PIC can be a very useful alternative to ion-exchange chromatography, and possesses distinct advantages, such as the inherent long lifetime of the reverse-phase packing material used. It is not necessary to regenerate the column being used with PIC, and repeated analyses can be run routinely in a matter of minutes [89].

At the present time, there are no reports in the literature describing the direct resolution of enantiomers by this particular method. Making use of an optically active counterion in the eluting solvent, and employing a recycling of the solution being used, should allow for the direct resolution of the desired enantiomers with reverse-phase LC. A suitable workup of the resolved materials in order to remove the counterion should then leave the desired enantiomers ready for subsequent studies [89]. There are now available preparative-scale reverse-phase packed columns, and these may eventually allow for the large-scale direct resolutions of various enantiomers [91]. Much remains to be learned regarding the direct application of ion-paired chromatography for the resolution of enantiomers by HPLC. There are a number of optically active counterions commercially available at the present time, making this method an attractive one for the future (Aldrich Chemical Company, Norse Chemical Company).

As an example of this suggestion, it should be possible to resolve various catecholamines and amino acids using either an optically active carboxylic or sulfonic acid as the counterion in solution. Detection could be by a number of the usual HPLC detectors, or perhaps more sensitively by the use of an electrochemical detector [92].

VII. CONCLUSIONS

There are many other approaches which remain to be studied with regard to direct liquid-chromatographic resolutions, and undoubtedly these will be described in the coming years. Although much has already been done in this area, the greatly improved equipment and methodology in HPLC that appear almost every day will lead to further improvements in both qualitative and preparative direct resolutions. In concluding this chapter, I wish to indicate some of the recent general literature in HPLC, as an aid to those interested in studying direct or indirect resolutions on their own. For many years, the standard textbook in HPLC has been that of Snyder and Kirkland [94], or the previous edition by Kirkland [95]. More recent texts are becoming somewhat specialized, such as that devoted to HPLC in clinical chemistry [96], or that by Brown [97] devoted to biochemical and biomedical applications to HPLC. A somewhat less expensive general text in this area appeared in 1973 [98].

There have been many review papers written on HPLC in the past decade, but space restricts mention to just a few of these. General summaries of this field have been presented by Knox [99] and Williams and Larmann [100]. The equipment for HPLC changes very rapidly, and newer injectors, pumps, or detectors are appearing on the market with regularity. General reviews of equipment are given by Berry and Karger [101], McNair and Chandler [102], and Knox [103]. Some of the packing materials available are discussed by Martin et al. [104], Rabel [105], and Laub [106]. Recent developments in the area of detectors have been summarized by Karasek [107, 108], Thomas [129], and Jupille [109]. Finally, the area of preparative HPLC has been discussed by a number of experts, such as Scott and Kucera [111], DeStefano and Kirkland [112], Pei et al. [113], Fallick [114], and Waters [115].

Throughout this discussion of direct resolutions by HPLC we have described different types of molecular interactions that occur between the solute to be resolved, the eluting solvent, and the packing material. Scott has defined the chromatographic process as follows:

"Chromatography has been classically defined as a separation that is achieved by the distribution of substances between two phases, a mobile phase and a stationary phase. Those substances distributed preferentially in the mobile phase will move through the system more rapidly than those distributed preferentially in the stationary phase" [116].

In principle at least, this is all that is involved in direct LC resolutions, but the design of the proper eluting solvent, the column coating material, or the packing material itself, is a difficult and very challenging problem for both organic and analytical chemists.

ACKNOWLEDGMENTS

Appreciation is expressed to many individuals for having made available to me their results prior to actual publication. I am especially grateful to Drs. Pirkle, Blaschke, Cram, Kirkland, Gil-Av, and Karger. Also appreciated were the encouraging comments of my colleagues at BTI, and the excellent translations of several papers by Dr. Jurg Bucher and Ms. Gaby Speyer. I am indebted to Dr. Renwick and Dr. Weinstein for their reviewing of the initial draft, as well as for their comments and suggestions. Finally, gratitude is expressed to Dr. Hawk and Dr. Bidlingmeyer of Waters Associates, Inc., for their technical contributions to the literature embodied within the paper. Any errors in omission or interpretation of the literature is the sole responsibility of the author.

REFERENCES

1. C. H. Lochmüller and R. W. Souter, J. Chromatogr., 113, 283 (1975).
2. E. Gil-Av and D. Nurok, in Advances in Chromatography, Vol. 10 (J. C. Giddings, E. Grushka, R. A. Keller, and J. Cazes, eds.), Marcel Dekker, New York, 1974, p. 99.
3. G. Losse and K. Kuntze, Z. Chem., 10, 22 (1970).
4. S. V. Rogozhin and V. A. Davankov, Russ. Chem. Rev., 37, 565 (1968).
5. D. R. Buss and T. Vermeulen, Ind. Eng. Chem., 60, 12 (1968).
6. L. H. Klemm and D. Reed, J. Chromatogr., 3, 364 (1960).
7. L. R. Snyder, J. Chromatogr. Sci., 10, 369, 200 (1972).

8. A. F. Michaelis, D. W. Cornish, and R. Vivilecchia, J. Pharm. Sci., 62, 1399 (1973).

9. H. Veening, J. Chem. Educ., 50, A429, A481, A529 (1973).

10. R. E. Majors, Am. Lab., October 1975, 13.

11. W. J. Haggerty, Jr. and E. A. Murrill, Res. Devel., August 1974, 30.

12. G. J. Fallick and J. L. Waters, Am. Lab., August 1972, 21.

13. W. F. Beyer and D. D. Gleason, J. Pharm. Sci., 64, 1557 (1975).

14. J. D. Roberts and M. C. Caserio, Basic Principles of Organic Chemistry, Benjamin, New York, 1964, pp. 569-608.

15. K. Mislow, Introduction to Stereochemistry, Benjamin, New York, 1965, Chaps. 3 and 4.

16. P. H. Boyle, Chem. Soc. (Lond.) Q. Rev., 25, 323 (1972).

17. K. Mislow and M. Raban, in Topics in Stereochemistry, Vol. I (N. L. Allinger and E. L. Eliel, eds.), Wiley-Interscience, New York, 1967, p. 1.

18. M. Raban and K. Mislow, in Topics in Stereochemistry, Vol. II (N. L. Allinger and E. L. Eliel, eds.), Wiley-Interscience, New York, 1967, p. 199.

19. S. A. Kaloustian and M. K. Kaloustian, J. Chem. Educ., 52, 56 (1975).

20. G. B. Kauffman and R. D. Myers, J. Chem. Educ., 52, 777 (1975).

21. S. Yamada, M. Yamamoto, and I. Chibata, J. Agr. Food Chem., 21, 889 (1973).

22. D. A. Labianca, J. Chem. Educ., 52, 156 (1975).

23. G. Paiaro, Organomet. Chem. Rev., Sec. A, 6, 319 (1970).

24. R. E. Ernst, M. J. O'Connor, and R. H. Holm, J. Am. Chem. Soc., 90, 5305 (1968).

25. A. C. Cope, W. R. Moore, R. D. Bach, and H. J. S. Winkler, J. Am. Chem. Soc., 92, 1243 (1970).

26. A. C. Cope, J. K. Hecht, H. W. Johnson, Jr., H. Keller, and H. J. S. Winkler, J. Am. Chem. Soc., 88, 761 (1966).

27. A. C. Cope, K. Banholzer, H. Keller, B. A. Pawson, J. J. Whang, and H. J. S. Winkler, J. Am. Chem. Soc., 87, 3644 (1965).

28. E. L. Eliel, private communication, 1976.

29. D. Valentine, Jr., K. K. Chan, C. G. Scott, K. K. Johnson, K. Toth, and G. Saucy, J. Org. Chem., 41, 62 (1976).

30. M. R. Peterson, Jr., and G. H. Wahl, J. Chem. Educ., $\underline{49}$, 790 (1972).

31. M. D. McCreary, D. W. Lewis, D. L. Wernick, and G. M. Whitesides, J. Am. Chem. Soc., $\underline{96}$, 1038 (1974).

32. B. C. Mayo, Chem. Soc. (Lond.) Rev., $\underline{2}$, 49 (1973)

33. S.-M. L. Chen, K. Nakanishi, N. Awata, M. Morisaki, N. Ikekawa, and Y. Shimizu, J. Am. Chem. Soc., $\underline{97}$, 5297 (1975).

34. M. Koreeda, G. Weiss, and K. Nakanishi, J. Am. Chem. Soc., $\underline{95}$, 239 (1973).

34a. G. Weiss, Ph.D. Thesis, Columbia University, New York, 1974; Diss. Abstr., 75-16, 148.

35. K. J. Judy, D. A. Schooley, R. G. Troetschler, R. C. Jennings, B. J. Bergot, and M. S. Hall, Life Sci., $\underline{16}$, 1059 (1975).

36. K. J. Judy, D. A. Schooley, M. S. Hall, B. J. Bergot, and J. B. Siddall, Life Sci., $\underline{13}$, 1511 (1973).

37. K. J. Judy, D. A. Schooley, L. L. Dunham, M. S. Hall, B. J. Bergot, and J. B. Siddall, Proc. Natl. Acad. Sci. U.S.A., $\underline{70}$, 1509 (1973).

38. P. Loew and W. S. Johnson, J. Am. Chem. Soc., $\underline{93}$, 3765 (1971).

39. Waters Associates Technical Bulletin B14, 1976.

40. M. Kainosho, K. Ajisaka, W. H. Pirkle, and S. D. Beare, J. Am. Chem. Soc., $\underline{94}$, 5924 (1972).

41. W. H. Pirkle, R. L. Muntz, and I. C. Paul, J. Am. Chem. Soc., $\underline{93}$, 2817 (1971).

42. W. H. Pirkle and S. D. Beare, J. Am. Chem. Soc., $\underline{91}$, 5150 (1969).

43. W. H. Pirkle and M. S. Hoekstra, J. Org. Chem., $\underline{39}$, 3904 (1974).

44. W. H. Pirkle and R. W. Anderson, J. Org. Chem., $\underline{39}$, 3901 (1974).

45. C. U. Pittman, Jr., and G. O. Evans, Chem. Tech., $\underline{1973}$, 560.

46. A. P. Schaap, A. L. Thayer, E. C. Blossey, and D. C. Neckers, J. Am. Chem. Soc., $\underline{97}$, 3741 (1975).

47. N. M. Weinshenker, G. A. Crosby, and J. Y. Wong, J. Org. Chem., $\underline{40}$, 1966 (1975).

48. D. C. Neckers, D. A. Kooistra, and G. W. Green, J. Am. Chem. Soc., $\underline{94}$, 9284 (1972).

49. E. C. Blossey and D. C. Neckers, Solid Phase Synthesis, Division of Wiley-Halsted Press, New York, 1975.

50. W. H. Pirkle and S. D. Beare, J. Am. Chem. Soc., **90**, 6250 (1968).

51. W. H. Pirkle and D. L. Sikkenga, J. Chromatogr., **123**, 400 (1976); private communication, 1976.

52. B. Siegel and R. Breslow, J. Am. Chem. Soc., **97**, 6869 (1975).

53. M. Mikolajczyk and J. Drabowicz, Tetrahedron Lett., **1972**, 2379.

54. I. Jonas and B. Norden, Nature (London), **258**, 597 (1975).

55. J. M. Timko, R. C. Helgeson, M. Newcomb, G. W. Gokel, and D. J. Cram, J. Am. Chem. Soc., **96**, 7097 (1974).

56. L. R. Sousa, D. H. Hoffman, L. Kaplan, and D. J. Cram, J. Am. Chem. Soc., **96**, 7100 (1974).

57. G. D. Y. Sogah, and D. J. Cram, J. Am. Chem. Soc., **97**, 1259 (1975).

58. H. Bartels and B. Prijs, in Advances in Chromatography, Vol. 10 (J. C. Giddings, E. Grushka, R. A. Keller, and J. Cazes, eds.), Marcel Dekker, New York, 1974, p. 115.

59. G. Blaschke and F. Donow, Chem. Ber., **108**, 2792 (1975).

60. A.-D. Schwanghart and G. Blaschke, Chem. Ber., in press.

61. G. Blaschke, private communication, 1976.

62. G. Blaschke, Chem. Ber., **107**, 237 (1974).

63. J. D. Navratil and H. F. Walton, Am. Lab., **January 1976**, 69.

64. R. E. Sievers, Nuclear Magnetic Resonance Shift Reagents, Academic Press, New York, 1973.

65. V. A. Davankov, S. V. Rogozhin, and A. V. Semechkin, J. Chromatogr., **91**, 493 (1974).

66. V. A. Davankov, S. V. Rogozhin, A. V. Semechkin, V. A. Baranov, and G. S. Sannikova, J. Chromatogr., **93**, 363 (1974).

67. V. A. Davankov, S. V. Rogozhin, A. V. Semechkin, and T. P. Sachkova, J. Chromatogr., **82**, 359 (1973).

68. V. A. Davankov and S. V. Rogozhin, J. Chromatogr., **60**, 280 (1971).

69. S. V. Rogozhin and V. A. Davankov, J. Chem. Soc. Chem. Commun., **1971**, 490.

70. R. V. Snyder, R. J. Angelici, and R. B. Meck, J. Am. Chem. Soc., **94**, 2660 (1972).

71. R. J. Baczuk, G. K. Landram, R. J. Dubois, and H. C. Dehm, J. Chromaotgr., **60**, 351 (1971).

72. J. Gaal and J. Inczedy, J. Chromatogr., 102, 375 (1974).
73. J. Gaal and J. Inczedy, Acta Chim. Acad. Sci. Hung., 81, 439 (1974).
74. K. Bernauer, M.-F. Jeanneret, and D. Vonderschmitt, Helv. Chim. Acta, 54, 297 (1971).
75. R. E. Leitch, H. L. Rothbart, and W. Rieman, III, J. Chromatogr., 28, 132 (1967).
76. M. I. Popova and C. G. Kratchanov, J. Chromatogr., 72, 192 (1972).
77. M. I. Popova, C. G. Kratchanov, and M. J. Kuntcheva, J. Chromatogr., 87, 581 (1973).
78. H. Guilford, Chem. Soc. (Lond.) Rev., 2, 249 (1973).
79. H. H. Weetall and C. C. Detar, Biotechnol. Bioeng., 16, 1537 (1974).
80. G. Wulff, A. Sarhan, and K. Zabrocki, Tetrahedron Lett., 1973, 4329.
81. K. K. Stewart and R. F. Doherty, Proc. Natl. Acad. Sci. U.S.A., 70, 2850 (1973).
82. Enzyme process to make L-malic detailed, Chem. Eng. News, August 25, 1975, p. 32.
83. I. Chibata, T. Tosa, T. Sato, T. Mori, and Y. Matsuo, Proc. IV, IFS: Ferment. Technol. Today, 1972, 383.
84. C. D. Scott, Science, 186, 226 (1974).
85. B.-A. Persson and B. L. Karger, J. Chromatogr. Sci., 12, 521 (1974).
86. B. L. Karger, S. C. Su, S. Marchese, and B.-A. Persson, J. Chromatogr. Sci., 12, 678 (1974).
87. W. Caldwell, paper presented at the North Jersey Chromatography Discussion Group of the American Chemical Society, Drew University, New Jersey, June 11, 1976.
88. P. D. Maestas and C. J. Morrow, Tetrahedron Lett., 1976, 1047.
89. Waters Associates, Technical Bulletin, Paired-Ion Chromatography. An Alternative to Ion Exchange, D61, December 1975.
90. R. K. Gilpin, J. A. Korpl, and C. A. Janicki, Anal. Chem., 47, 1498 (1975).
91. Waters Associates, Technical Bulletins B14 and P25, Prep LC/System 500 for Preparative Liquid Chromatography, 1976.
92. R. C. Buchta and L. J. Papa, J. Chromatogr. Sci., 14, 213 (1976).
93. D. C. Locke, J. Chromatogr. Sci., 12, 433 (1974).

94. L. R. Snyder and J. J. Kirkland, Introduction to Modern Liquid Chromatography, Wiley-Interscience, New York, 1974.

95. J. J. Kirkland (ed.), Modern Practice of Liquid Chromatography, Wiley-Interscience, New York, 1971.

96. P. H. Dixon, C. H. Gray, C. K. Lim, and M. S. Stoll (eds.), High Pressure Liquid Chromatography in Clinical Chemistry, Academic Press, New York, 1976.

97. P. R. Brown, High Pressure Liquid Chromatography—Biochemical and Biomedical Applications, Academic Press, New York, 1973.

98. S. G. Perry, R. Amos, and P. I. Brewer, Practical Liquid Chromatography, Plenum-Rosetta, New York, 1973.

99. J. H. Knox, Annu. Rev. Phys. Chem., 24, 29 (1973).

100. R. C. Williams and J. P. Larmann, Ind. Res., August 1975, 62.

101. L. Berry and B. L. Karger, Anal. Chem., 45, 819A (1973).

102. H. M. McNair and C. D. Chandler, J. Chromatogr. Sci., 12, 425 (1974), and preceding papers in this series.

103. J. H. Knox, Lab. Prac., 22, 52 (1973).

104. M. Martin, C. Eon, and G. Guiochon, Res. Devel., April 1975, 24.

105. F. M. Rabel, Am. Lab., May 1975, 53.

106. R. J. Laub, Res. Devel., July 1974, 24.

107. F. W. Karasek, Res. Devel., March 1975, 34.

108. F. W. Karasek, Res. Devel., January 1976, 28.

109. T. H. Jupille, Am. Lab., May 1976, 85.

110. K. Karch, I. Sebestian, I. Halasz, and H. Engelhardt, J. Chromatogr., 122, 171 (1976).

111. R. P. W. Scott and P. Kucera, J. Chromatogr., 119, 467 (1976).

112. J. J. DeStefano and J. J. Kirkland, Anal. Chem., 47, 1103A (1975).

113. P. Pei, S. Ramachandran, and R. S. Henly, Am. Lab., January 1975.

114. G. Fallick, Am. Lab., August 1973, 19.

115. J. L. Waters, J. Chromatogr., 55, 213 (1971).

116. R. P. W. Scott, J. Chromatogr., 122, 35 (1976).

117. C. C. Leznoff, Chem. Soc. (Lond.) Rev., 3, 65 (1974).

118. C. G. Overberger and K. N. Sannes, Angew. Chem. Int. Ed., **13**, 99 (1974).
119. G. A. Crosby, Aldrichim. Acta, **9**, 15 (1976).
120. Eastman Org. Chem. Bull., **48**, 1 (1976).
121. E. K. Barefield, Strem Chem., **3**, 3 (1975).
122. G. W. Gokel and H. D. Durst, Aldrichim. Acta, **9**, 3 (1976).
123. EM Laboratories Technical Bulletin, Kryptofix-Polyoxadiazamacrobicycles.
124. W. D. Curtis, D. A. Laidler, J. F. Stoddart, and G. H. Jones, J. Chem. Soc. Chem. Commun., **1975**, 833.
125. W. D. Curtis, D. A. Laidler, J. F. Stoddart, and G. H. Jones, J. Chem. Soc. Chem. Commun., **1975**, 835.
126. D. J. Cram, paper presented at the American Chemical Society National Meeting, New York, April 4-9, 1976. Abstracts ORGN 007.
127. G. D. Y. Sogah and D. J. Cram, J. Am. Chem. Soc., **98**, 3038 (1976).
128. H. Takahagi and S. Seno, J. Chromatogr., **108**, 354 (1975).
129. H. L. Thomas, Ind. Res., **August 1976**, 51.
130. F. Mikes, G. Boshart, and E. Gil-Av, J. Chem. Soc. Chem. Commun., **1976**, 99.
131. F. Mikes, G. Boshart, and E. Gil-Av, J. Chromatogr., **122**, 205 (1976).
132. D. I. Kingston and B. B. Gerhart, J. Chromatogr., **116**, 182 (1976).
133. B. B. Wheals, J. Chromatogr., **107**, 402 (1975).
134. B. B. Wheals, C. G. Vaughan, and M. J. Whitehouse, J. Chromatogr., **106**, 109 (1975).
135. S. H. Chang, K. M. Gooding, and F. E. Regnier, J. Chromatogr., **120**, 321 (1976).
136. Realco Chemical Company, New Brunswick, New Jersey, Technical bulletins on Spheron and Photox, private communication.
137. J. Lesec, F. Lafuma, and C. Quivoron, J. Chromatogr. Sci., **12**, 683 (1974).
138. J. Lesec and C. Quivoron, Analysis, **4**, 120 (1976).

Chapter 7

THE USE OF HIGH-PRESSURE LIQUID CHROMATOGRAPHY
IN RESEARCH ON PURINE NUCLEOSIDE ANALOGS

William Plunkett

Department of Developmental Therapeutics
M. D. Anderson Hospital and Tumor Institute
The University of Texas System Cancer Center
Houston, Texas

I. INTRODUCTION 211
II. METHODS 212
III. RESULTS AND DISCUSSION 215
    A. Metabolism of 9-β-D-Arabinofuranosyladenine 215
    B. Metabolism of 9-β-D-Xylofuranosyladenine 236
IV. CONCLUSIONS 245
    ACKNOWLEDGMENTS 246
    REFERENCES 246

I. INTRODUCTION

    An understanding of the biological disposition of a potentially useful chemotherapeutic compound is helpful in gaining insight into the mechanism by which the compound exerts its activity. After observing the metabolism of a drug by an organism over a period of time, the investigator may be able to deduce the enzymatic steps which could be limiting to the efficacy of the compound. The results of such pharmacokinetic studies may suggest ways by which the toxicity of drugs may be increased and thereby aid in the design of more effective modes and schedules of drug administration.

    High-pressure liquid chromatography, a relatively new technique, is proving to be valuable in pharmacologic studies on a wide spectrum of compounds. The theory governing the use of this method of analysis and its application to the separation of a variety of classes of compounds is highly developed and has been extensively reviewed in recent years [1-7]. The purpose of this chapter is to illustrate the utility of high-pressure liquid chromatography as an analytical technique, when integrated with more traditional biochemical methods, in the investigation of the metabolism of purine nucleoside analogs which exhibit chemotherapeutic activity.

Many nucleoside analogs are not biologically active until they are phosphorylated to the respective nucleotides. Thus, much of the emphasis in this line of research is placed on determining the rate of formation and cellular concentration of nucleotide analogs. The recent reviews by Horvath [4] and Brown [5] have documented much of the methodology for the analysis of complex mixtures of nucleotides in cell and tissue extracts by high-pressure liquid chromatography. Some advantages that recommend this technique include the ability to obtain good resolution of chemically related compounds in relatively short times. Also, both qualitative and quantitative information may be derived from small samples of biological materials, since the latter are not destroyed by high-pressure liquid-chromatographic analysis and may be further processed. In addition, the cost of the analyses in terms of both time and money is decreased since the columns may be rapidly regenerated and generally have a long lifetime.

At the editor's suggestion, this chapter will not deal with all the applications of high-pressure liquid chromatography to chemotherapeutic nucleoside analogs. Rather, the material covered reflects the author's own experience with the technique in investigations on the cellular metabolism of the purine nucleoside analogs, arabinosyladenine and xylosyladenine. It is hoped that this approach will illustrate the variety of ways that high-pressure liquid chromatography may be usefully integrated into a biochemical research program.

II. METHODS

The theory underlying both the operation of the individual components used in a high-pressure liquid chromatograph and the optimization of the various operating parameters involved has been covered in several recent comprehensive reviews, and will be dealt with sparingly here [2-5]. The instrument used for the work to be presented was the Varian Aerograph dual-column high-pressure liquid chromatograph, Model LCS-1000. A diagram of this instrument is shown in Fig. 1. The basic features of this chromatograph are representative of a variety of instruments currently on the market and also serve as an outline for the investigator interested in assembling his own chromatograph from components.

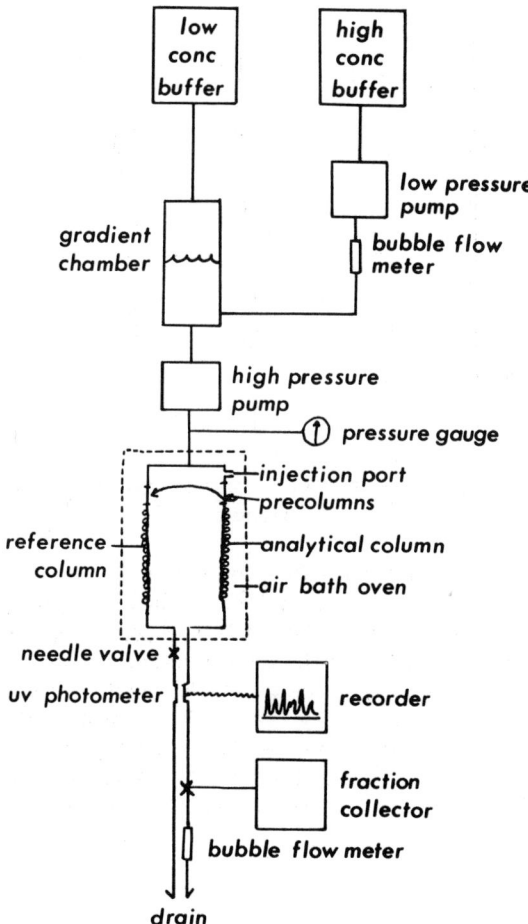

FIG. 1. A schematic diagram of the Varian Aerograph dual-column liquid chromatograph, Model LCS-1000.

The work to be described has employed the pellicular anion-exchange resins PA-39 (Varian Aerograph) and AS-Pellionex-SAX (Whatman, Inc.), which were supplied prepacked in stainless steel columns (1 mm × 3 m). The presence of uv-absorbing contaminants in analytical grade KH_2PO_4 results in an increase in the baseline during gradient elutions at high sensitivities if the reference cell of the detector contains only the dilute buffer (Fig. 2B). This difficulty has been eliminated by connecting two identical columns in parallel to the analytical and reference sides of the detector

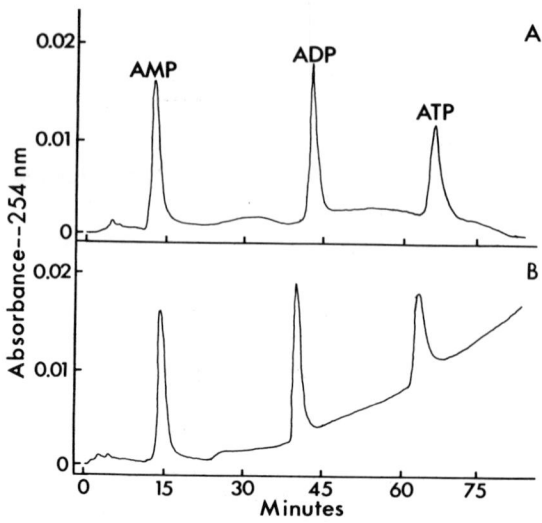

FIG. 2. The effect of a reference column on the background absorption in high-pressure liquid-chromatographic separations. Two portions of a mixture of 1 nmol each of AMP, ADP, and ATP were separated serially, under identical conditions, except in (A) the reference column was connected to the uv detector flow cell, whereas in (B) the eluate of the reference column was passed immediately to the drain without entering the detector. Anion-exchange resin: AS-Pellionex-SAX. Temperature: 60°C. Pressure: 775 psi. Starting volume: 58 ml. Eluents: 0.002 M KH_2PO_4, pH 3.55, and 1.00 M KH_2PO_4, pH 4.15. Flow rates: 24 ml/hr each. Gradient delay: 15 min.

so that absorbance due to contaminants in the eluent is cancelled (Fig. 2A). Use of the needle valve placed on the reference side to adjust for anticipated differences in the backpressure properties of the two columns has not been necessary to date.

Nucleotide samples from cells and tissues are generally prepared by acid extraction with either perchloric acid [8, 9] or trichloroacetic acid [5] followed by neutralization of the sample. Alternatively, extraction with 60% methanol results in the efficient removal of cellular nucleotides [10] and provides a sample that is easily concentrated for analysis by high-pressure liquid chromatography [11].

III. RESULTS AND DISCUSSION

High-pressure liquid chromatography has played an integral role in our studies on the cellular disposition of analogs of adenine nucleosides. It is hoped that a review of selected portions of this research will illustrate the utility of this technique and perhaps suggest further applications in other investigations.

A. Metabolism of 9-β-D-Arabinofuranosyladenine

9-β-D-Arabinofuranosyladenine (araA) is an adenine nucleoside analog of the normal DNA constituent, 2'-deoxyadenosine, differing from that molecule by the presence of a hydroxyl group on the 2'-carbon (Fig. 3). This hydroxyl is positioned trans to the 3'-hydroxyl in contrast to the cis arrangement seen in adenosine (Fig. 3). Currently, araA is recognized as an important antiviral agent [12] and is undergoing tests as an antineoplastic compound [13, 14]. Incubation of mouse fibroblasts (L cells) with araA results in the inhibition of DNA synthesis and the subsequent loss of cell viability [15]. Enzymatic studies have suggested that these effects may be due to the inhibition of DNA polymerase by araATP [16] and possibly from the inhibition of ribonucleotide reductase by araADP and araATP [17]. Since some nucleosides may be toxic in their own right as nonphosphorylated compounds, e.g., psicofuranine [18], arabinosyl-6-mercaptopurine [19], and decoyinine [20], it was essential to demonstrate that exogenously supplied araA was converted to inhibitory concentrations of arabinosyl nucleotides in the cells.

Exponentially growing L cells were incubated 4 hr with 0.1 mM [2-^{3}H]-araA, washed thoroughly, and extracted with cold PCA. In several identical experiments, an average of only 0.14% of the total tritium was associated

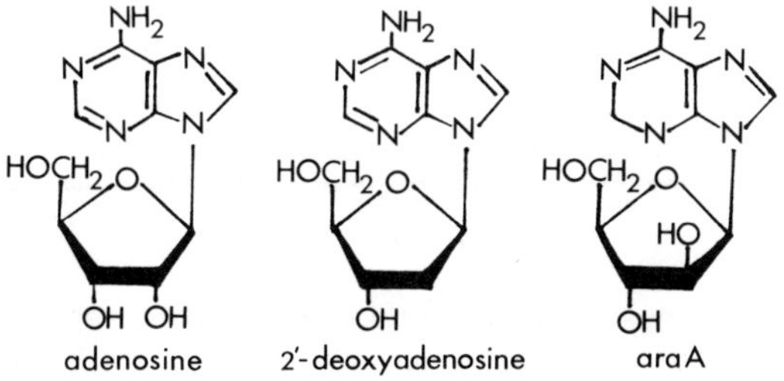

FIG. 3. Structures of adenosine, 2'-deoxyadenosine, and araA.

with the cell fraction, whereas the cells had deaminated about half of the araA in the medium to 9-β-D-arabinofuranosylhypoxanthine (araHx). Of the cellular tritium, the acid-soluble material contained 82 to 94% of the cellular radioactivity, whereas the RNA and DNA fractions consisted of 5.4 to 16.1% and 0.6 to 3.2%, respectively [9]. The following procedures were carried out to identify the radioactive components in each cellular fraction.

A portion of the acid-soluble fraction was concentrated under vacuum at 33°C, and after chilling, the insoluble $KClO_4$ was removed by centrifugation. An aliquot of the concentrated supernatant equivalent to the acid-soluble material extracted from 9.2×10^6 cells was analyzed by high-pressure liquid

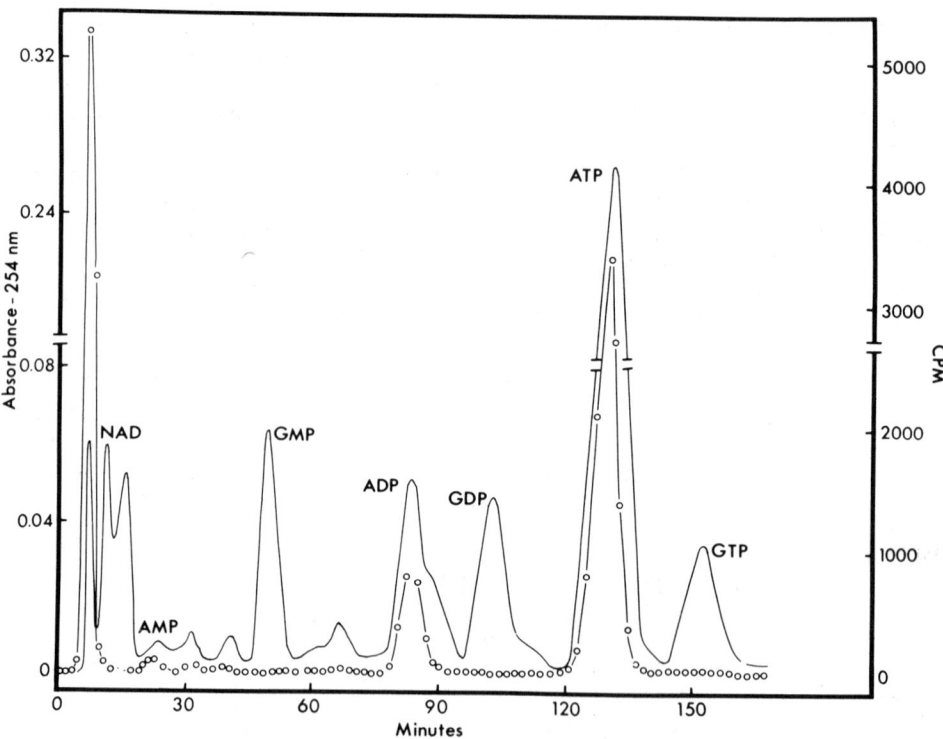

FIG. 4. High-pressure liquid-chromatographic separation of the acid-soluble components extracted from the equivalent of 9.2×10^6 L cells following 4 hr of incubation with 1×10^{-4} M [^{3}H]araA (specific activity, 1.07×10^7 cpm/μmol). Anion-exchange resin: PA-39. Temperature: 70°C. Pressure: 1900 psi. Starting volume: 63 ml. Eluents: 0.04 M KH_2PO_4, pH 3.55, and 1.00 M KH_2PO_4, pH 4.00. Flow rates: 20 ml/hr each. Gradient delay: 20 min [9]. Solid line, absorbance at 254 nm; open circles, cpm/2 min fraction.

chromatography and fractions of 2-min duration were collected directly into scintillation vials (Fig. 4). The identity of individual peaks was generally established by comparison of retention times with those of standards. For convenience, labels on individual peaks indicate the major constituent of that peak, not that the designated nucleotide is the exclusive constituent. Tritium activity from the acid-soluble material from araA-treated cells was associated predominantly with the first-eluted nucleoside peak and with the ADP and ATP peaks.

High-pressure ion-exchange chromatography undertaken as just described results in the separation of nucleotides on the basis of the identity of the base and on the degree of phosphorylation, but is unable to discriminate between nucleotides differing only in the carbohydrate moiety. Accordingly, a mixture of ATP, dATP, and araATP is eluted as a single peak. For this reason, additional processing of the adenine nucleoside triphosphate fraction was necessary to identify the radioactive components. Another portion of the acid-soluble material equivalent to that extracted from 2.2×10^7 cells was fractionated by high-pressure liquid chromatography as described in Fig. 4. The adenine nucleoside triphosphate fraction, eluted between 120 and 140 min, was collected, extracted on Norit to remove the phosphate buffer, and dephosphorylated with alkaline phosphatase. The resulting nucleosides were separated by two-dimensional thin-layer chromatography (Table 1). Adenosine and deoxyadenosine contain 3.8 and 4.9%

TABLE 1

Distribution of ^{3}H from [^{3}H]AraA in the Adenine Nucleoside Triphosphate Fraction[a]

Fraction	AraA equivalents (pmol)
AraA	1032.0 (88.7)[b]
AraHx	30.3 (2.6)
Deoxyadenosine	57.2 (4.5)
Adenosine	44.4 (3.5)

[a]The acid-soluble fraction equivalent to that extracted from 2.2×10^7 cells was fractionated by high-pressure liquid chromatography as described in Fig. 6. The adenine nucleoside triphosphate peak was collected, extracted on Norit, and dephosphorylated with alkaline phosphatase; the resulting nucleosides were separated by two-dimensional TLC [9].

[b]Numbers in parentheses, percentage of total radioactivity in nucleosides eluted from the plate.

of the radioactive nucleoside, respectively. These most likely were derived from araA after deamination to araHx by adenosine deaminase, followed by cleavage to free hypoxanthine which was subsequently reutilized to form [^{3}H]IMP with ensuing anabolic metabolism. In contrast, arabinose-containing nucleosides constituted about 90% of the radioactive nucleoside, indicating that percentage of radioactivity in the adenine nucleoside triphosphates had been associated with araATP. This knowledge, together with that of the average cell volume and the specific activity of [^{3}H]araA, permitted us to calculate that after 4 hr of incubation, the cellular araATP concentration exceeded 20 μM [9]. This value is about 20-fold greater than the K_i of araATP for a mammalian DNA polymerase [16] and thus may be presumed to be inhibitory to DNA synthesis.

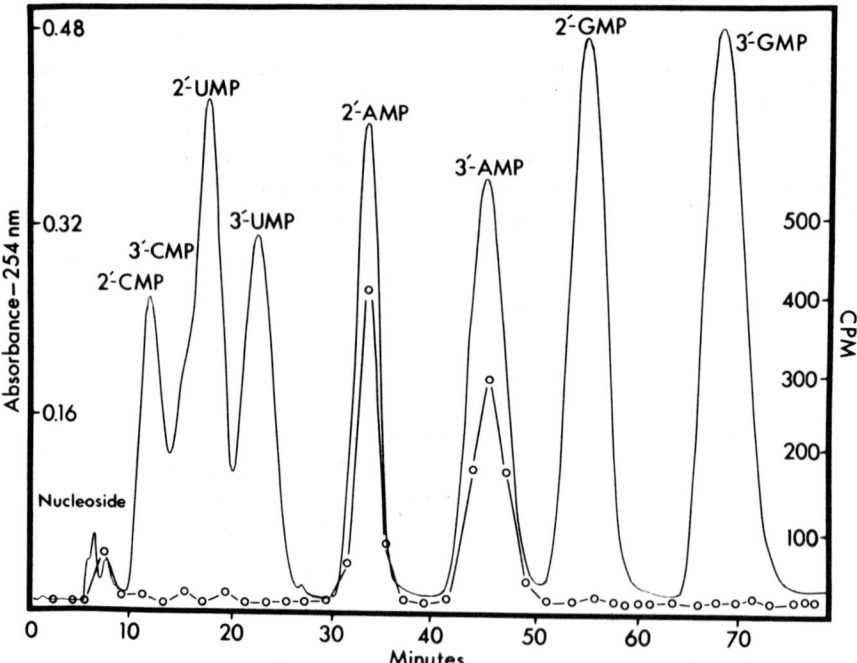

FIG. 5. High-pressure liquid-chromatographic separation of the alkali-hydrolyzed RNA fraction from the equivalent of 1.3×10^7 L cells following 4 hr of incubation with 1×10^{-4} M [^{3}H]araA (specific activity, 1.07×10^7 cpm/μmol). Anion-exchange resin: PA-39. Temperature: 70°C. Pressure: 1800 psi. Starting volume: 60 ml. Eluents: 0.04 M KH_2PO_4, pH 3.55, and 1.00 M KH_2PO_4, pH 4.00. Flow rates: 20 and 10 ml/hr, respectively. Gradient delay: 20 min [9]. Solid line, absorbance at 254 nm; open circles, cpm/2 min fraction.

An RNA hydrolysate had been obtained by treating the PCA-insoluble material from [^{3}H]araA-treated cells with 0.3 N KOH for 18 hr. The resulting mixture of 2'(3')-nucleoside monophosphates was fractionated by high-pressure liquid chromatography (Fig. 5). The major radioactive components coelute with 2'-AMP and 3'-AMP, and a minor portion of the tritium is associated with the nucleoside fraction. The other nucleotide peaks do not contain significant amounts of radioactivity and further elution with buffer concentrations sufficient to elute dinucleoside monophosphates failed to produce additional tritium-containing peaks. To establish the identity of the radioactive compounds that coelute with the AMP isomers, a similar portion of KOH hydrolysate was fractionated by high-pressure liquid chromatography as before and the nucleoside and AMP peaks were collected.

The AMP fractions were pooled, extracted on Norit, and dephosphorylated with alkaline phosphatase. Complete degradation of the nucleotide material to nucleosides was verified by high-pressure liquid chromatography. The resulting nucleosides were separated by two-dimensional thin-layer chromatography (Table 2, column A). About 70% of the radioactivity was

TABLE 2

Distribution of the ^{3}H from [^{3}H]AraA in the RNA Fraction[a]

Fraction	AraA equivalent (pmol)	
	A	B
AraA	14.6 (25.6)[b]	4.3 (61.4)
AraHx	3.1 (5.4)	1.0 (14.3)
Deoxyadenosine	0.6 (1.0)	0.5 (7.1)
Adenosine	38.7 (67.9)	1.2 (17.1)

[a] Exponentially growing L cells were incubated for 4 hr with 0.1 mM [^{3}H]araA (specific activity, 6.19×10^7 cpm/μmol), and the acid-insoluble material was hydrolyzed with KOH. The RNA hydrolysate equivalent to that extracted from 1.0×10^7 cells was fractionated as described in Fig. 7 and the nucleoside peak and the 2'- and 3'-AMP peaks were collected. After pooling, the contents of the AMP peaks were extracted on Norit and dephosphorylated, and the reaction products were separated by two-dimensional TLC (A). The contents of the nucleoside peak were extracted on Norit and fractionated by two-dimensional TLC (B) [9].

[b] Numbers in parentheses, percentage of the total radioactivity eluted from the plate.

associated with adenosine, suggesting extensive cellular metabolism of araA. The nucleoside portion of the KOH hydrolysate is representative of the 3'-terminal nucleotides of RNA chains. In contrast with the AMP fractions, separation of the nucleoside fraction by two-dimensional thin-layer chromatography indicated that most of the 3'-terminal radioactivity in RNA was associated with arabinosyl nucleotides (Table 2, column B). These data are consistent with the suggestion [21] that araAMP may be incorporated on the 3'-terminus of t-RNA.

For identification of the radioactive components of the DNA fraction, a portion of that material which remained acid insoluble following KOH hydrolysis was degraded to 5'-mononucleotides by the sequential action of DNase I and venom phosphodiesterase. Figure 6 illustrates the separation of the 5'-deoxynucleotide peaks by conventional anion-exchange chromatography. The single peak of radioactivity is seen to elute after dCMP and to trail slightly into dAMP. When a control mixture of the 5'-deoxynucleotide monophosphates and authentic [^{3}H]araAMP was chromatographed, the araAMP eluted immediately after dCMP without any trailing into the 5'-dAMP

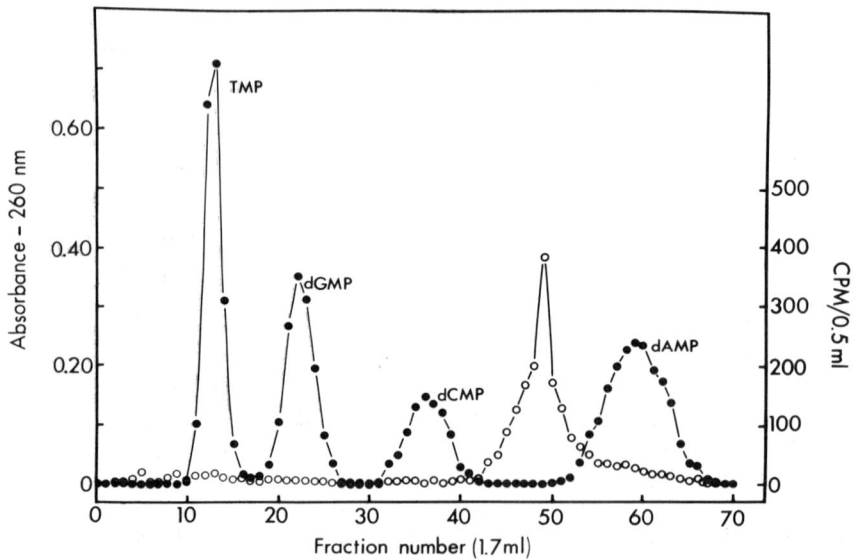

FIG. 6. Fractionation of the 5'-nucleotides from the DNA from the equivalent of 7.3×10^7 L cells following 4 hr of incubation with 1×10^{-4} M [^{3}H]araA (specific activity, 6.19×10^7 cpm/μmol). Anion-exchange resin: AG 50W-X4 (NH$_4^+$) (200-400 mesh). Eluent: 0.10 M ammonium formate, pH 3.2 [9].

fraction. The remainder of fractions 44 to 53 was pooled, dephosphorylated, and separated by two-dimensional thin-layer chromatography. The results indicated that over 95% of the radioactivity was associated with araAMP [9].

Since araATP had been cited as a possible DNA chain-terminating nucleotide [22], it was important to determine the position of araAMP in DNA. Another portion of the DNA fraction was degraded to 3'-nucleoside monophosphates by the sequential action of micrococcal nuclease and spleen phosphodiesterase. Fractionation of 2 to 10 nmol of the 3'-mononucleotide mixture by high-pressure liquid chromatography resulted in the separation of the four 3'-deoxynucleoside monophosphates and a nucleoside peak (Fig. 7). However, when the 200-fold larger quantities of the 3'-nucleotide mixture necessary for radioactive analysis were fractionated by this procedure, the resolution between 3'-dCMP and 3'-TMP was lost (Fig. 8). Nevertheless, the fractionation proved useful since the radioactivity was associated only with the nucleoside fraction (32%) and the 3'-dAMP peak (68%) (Fig. 8). To make definite identification of the radioactive compounds associated with the 3'-dAMP peak, the material was collected from the analytical column of the chromatograph, extracted on Norit, dephosphorylated, and separated by two-dimensional thin-layer chromatography. The results (Table 3, column A) demonstrated that 80% of the radioactivity of the 3'-dAMP peak was associated with araA and araHx, the latter probably arising from a slight

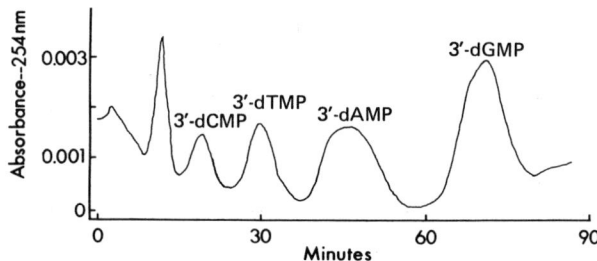

FIG. 7. High-pressure liquid-chromatographic separation of 3'-deoxynucleotides. The DNA fraction from cells incubated 4 hr with 1×10^{-4} M [^{3}H]araA was degraded to 3'-mononucleotides by the sequential action of micrococcal nuclease and spleen phosphodiesterase. A portion of the hydrolysate equivalent to the material from 9×10^4 cells was fractionated by high-pressure liquid chromatography. Anion-exchange resin: PA-39. Temperature: 70°C. Pressure: 800 psi. Starting volume: 58 ml. Eluents: 0.04 M KH_2PO_4, pH 3.55, and 1.00 M KH_2PO_4, pH 4.00. Flow rates: 12 ml/hr each. Gradient delay: 20 min. (Courtesy of W. Plunkett and S. S. Cohen.)

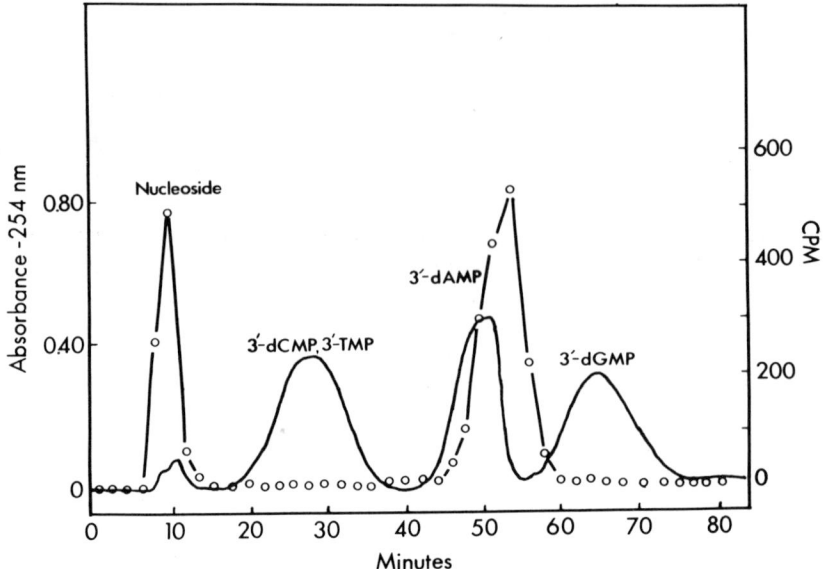

FIG. 8. High-pressure liquid-chromatographic fractionation of the 3'-nucleotides from the DNA fraction from the equivalent of 7.3×10^7 L cells following a 4-hr incubation with 1×10^{-4} M [^{3}H]araA (specific activity, 6.19×10^7 cpm/μmol). Conditions of the fractionating were identical to those described in Fig. 7 [9]. Solid line, absorbance 254 nm; open circles, cpm/2 min fraction.

TABLE 3

Distribution of ^{3}H from [^{3}H]AraA in DNA Following Degradation to 3'-Mononucleotides[a]

Fraction	AraA equivalents (pmol)	
	A	B
AraA	22.9 (76.8)[b]	7.1 (71.7)
AraHx	1.2 (4.0)	1.4 (14.1)
Deoxyadenosine	4.2 (14.1)	0.4 (4.0)
Adenosine	1.5 (5.0)	1.0 (10.1)

[a]The DNA fraction equivalent to that extracted from 7.3×10^7 cells was degraded to 3'-mononucleotides and fractionated as described in Fig. 7, and the nucleoside and 3'-dAMP fractions were collected. After extraction on Norit, the 3'-dAMP fraction was dephosphorylated with alkaline phosphatase and the reaction products were separated by two-dimensional TLC (A). The nucleoside fraction was extracted on Norit and separated by two-dimensional TLC (B) [9].

[b]Numbers in parentheses, percentage of the total radioactivity eluted from the plate in each fraction.

deaminase contamination of the alkaline phosphatase. This indicates that the majority of the arabinosyl nucleotides in DNA had been incorporated internally in DNA in 3'-5' phosphodiester linkage. The radioactivity associated with the nucleoside peak, representative of the 3'-terminus of the DNA chain, was also seen to be predominantly in araA and araHx (Table 3, column B).

These results suggest that after incubation of mouse fibroblasts with araA, a small amount is phosphorylated to arabinosyl nucleotides, which attain inhibitory cellular concentrations, and some araATP is incorporated in internucleotide linkage into DNA. However, analysis of the growth medium in these incubations showed that most of the exogenous araA was rapidly deaminated to araHx, a compound that has long been recognized as possessing greatly reduced toxicity relative to araA [15, 21, 23]. For example, when L cells were incubated with 0.2 mM araA, the viability in the culture decreased rapidly for 24 hr followed thereafter by the multiplication of surviving cells (Fig. 9). Analysis of the growth medium indicated that essentially all the araA had been deaminated to araHx after 24 hr. In several identical experiments, the growth of cells treated with 0.2 mM araHx was the same as the control [24]. Therefore, the arrest of the decrease in viable cells in the araA-treated culture appears to be due to complete deamination of araA followed by turnover and decrease of the cellular araATP. Surviving cells then go on to multiply. Therefore, it was reasoned that methods of administering araA that protected the nucleoside from deamination would result in increased toxicity of the compound. Ortiz [25] first observed that araAMP, the 5'-monophosphate of araA, was more toxic to L cells than was araA. As shown in Fig. 9, cells treated with araAMP continue to multiply for 12 hr after the start of the incubation. Thereafter, the culture undergoes a rapid decrease in viability equal in rate to that experienced by araA-treated cells but more sustained in duration.

There were at least three possibilities that could explain why the toxicity of araAMP was more sustained than that of araA. First, the nucleotide, protected from nucleoside deaminases by the phosphate and acting as a depot form of araA, might be gradually dephosphorylated and slowly enter the cell as araA over a prolonged period of time, thus leading to sustained toxicity. Second, araAMP could only enter the cell after dephosphorylation, but the cells might metabolize a nucleoside derived from a nucleotide differently, perhaps in a way that made the nucleoside product more toxic than the exogenously added nucleoside itself. Such differential metabolism had previously been found in bacteria [26]. A third possibility was that araAMP had penetrated the cell intact and had been directly phosphorylated to the active triphosphate. Experiments utilizing doubly labeled araAMP were carried out to choose between these possibilities.

Exponentially growing L cells were incubated with 0.1 mM $[2-^3H, ^{32}P]$-araAMP. Samples were taken at hourly intervals, washed thoroughly, and

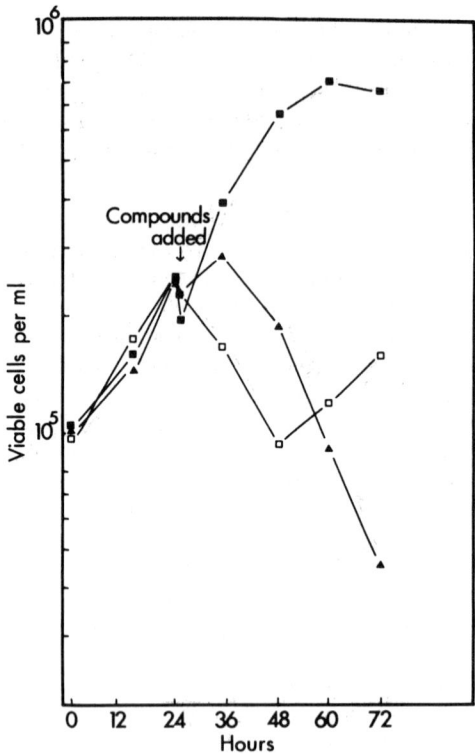

FIG. 9. Viability of L cells during incubation with arabinosyl compounds. ■, Control; □, 2×10^{-4} M araA; ▲, 2×10^{-4} M araAMP [11].

extracted with PCA. Both the ^{32}P and ^{3}H of the nucleotide were accumulated in the acid-soluble material at similar constant rates for 3 hr (Fig. 10A). In a parallel experiment, [^{3}H]araA was incorporated into the acid-soluble pool at an initial rate of about 20 times the rate of the nucleotide (Fig. 10A). In contrast to the incorporation from araAMP, ^{3}H incorporation from araA appeared to reach a maximum rate after 120 min. The fact that the ^{3}H from araAMP was taken up in a linear fashion over 4 hr argues that this activity was not derived from an [^{3}H]araA contamination in the doubly labeled nucleotide. Both labels from the nucleotide also accumulated in the acid-insoluble material at a rate only two or three times less than that of [^{3}H]araA (Fig. 10B). During the initial incubation, the ratio of ^{32}P to ^{3}H remained close to that of the exogenous nucleotide, suggesting that extensive extracellular dephosphorylation with subsequent incorporation of the nucleoside did not occur.

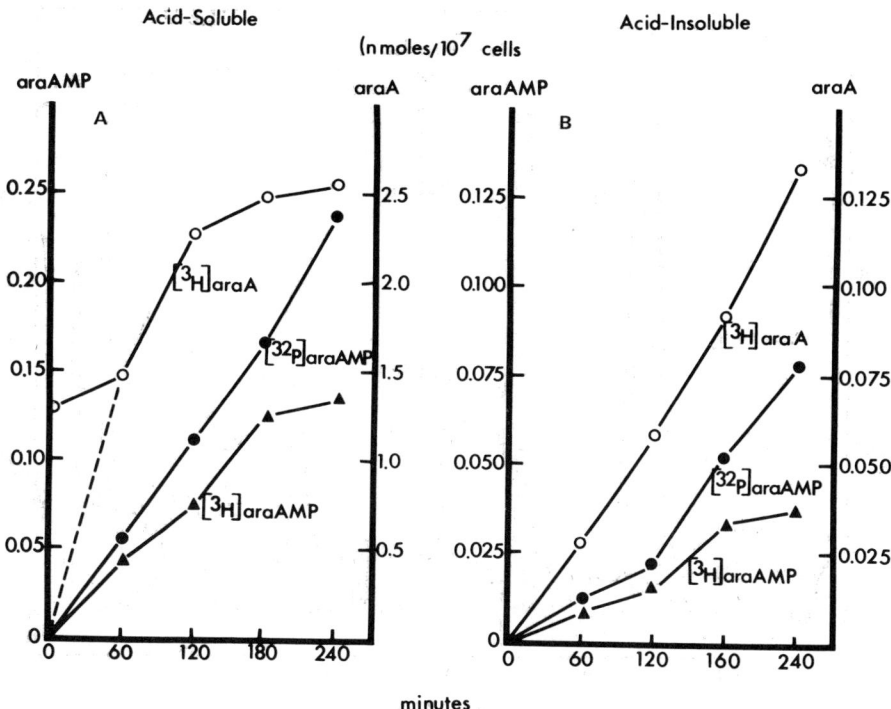

FIG. 10. Uptake and incorporation of [^{3}H, ^{32}P]araAMP and [^{3}H]araA by L cells. (A) Acid-soluble fraction; (B) acid-insoluble fraction [8].

To establish the identity of the radioactive compounds in the acid-soluble material, PCA-soluble extracts from cells incubated 4 hr with double-labeled araAMP were fractionated by high-pressure liquid chromatography and samples of 2-min duration were analyzed for radioactivity (Fig. 11). Tritium was found in the nonretained nucleoside peak and in increasing amounts in peaks containing AMP, ADP, and ATP. Some ^{32}P was associated with the nonretained compounds and also with the NAD-containing peak. The latter was probably ^{32}P$_i$. Otherwise, most of the ^{32}P eluted with guanine nucleotides, indicating some metabolism of araAMP by the cells. However, the ratios of ^{32}P to ^{3}H found in the adenine nucleotides (Fig. 11, insert) show only relatively small deviations from that of the initial ratio of the exogenously added compound. This again suggests that araAMP may have entered the cells with relatively little dephosphorylation.

Since high-pressure ion-exchange chromatography, as carried out in Fig. 11, is unable to separate araATP from other adenine nucleoside

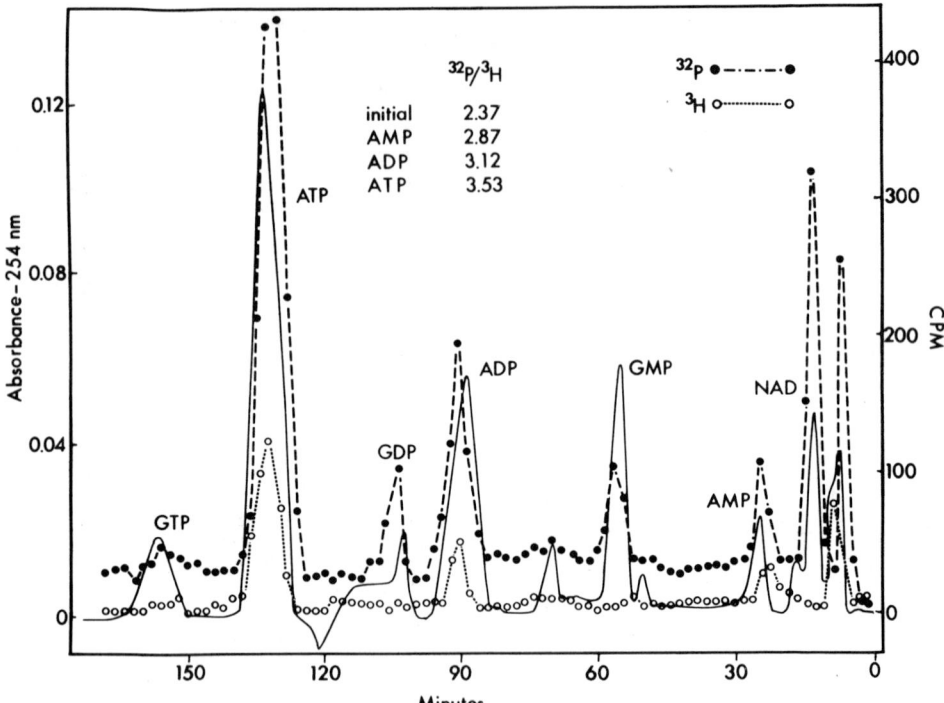

FIG. 11. High-pressure liquid-chromatographic fractionation of acid-soluble components extracted from the equivalent of 5.6×10^6 L cells following 4 hr of incubation with 1×10^{-4} M [^{3}H, ^{32}P]araAMP (specific activities, [^{3}H]araAMP 1.11×10^7 cpm/μmol; [^{32}P]araAMP 2.63×10^7 cpm/μmol). Anion-exchange resin: PA-39. Temperature: 70°C. Pressure: 1800 psi. Starting volume: 63 ml. Eluents: 0.04 M KH_2PO_4, pH 3.55, and 1.00 M KH_2PO_4, pH 4.00. Flow rates: 20 ml/hr each. Gradient delay: 20 min [8].

triphosphates, an additional analysis was required to verify that the ^{32}P was actually associated with arabinosyl nucleotides. The nucleoside triphosphate fraction was isolated, extracted on Norit to remove the phosphate buffer, and degraded to monophosphates by the enzyme, apyrase. The resulting monophosphates and nucleosides (due to a slight phosphatase contamination) were subjected first to thin-layer electrophoresis followed by thin-layer chromatography in the second dimension. The results (Table 4) showed that all of the ^{32}P and 97% of the ^{3}H was associated with araAMP, whereas no ^{32}P and only 3% of the nucleotide-bound ^{3}H were associated with AMP. The ratio of ^{32}P to ^{3}H for araAMP was reasonably close to that

TABLE 4

Fractionation of Apyrase Reaction Products[a]

	^{3}H (cpm)	^{32}P (cpm)	^{32}P/^{3}H [b]
AraA	141	0	—
Adenosine	19	0	—
AraAMP	1875	179	0.096
AMP	66	0	—
Exogenous araAMP			0.118

[a] Exponentially growing L cells (1.2×10^8 cells) were incubated 3 hr with 1×10^{-4} M [^{3}H,^{32}P]araAMP (specific activities: [^{3}H]araAMP 6.09×10^6 cpm/μmol; [^{32}P]araAMP 7.07×10^5 cpm/μmol) in 250 ml of growth medium. The nucleoside triphosphates, isolated from the acid-soluble fraction, were treated with apyrase and the reaction products were separated by thin-layer electrophoresis and TLC [8].

[b] Ratio of cpm.

of the exogenous nucleotide. These results indicate that araAMP entered the cells intact and was subsequently phosphorylated to araATP. Assuming that 97% of the ^{3}H in the adenine nucleoside triphosphate peak was associated with araATP, it is possible to calculate that the cellular araATP concentration after 4 hr incubation with araAMP was 2 μM. This is only one-tenth the concentration found after similar incubation with araA, but still exceeds the K_i of araATP for a mammalian DNA polymerase [16]. In the absence of subsequent time points, it can only be assumed that the araAMP in the medium, bathing the cells over a long time and slowly penetrating the cells, eventually contributed to intracellular araATP concentrations sufficiently high to evoke the observed sustained lethality.

A portion of the 2'(3')-ribonucleotide mixture produced by alkaline hydrolysis of the acid-insoluble material from cells treated with [^{3}H, ^{32}P]-araAMP was fractionated by high-pressure liquid chromatography as described in Fig. 5. The small quantities of ^{3}H in the nucleoside fraction and similar small amounts of ^{3}H and ^{32}P that were observed in the 2'- and 3'-AMP peaks were not sufficient to permit identification of the radioactive compounds by methods comparable to those used on the RNA fraction from [^{3}H]araA-treated cells (Fig. 5). However, Table 5 shows a comparison of total ^{3}H from the nucleoside peak and the total pooled 2'- and 3'-AMP peaks obtained by high-pressure liquid chromatography of the RNA fraction from

TABLE 5

Incorporation of ^{3}H from [^{3}H,^{32}P]AraAMP or [^{3}H]AraA into RNA[a]

Precursor	No.	Specific activity (cpm/μmol)	pmol ^{3}H/10^7 cells	
			Nucleoside	2'-AMP + 3'-AMP
[^{3}H,^{32}P]AraAMP	1[b]	1.43×10^7	2.49	5.54
	2	0.97×10^7	2.67	6.02
			Average 2.56	5.70
[^{3}H]AraA	1	1.70×10^7	3.92	51.3
	2	6.45×10^7	5.95	136.0
			Average 4.93	93.6

[a] Exponentially growing cells were incubated 4 hr with either 1×10^{-4} M [^{3}H,^{32}P]araAMP or [^{3}H]araA. Alkaline hydrolysates were fractionated by high-pressure liquid chromatography as described in Fig. 5 and the ^{3}H incorporation into the nucleoside fraction and the pooled 2'- and 3'-AMP fraction was determined [27].
[b] Average of two determinations.

[^{3}H, ^{32}P]araAMP-treated cells with those of [^{3}H]araA-treated cells. The results indicate that ^{3}H incorporation into the nucleoside fraction, i.e., the 3'-terminus of RNA, occurs to a similar extent in cells incubated with either precursor, but that ^{3}H incorporation into the nucleotide fraction is nearly 20-fold greater in araA-treated cells [27]. Since most of the ^{3}H label in the AMP fractions of the RNA hydrolysate from [^{3}H]araA-treated cells was associated with ribosyladenine nucleotide (Table 2), the lack of similar labeling in araAMP-treated cells is consistent with the findings that the adenine moiety of araAMP is not extensively metabolized to the ribosyl series (Table 4) and that araATP is a poor substrate for a mammalian RNA polymerase [28, 29]. Thus, conversion to the ribosyl series appears to be the main pathway by which [^{3}]araA may label the internal nucleotides of RNA.

Both araA and araAMP label the 3'-terminus of the RNA to similar extents (Table 5). This may be due to arabinosyl nucleotide addition to the 3'-terminus of t-RNA, a reaction of araATP that has been observed to be catalyzed by an E. coli enzyme (I. Fukuma and S. S. Cohen, cited in Ref. 9). The quantitative differences in this terminal labeling by araA and araAMP may possibly be explained by the ten-fold difference in the araATP concentration present in the araA-treated versus the araAMP-treated cells after the 4-hr incubation.

To identify the radioactive compounds in the DNA of cells incubated 4 hr with [^{3}H, ^{32}P]araAMP, a portion of the DNA fraction was enzymatically degraded to 5'-monophosphates. The mixture was separated on AG50W-X2, a cation-exchange resin on which dAMP and araAMP nearly coelute. Owing to the small amount of total radioactivity, the contents of each deoxynucleotide peak were pooled and the total radioactivity in each pooled fraction was determined (Table 6). The bulk of both the ^{32}P and the ^{3}H were associated with the dAMP fraction in a ratio similar to that of the exogenous araAMP. The ^{3}H (and up to half the ^{32}P) seen in the dCMP fraction was apparently carryover contamination from the closely adjacent dAMP peak, since no significant ^{3}H was present in deoxcytidine, thymidine, or deoxyguanosine following dephosphorylation of the nucleotides. However, 78.4 and 14.5% of the ^{3}H was contained in araA and araHx, respectively, the latter probably arising from a slight deaminase activity contamination in the alkaline phosphatase.

The nearest neighbor exchange experiment was completed by enzymatic degradation of another portion of the DNA fraction to 3'-mononucleotides. The mixture was separated by high-pressure liquid chromatography as described in Fig. 8, the eluate in each peak was pooled, and the total radioactivity was determined. As seen in Table 7, the ^{3}H was localized in the 3'-dAMP peak, but ^{32}P activity was distributed among all the deoxynucleotides.

TABLE 6

Fractionation of the 5'-Monophosphates from the DNA of [^{3}H, ^{32}P]AraAMP-Treated Cells[a]

	^{32}P (cpm)	^{3}H (cpm)	^{32}P/^{3}H [b]
5'-TMP	138	35	
5'-dGMP	99	15	
5'-dCMP	311	100	
5'-dAMP	1645	1374	1.20
Exogenous araAMP			1.75

[a] Exponentially growing L cells were incubated 4 hr with 1×10^{-4} M [^{3}H, ^{32}P]araAMP (specific activities: [^{3}H]araAMP 1.25×10^7 cpm/µmol; [^{32}P]araAMP 2.18×10^7 cpm/µmol). A portion of the DNA fraction equivalent to 6.8×10^7 cells was degraded to 5'-monophosphates with DNase I and venom phosphodiesterase, fractionated on AG50W-X2 (NH$_4^+$), and each nucleotide peak was pooled and the radioactivity determined [8].
[b] Ratio of cpm.

TABLE 7

Fractionation of the 3'-Monophosphates from the DNA of [^{3}H, ^{32}P]AraAMP-Treated Cells[a]

	^{3}H (cpm)	^{32}P (cpm)	^{32}P/^{3}H [b]
Nucleoside	22	—	
3'-dCMP	21	144	
3'-TMP	56	86	
3'-dAMP	637	60	
3'-dGMP	15	76	
Total	751	366	0.49
Exogenous araAMP			0.48

[a] A portion of the DNA fraction from the equivalent of 3.4×10^7 cells incubated as described in the legend to Table 6, was degraded to 3-monophosphates by the sequential action of micrococcal nuclease and spleen phosphodiesterase. The nucleotides were fractionated by high-pressure liquid chromatography as described in Fig. 7 [8].
[b] Ratio of cpm.

The ratio of the sum of the ^{32}P to the ^{3}H isolated was that of the exogenous [^{3}H, ^{32}P]araAMP. Moreover, only 3% of the total ^{3}H was found in the terminating nucleoside fraction, indicating again that most of the arabinosyl nucleotide had been incorporated into internucleotide linkage.

In sum, the evidence indicates that the increased toxicity of araAMP over araA to L cells is due to the resistance of araAMP to detoxifying deamination by virtue of protection by the 5'-phosphate. The continued presence of araAMP in the growth medium serves as a source of replenishment for the toxic cellular concentrations of araATP accumulated after penetration of the nucleotide.

Consistent with previous suggestions [15, 30, 31], these data predict that the biological activity of araA may be potentiated by the simultaneous administration of an inhibitor of adenosine deaminase. Recently, Schaeffer and Schwender [32] synthesized erythro-9-(2-hydroxy-3-nonyl)adenine (EHNA), an extremely potent inhibitor of adenosine deaminase. When added to deaminase assays at a final concentration of 1×10^{-5} M, the inhibitor blocked all detectable deamination of adenosine, araA, and other adenine nucleosides by extracts of L cells and Ehrlich ascites carcinoma cells [11, 24]. L cells growing exponentially in suspension culture at 4.6×10^5 cells/ml deaminated araA at a rate of 11 nmol/ml/hr. However, L cells incubated with 1×10^{-5} M EHNA deaminated only 6 nmol araA/ml in 24 hr. Since the deaminase inhibitor seemed quite potent in the L-cell system, we decided to test its effect on the toxicity of araA [11, 24].

Tested alone, concentrations of EHNA up to 1×10^{-5} M showed no effect on L-cell viability (Fig. 12). When cells were incubated with 1×10^{-5} M deaminase inhibitor and 2×10^{-5} M araA, a concentration which evokes little or no toxicity alone, the increase in viable cells stopped, and after a lag of 22 hr, viability in the culture decreased exponentially for the duration of the experiment. AraA at 1×10^{-5} M was also very toxic in the

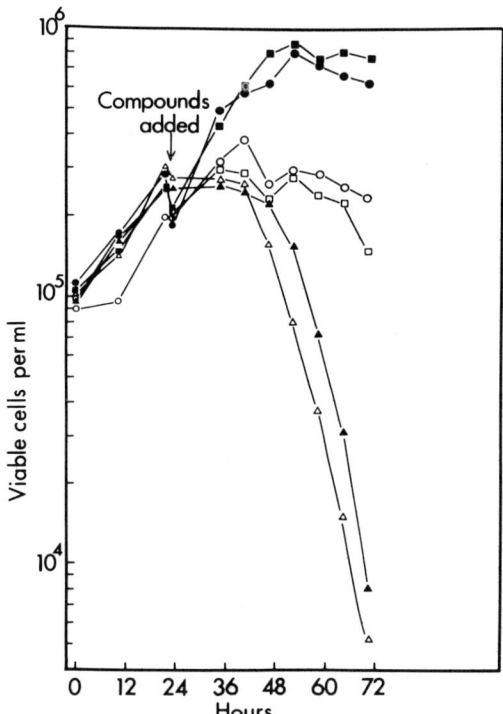

FIG. 12. Viability of L cells during incubation with EHNA and various concentrations of araA. ■, Control; ●, 1×10^{-5} M EHNA; ○, 2.5×10^{-6} M araA + 1×10^{-5} M EHNA; □, 5.0×10^{-6} M araA + 1×10^{-5} M EHNA; ▲, 1×10^{-5} M araA + 1×10^{-5} M EHNA; △, 2×10^{-5} M araA + 1×10^{-5} M EHNA [11].

presence of EHNA, whereas concentrations of araA that were less than 1×10^{-5} M were only slightly toxic with the deaminase inhibitor. As the araA concentration was lowered, the lag time prior to the onset of toxicity increased, suggesting that the lag may represent the time necessary to accumulate toxic concentrations of araATP in the cells.

The synergistic effect of EHNA on araA toxicity was also observed in the protection of mice bearing tumors. The data in Table 8 show that the average survival time of BD_2F_1 mice bearing Ehrlich ascites carcinoma cells was extended beyond that of control mice given daily intraperitoneal injections of araA (50 mg/kg) over 5 days. However, tumor-bearing mice treated simultaneously with araA and EHNA survived significantly longer than did those mice treated with araA alone. Recently, 2'-deoxycoformycin, another potent inhibitor of adenosine deaminase, has also been shown to potentiate the antitumor activity of araA [33, 34].

The finding that normally ineffective concentrations of araA (5-10 μM) were highly lethal to L cells in the presence of a deaminase inhibitor confirmed the hypothesis that the low toxicity of araA alone was indeed the result of deamination of this nucleoside. Therefore, it was important to study the metabolism of araA in the presence of the deaminase inhibitor for comparison with the studies of the metabolism of araA alone and also with that of araAMP.

Some work has been carried out in that direction. Exponentially growing L cells were incubated for 4 hr with either 1×10^{-4} M [^{3}H]araA or 1×10^{-5} M EHNA, or both. Although 1×10^{-4} M araA is essentially

TABLE 8

Effect of AraA and Erythro-9-(2-Hydroxy-3-nonyl)adenine (EHNA) on the Survival of BD_2F_1 Mice Bearing Ehrlich Ascites Carcinoma [11]

	Days surviving	
Treatment	I	II
PBS	14.1 ± 2.1[a] (10)[b]	13.8 ± 2.1 (10)
AraA (50 mg/kg)	20.3 ± 3.8 (9)	16.2 ± 2.0 (8)
AraA (50 mg/kg) + EHNA (3.1 mg/kg)	31.8 ± 6.9 (10)	29.4 ± 11.2 (9)
EHNA (3.1 mg/kg)	15.2 ± 1.9 (5)	

[a]Mean ± SD.
[b]Numbers in parentheses, number of mice tested.

cytostatic [25] and 1×10^{-5} M EHNA has no effect on cell growth (Fig. 12), the two agents in combination produce a concentration-dependent exponential loss of cell viability with little detectable lag time [11]. The nucleotide pools from cells grown under each condition were extracted and fractionated by high-pressure liquid chromatography (Fig. 13). With minor exceptions, the qualitative and quantitative aspects of each chromatogram appear to be identical. Thus, inhibition of adenosine deaminase by EHNA for 4 hr appears to have little effect on the amounts of adenine nucleotides present in the cells. As shown previously [9], incubation of cells with 1×10^{-4} M araA neither affects the cellular nucleotide concentrations nor results in the appearance of new arabinosyl nucleotide peaks. The latter finding arises because only relatively small quantities of arabinosyl nucleotides are produced in this time, and these coelute with ADP and ATP, nucleotides that are present in much higher cellular concentrations. Therefore, it is perhaps not surprising that the combination of EHNA and araA produces little apparent change in the nucleotide chromatogram. For instance, if the presence of the deaminase inhibitor resulted in, say, a five-fold increase in the cellular araATP concentration (to 100 μM) over that in cells treated with araA alone (20 μM), this quantity would still represent only about 10% of the total adenine nucleoside triphosphates with which it coelutes. As such, it would be difficult to discern this change in peak area accurately without multiple analyses of the nucleotide pools from treated and control cells. Therefore, the radioactivity derived from [^{3}H]araA in the nucleotide pools was analyzed to search for biochemical differences between araA-treated cells incubated with or without the deaminase inhibitor.

Fractions of 2-min duration were collected directly into scintillation vials during the fractionations represented by the chromatograms in Figs. 13C and 13D. In both samples, greater than 90% of the radioactivity eluted with either the first-eluted peak, containing nucleosides and other nonretained compounds, ADP, or ATP. The total radioactivity in each nucleotide fraction was summed and is presented in Table 9. Although the fractions from the cells treated with both araA and EHNA contained about 30% fewer total counts than did the fractions from araA-treated cells, the percentage distribution of the counts in the fractions from each of the incubations was nearly identical. Since these data provided no clues that could account for the observed differences in toxicity, the acid-soluble extracts were analyzed chemically.

These analyses revealed major differences in the nature of the tritiated components in the two systems. In particular, calculations made after thin-layer chromatographic analysis of the acid-soluble material from cells treated with 1×10^{-4} M [^{3}H]araA indicated that arabinosyl nucleotides totaled 50.4 and 92.2% of the radioactivity for cells incubated in the absence and presence of 1×10^{-5} M EHNA, respectively [24]. Assuming these percentages reflect the portion of araATP-associated ^{3}H in the adenine triphosphate

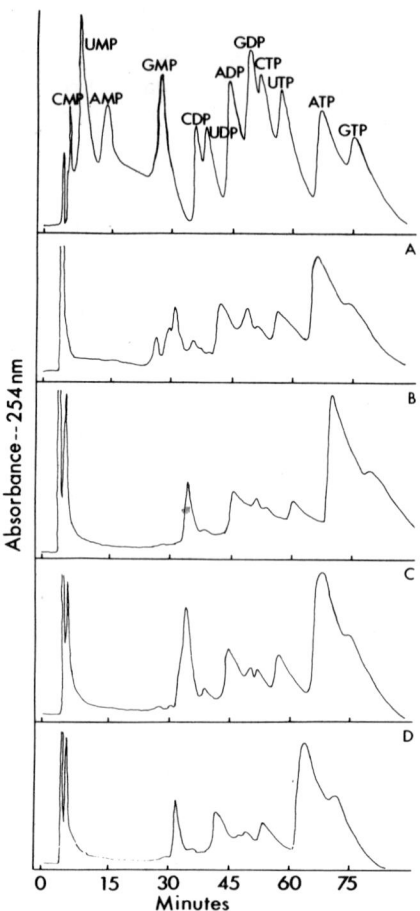

FIG. 13. High-pressure liquid-chromatographic fractionation of the acid-soluble material from the equivalent of 2×10^6 L cells after incubation with [^{3}H]araA in the absence or presence of EHNA. Anion-exchange resin: AS-Pellionex-SAX. Temperature: 60°C. Pressure: 775 psi. Starting volume: 58 ml. Eluents: 0.002 M KH_2PO_4, pH 3.55, and 1.00 M KH_2PO_4, pH 4.15. Flow rates: 24 ml/hr each. Gradient delay: 15 min. (A) Control cells; (B) 1×10^{-5} M EHNA; (C) 1×10^{-4} M araA; (D) 1×10^{-4} M araA + 1×10^{-5} M EHNA. Full-scale absorbance: 0.08 [24].

TABLE 9
Distribution of ^{3}H from [^{3}H]AraA in the Acid-Soluble Fraction[a]

Fraction	−EHNA (cpm)	+EHNA (cpm)
Nucleoside	3067 (31.2)[b]	2492 (29.4)
ADP	1086 (11.0)	555 (6.5)
ATP	5657 (57.8)	5435 (64.1)

[a]Cells were incubated for 4 hr with 0.1 μmol [^{3}H]araA/ml (specific activity, 1.55×10^7 cpm/μmol) in the presence and absence of 0.01 μmol EHNA/ml. Portions of the PCA-soluble material equivalent to 2.5×10^6 cells were fractionated by high-pressure liquid chromatography as described in Fig. 13 and the radioactivity in each fraction was determined [24].

[b]Numbers in parentheses, percentage of radioactivity in the three fractions.

fraction (Table 9), and knowing the number of cells analyzed (2.5×10^6) and the average cell volume (3×10^3 μm^3), it was possible to calculate the araATP concentration in cells incubated with araA in the absence and presence of EHNA to have been 24 and 43 μM, respectively. The former value is in good agreement with that previously determined by isolation of araATP from the acid-soluble fraction after an identical incubation (20 μM) [9]. The estimate of 43 μM araATP in cells incubated with araA in the presence of EHNA is nearly twice that of cells similarly incubated but in the absence of EHNA, and greatly exceeds the K_i of araATP for a mammalian DNA polymerase (K_i = 1 μM) [16]. It may be expected that as more araA is deaminated by the cells incubated in the absence of EHNA, the cellular araATP concentration would maximize and then decline, whereas araATP might continue to accumulate in cells incubated in the presence of EHNA.

These results suggest a biochemical basis for the increased toxicity of araA in the presence of EHNA. However, an in-depth understanding of the relationships among cell viability, the cellular araATP concentration, inhibition of DNA synthesis, and incorporation of arabinosyl nucleotides into DNA will require detailed pharmacokinetic analysis not only covering the early times of lethality, but also extending for the duration of the killing phenomenon. It is clear that high-pressure liquid chromatography will be an important technique in conducting those studies.

B. Metabolism of 9-β-D-Xylofuranosyladenine

9-β-D-Xylofuranosyladenine (xylA) is an analog of adenosine, differing from that nucleoside in that its 3'-hydroxyl group is arranged <u>trans</u> to the 2'-hydroxyl (Fig. 14). The analog exhibits selected antitumor activity against those rat and mouse tumors known to possess low adenosine deaminase activities [35], suggesting that the deamination product, xylosylhypoxanthine, is not toxic. XylA is thought to be active only after phosphorylation to the 5'-triphosphate, xylATP, a compound that is synthesized by biological systems both in vivo and in vitro, and that has been shown to inhibit the synthesis of phosphoribosyl pyrophosphate (PRPP) [36]. Since PRPP is required for both purine and pyrimidine nucleotide synthesis by de novo pathways, it was suggested that the observed inhibition of adenine incorporation into RNA and DNA in xylA-treated cells was due to depletion of the nucleotide pools of the precursors of nucleic acid synthesis [36].

We have employed the L-cell viability system to quantitate further the toxicity of xylA. In addition, high-pressure liquid-chromatographic analysis has been utilized to study the metabolism of xylA by L cells in the presence and the absence of an inhibitor of adenosine deaminase, and to visualize the effect of this combination on the levels of nucleotide pools.

Exponentially growing L cells were incubated with 1×10^{-4} M xylA in the presence or absence of 1×10^{-5} M EHNA (Fig. 15). The viability of those cells treated with xylA alone decreased by 90 to 95% in 5 hr, after which little additional loss of viability was seen, even if the experiment was continued an additional 30 hr. However, when cells were incubated with xylA + EHNA, the loss of viability was not only more rapid than that produced by xylA alone, but also continued until the culture was essentially sterilized. Greater than four decades of cells were rendered unable to form colonies after only 7 hr of incubation.

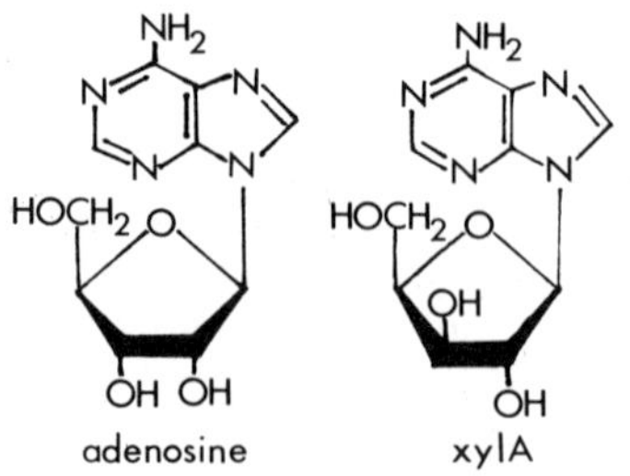

FIG. 14. Structures of adenosine and xylA.

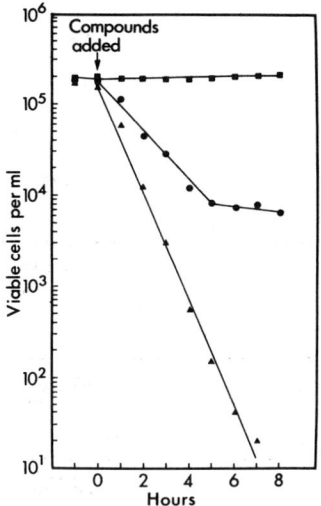

FIG. 15. Viability of L cells during incubation with xylA in the presence and absence of EHNA. ■, Control; ●, 1×10^{-4} M xylA; ▲, 1×10^{-4} M xylA + 1×10^{-5} M EHNA [37].

The great toxicity of xylA + EHNA in this system suggested that the inhibitory effect on nucleic acid synthesis might be more severe than previously observed [36]. L cells in exponential growth were incubated with xylA + EHNA and the incorporation of [^{3}H]uridine into acid-insoluble material was measured. After a 10-min incubation, uridine incorporation by cells treated with xylA + EHNA was inhibited by 90% (Plunkett, unpublished observation). Although control cells continued to accumulate uridine at a linear rate, cells treated with the combination had accumulated only 3% of control values by 60 min. Incorporation of uridine into acid-soluble material was inhibited only 43% by 10 min, a value which rose to 74% after 60 min. Thymidine incorporation into acid-insoluble material by cells treated with 1×10^{-4} M xylA + 1×10^{-5} M EHNA was entirely blocked at 2% of the control value by 1 hr. At this time, radioactivity from thymidine in the acid-soluble pool of the treated cells was 118% of that in control cells. As previously demonstrated [11], 1×10^{-5} M EHNA alone has little effect on either uridine or thymidine incorporation.

Possibly, visualization of the contents of the nucleotide pools by high-pressure liquid chromatography might provide an explanation for the severe inhibition of nucleic acid synthesis at times when incorporation of the labeled precursors into the nucleotide pools was not proportionately inhibited. Exponentially growing cells were treated with 1×10^{-4} M xylA + 1×10^{-5} M EHNA and the acid-soluble material was prepared from samples taken at

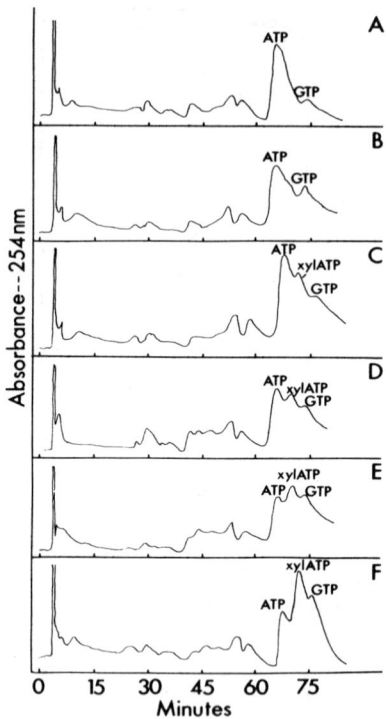

FIG. 16. High-pressure liquid-chromatographic analysis of PCA-soluble material extracted from the equivalent of 1×10^6 L cells incubated for various times with 1×10^{-4} M xylA + 1×10^{-5} M EHNA. Anion-exchange resin: AS-Pellionex-SAX. Temperature: 60°C. Pressure: 725 psi. Starting volume: 58 ml. Eluents: 0.002 M KH_2PO_4, pH 3.55, and 1.00 M KH_2PO_4, pH 4.15. Flow rates: 24 ml/hr each. Gradient delay: 15 min. (A) Zero time; (B) 10 min; (C) 20 min; (D) 30 min; (E) 45 min; (F) 60 min.

times up to 1 hr, i.e., a time when 60% of the population was nonviable. Analyses of the samples are shown in Fig. 16. The accumulation of xylATP in the nucleotide pool is first seen as a distinct peak at 20 min. Unlike araATP, which is synthesized in small quantities and coelutes with ATP, xylATP becomes the predominant adenine nucleoside triphosphate in the cell after 45 min of incubation, and elutes several minutes following ATP but prior to GTP. Still, quantitation of ATP and xylATP was difficult since the two peaks were not well resolved. The ATP concentration in exponentially growing L cells had previously been calculated at about 1 mM [8, 11]. By comparison of the area of the ATP from the zero-time cells with that of xylATP in cells after 60 min of incubation with xylA + EHNA, it may be estimated that the xylATP concentration approached 1 mM in cells after this time. XylADP also accumulated in the cells, although its presence is not clearly detectable in the chromatograms in Fig. 16.

To provide further evidence of the identity of the new peaks, acid-soluble extracts were prepared from control cells and cells incubated 1 hr with xylA + EHNA. After a portion of each extract was treated with periodate to destroy ribonucleotides [38], the extracts were analyzed by high-pressure liquid chromatography. As seen in Fig. 17, most of the acid-soluble

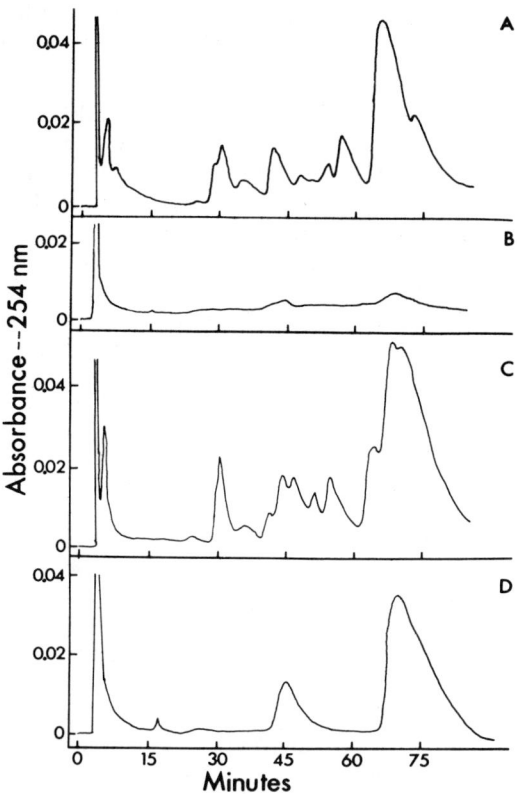

FIG. 17. High-pressure liquid-chromatographic analysis of PCA-soluble material extracted from the equivalent of 1×10^6 L cells. (A) Control; (B) control after treatment with periodate; (C) cells incubated for 1 hr with 1×10^{-4} M xylA + 1×10^{-5} M EHNA; (D) cells incubated as in (C) but treated with periodate. Periodate treatment was conducted as described by Neu and Heppel [38]. The chromatographic analyses were carried out under the conditions described in Fig. 16.

material from control cells was eliminated by treatment with periodate. However, degradation of the extract from (xylA + EHNA)-treated cells left two major peaks remaining which eluted at the same times as the two new peaks described in Fig. 16. This would be the expected result if those peaks contained no <u>cis</u>-hydroxyls, as is the case for xylosyladenine nucleotides.

During the first 30 min of the incubation, when synthesis of both RNA and DNA was strongly inhibited, the major identifiable nucleoside triphosphates, i.e., UTP, ATP, and GTP, did not appear to undergo quantitative changes (Fig. 16). After 30 min there was a decrease in the content of the ATP fraction, although through 60 min of incubation, neither GTP, UTP, nor many other unidentified constituents of the nucleotide pool appeared to have undergone significant quantitative changes.

To gain a long-term view of the effect of xylA + EHNA on the cellular nucleotides, the incubation was repeated but samples were taken at hourly intervals for 5 hr, a time at which 99.9% of the population was nonviable (Fig. 15). Again, after 1 hr of treatment, xylATP was the major nucleotide in the cells (Fig. 18). However, the xylATP concentration became progressively diminished between 1 and 2 hr, apparently shifting to xylADP, which was seen to have increased by 2 hr. This shift may reflect both the cellular utilization of the γ-PO_4 of xylATP and the inability of the cells to rephosphorylate xylADP, perhaps because of the reduced cellular ATP level. Most of the xylATP was gone by 3 hr. After 1 hr, the normal constituents of the nucleotide pool showed a general decrease until at 3 hr, the nucleotide pool had been virtually eliminated.

The fact that 40% of the population was able to form colonies when washed free of xylA and EHNA following 1 hr of incubation (Fig. 15) suggested that the xylATP concentration may have been differentially decreased with respect to the normal nucleotides after washing and plating. To test this possibility, exponentially growing L cells were incubated 1 hr in the presence of xylA and EHNA, washed out of the drug-containing medium, and allowed to recover in fresh medium for 5 hr. Acid-soluble extracts from nucleotide pools were prepared at various times and analyzed by high-pressure liquid chromatography (Fig. 19). As shown previously, after 1 hr of incubation (Fig. 19) [13], a large xylATP peak is present as well as a smaller peak of xylADP. One hour after washing (Fig. 19C), a reduction in the size of the xylATP peak was evident. The reduction of the ATP peak appeared roughly proportional to that of the xylATP peak, but other nucleotides remained undiminished for the most part. Three hours after washing (Fig. 19D), cellular xylATP levels appeared to have decayed to less than the quantities of either ATP or GTP in the cells. Reduction of xylADP was also evident. By 5 hr after washing (Fig. 19E), the decay of xylATP and xylADP from the nucleotide pool was nearly complete. Normal nucleotide triphosphates were evident at 3 and 5 hr after washing in contrast to their

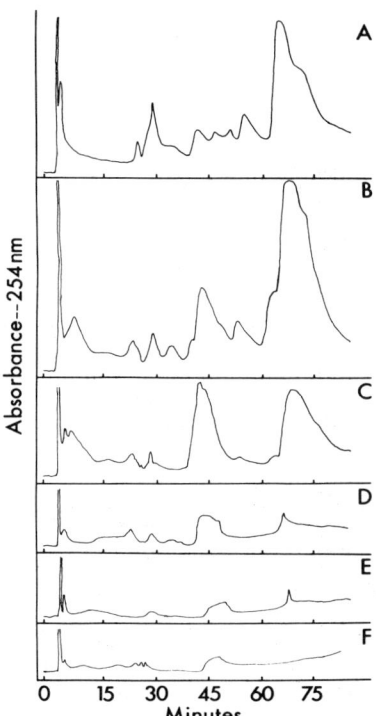

FIG. 18. High-pressure liquid-chromatographic analysis of PCA-soluble extracts from the equivalent of 1×10^6 L cells after various times of incubation with 1×10^{-4} M xylA + 1×10^{-5} M EHNA. (A) Zero time; (B) 1 hr; (C) 2 hr; (D) 3 hr; (E) 4 hr; (F) 5 hr. The chromatographic analyses were carried out under the conditions described in Fig. 16.

total absence after 3 hr of continuous incubation with xylA and EHNA (Fig. 18).

These results demonstrate that the concentration of xylosyl nucleotides in the cells after 1 hr of incubation with xylA and EHNA was sufficient to cause the partial depletion of the normal cellular nucleotides. The xylATP concentration, which was several-fold greater than that of ATP and much greater than that of GTP after 1 hr of incubation, was reduced to less than that of either ATP or GTP following 5 hr of incubation in xylA-free medium. This differential loss of xylosyl nucleotides suggests that xylosyl nucleotides were either more extensively utilized or less efficiently replenished than normal nucleotides at a time when the concentration of normal nucleotides was markedly depressed.

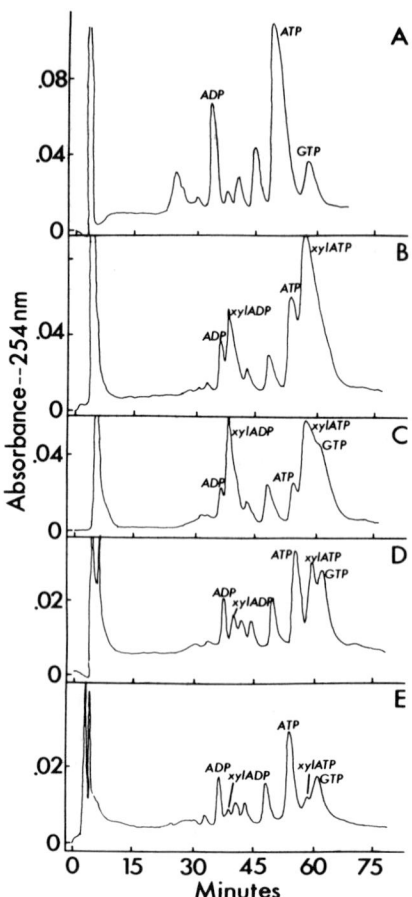

FIG. 19. High-pressure liquid-chromatographic analysis of PCA-soluble material extracted from the equivalent of 1×10^6 L cells after incubation with 1×10^{-4} M xylA + 1×10^{-5} M EHNA for 1 hr, followed by washing and resuspending in drug-free media. (A) Zero time; (B) 1 hr of incubation with xylA + EHNA; (C) 1 hr after washing into drug-free media; (D) 3 hr after washing into drug-free media; (E) 5 hr after washing into drug-free media. The chromatographic analyses were carried out under the conditions described in Fig. 16.

To what extent could the inhibition of PRPP synthesis have contributed to the loss of all the normal elements of the nucleotide pool as well as to the diminution of high concentration of xylosyl nucleotides? Inhibition of PRPP synthesis by xylATP might be expected to produce cellular deficiencies in compounds whose synthesis requires PRPP. Of those compounds, both histidine and tryptophan are essential amino acids and are supplied in the growth medium. Nicotinamide mononucleotide, from which pyridine nucleotides are produced, also requires PRPP. However, since pyridine nucleotides are required only as cofactors, it is unlikely that inhibition of their synthesis for periods as short as 1 hr, about 5% of the population doubling time, could lead to the observed death of 60% of the population. Purine and pyrimidine nucleotides constitute the remaining compounds whose synthesis requires PRPP. One could predict then, that if the major action of xylosyl nucleotides that leads to cellular toxicity was the inhibition of PRPP synthesis, addition of the appropriate nucleosides in the presence of xylA might spare the nucleotide deficiency. This sparing action would be expected to result in the less rapid onset of toxicity and the maintenance of the nucleotide pools.

To test this possibility, cells were incubated with 1×10^{-4} M xylA + 1×10^{-5} M EHNA with or without the nucleosides adenosine, guanosine, uridine, and thymidine, each at 1×10^{-5} M. The number of viable cells was determined at hourly intervals (Fig. 20). Cells treated with xylA and the nucleosides showed slightly greater viability than those incubated without the nucleosides after 1 hr. Thereafter, the extent of lethality in each culture was identical. Cells incubated with EHNA and nucleosides showed essentially the same viability as the control.

To determine the effect of added nucleosides on the nucleotide pool levels of cells treated with xylA + EHNA, cells were treated as described in Fig. 20 for 4 hr and then nucleotide pools were analyzed by high-pressure chromatography (Fig. 21). The nucleotide profile from cells treated with EHNA and the nucleosides (Fig. 21B) is qualitatively similar to that of control cells (Fig. 21A), although the ATP peak appears smaller than that of the control. As seen previously (Fig. 18), little remains in the nucleotide pool of cells treated with xylA + EHNA for 4 hr (Fig. 21C). The pools of those cells treated with xylA, EHNA, and the nucleosides give evidence of both diphosphate and triphosphate components as well as several other peaks not present in cells treated with only xylA + EHNA (Fig. 21D). However, it is evident that the presence of the nucleosides was not sufficient to maintain either the normal pool constituents or the xylosyl nucleotides during the 4 hr of incubation with xylA + EHNA.

The effect of xylA on exponentially growing L cells is a complex phenomenon. At times up to 30 min of incubation, as xylATP was accumulating, the synthesis of both RNA and DNA was severely inhibited. Yet, no substantial quantitative changes were seen among the normal nucleoside

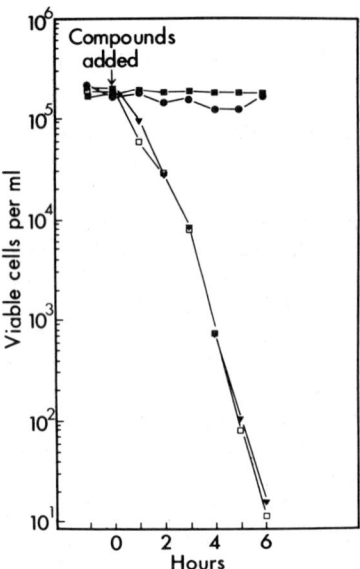

FIG. 20. Viability of L cells during incubation with xylA + EHNA and with nucleosides. ■, Control cells; ●, EHNA, adenosine, guanosine, uridine, and thymidine, 1×10^{-5} M each; ▼, 1×10^{-4} M xylA + 1×10^{-5} M EHNA; □, 1×10^{-4} M xylaA, and EHNA, adenosine, guanosine, uridine, and thymidine, 1×10^{-5} M each.

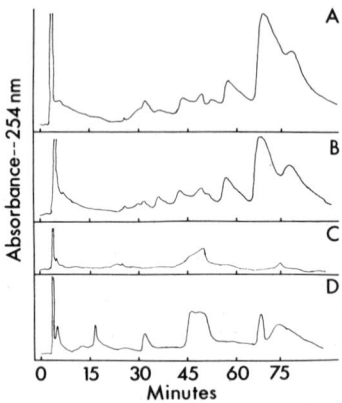

FIG. 21. High-pressure liquid-chromatographic analyses of PCA-soluble extracts from the equivalent of 2.5×10^6 L cells. (A) Control cells; (B) 4 hr incubation with 1×10^{-5} M EHNA + 1×10^{-5} M nucleosides; (C) 4 hr incubation with 1×10^{-4} M xylA + 1×10^{-5} M EHNA; (D) 4 hr incubation with 1×10^{-4} M xylA, 1×10^{-5} M EHNA, and 1×10^{-5} M nucleosides. The chromatographic analyses were carried out under the conditions described in Fig. 16.

triphosphate constituents of the cells. These observations argue against the possibility that inhibition of nucleic acid synthesis at the early stages of the incubation was due to diminution of the nucleoside triphosphate pools after xylATP inhibition of PRPP synthesis. Rather, xylATP may be playing a more direct role in the inhibition of nucleic acid synthesis than has previously been suggested. It is possible that xylATP might have been incorporated into RNA, where, because the configuration of the 3'-hydroxyl group would be sterically unfavorable to the addition of a subsequent nucleotide, it could have acted as a chain-terminating nucleotide. Furthermore, such premature termination of an RNA required for the initiation of DNA synthesis [39] could account for the observed abrupt termination of DNA synthesis. This has recently been suggested to be the mechanism by which 3'-deoxyadenosine (cordycepin), a true RNA chain terminator, inhibits DNA synthesis in L cells, and by which it causes lethality which is kinetically similar to that produced by xylA [11].

The mechanism by which the cells are depleted of the normal and xylosyl nucleotides in the later stages of the incubation with xylA and EHNA is not clear. After the first hour, RNA and DNA synthesis was at the limit of detection (50 pmol/2.5×10^7 cells) and might not have been a major factor in the nucleotide attrition. The high concentrations of xylATP which were present could have interfered with numerous cellular reactions that require ATP. One obvious effect involved the pathways which function to replenish the ATP pool, which fell dramatically between 30 min and 2 hr. Such an upset of the energy charge in the cell may have led to a loss in the integrity of the cell membrane and the ensuing efflux of the cellular nucleotides. Although in vivo, liver parenchyma cells are known to survive prolonged depression of ATP levels [40], the absence of either diphosphates or monophosphates after incubation with xylA + EHNA favors the idea that such treatment increases the permeability of the plasma membrane to nucleotides. The preparation of [2-^{3}H]xylA will greatly facilitate further investigation of these possibilities.

IV. CONCLUSIONS

The technique of high-pressure liquid chromatography has been essential to the success of our work with purine nucleoside analogs. Hopefully, this chapter has illustrated how this powerful new technique can be integrated into a research program dealing with the biochemistry and pharmacology of drug development. Although the use of high-pressure liquid chromatography is largely confined to the basic research laboratory at the present, the time when it will become an important clinical tool for the investigation of drug metabolism is close at hand.

ACKNOWLEDGMENTS

The work described herein was supported in part by Grant No. NP-60B from the American Cancer Society and by Grant No. 11636 from the National Institute of Allergy and Infectious Diseases awarded to Dr. Seymour S. Cohen, and was conducted in Dr. Cohen's laboratory at the University of Colorado Medical Center, Denver, Colorado.

REFERENCES

1. P. R. Brown, J. Chromatogr., 52, 272 (1970).
2. N. Hadden, F. Baumann, F. MacDonald, M. Munk, R. Stevensen, D. Gere, and F. Zamaroni, Basic Liquid Chromatography, Varian Aerograph, Walnut Creek, California, 1971.
3. J. J. Kirkland (ed.), Modern Practice of Liquid Chromatography, Wiley-Interscience, New York, 1971.
4. C. Horvath, Methods Biochem. Anal., 21, 79 (1972).
5. P. R. Brown, High Pressure Liquid Chromatography, Biochemical and Biomedical Applications, Academic Press, New York, 1973.
6. J. N. Done, J. H. Knox, and J. Loheac, Applications of High-Speed Liquid Chromatography, Wiley, New York, 1974.
7. P. R. Brown, Adv. Chromatogr., 13, 1 (1975).
8. W. Plunkett, L. Lapi, P. J. Ortiz, and S. S. Cohen, Proc. Natl. Acad. Sci. U.S.A., 71, 73 (1974).
9. W. Plunkett and S. S. Cohen, Cancer Res., 35, 415 (1975).
10. B. Munch-Petersen, G. Tyrsted, and B. DuPont, Exp. Cell Res., 79, 249 (1972).
11. W. Plunkett and S. S. Cohen, Cancer Res., 35, 1547 (1975).
12. D. Pavan-Langston, R. A. Buchanan, and C. A. Alford, Jr. (eds.), Adenine Arabinoside: An Antiviral Agent, Raven Press, New York, 1975.
13. G. P. Bodey, J. A. Gottlieb, K. B. McCredie, and E. J. Freireich, Adenine Arabinoside: An Antiviral Agent (D. Pavan-Langston, R. A. Buchanan, and C. A. Alford, Jr., eds.), Raven Press, New York, 1975, p. 281.
14. G. A. LePage, A. Khaliq, and J. A. Gottlieb, Drug Metab. Disposit., 1, 759 (1973).

15. A. Doering, J. Keller, and S. S. Cohen, Cancer Res., 26, 2444 (1966).
16. J. J. Furth and S. S. Cohen, Cancer Res., 28, 2061 (1968).
17. E. C. Moore and S. S. Cohen, J. Biol. Chem., 242, 2116 (1968).
18. H. S. Moyed, Cold Spring Harbor Symp. Quant. Biol., 26, 323 (1961).
19. A. P. Kimball, G. A. LePage, and B. Bowman, Can. J. Biochem., 42, 1753 (1964).
20. A. Bloch and C. A. Nichol, Biochem. Biophys. Res. Commun., 16, 400 (1964).
21. M. Hubert-Habart and S. S. Cohen, Biochim. Biophys. Acta, 59, 468 (1962).
22. M. A. Waqar, L. A. Burgoyne, and M. R. Atkinson, Biochem. J., 121, 803 (1971).
23. J. J. Brink and G. A. LePage, Cancer Res., 24, 312 (1964).
24. W. Plunkett and S. S. Cohen, Ann. N.Y. Acad. Sci., 284, 91-102 (1977).
25. P. J. Ortiz, M. J. Manduka, and S. S. Cohen, Cancer Res., 32, 1512 (1972).
26. J. Lichtenstein, H. D. Barner, and S. S. Cohen, J. Biol. Chem., 235, 457 (1960).
27. S. S. Cohen and W. Plunkett, Ann. N.Y. Acad. Sci., 255, 269 (1975).
28. J. J. Furth and S. S. Cohen, Cancer Res., 27, 1528 (1967).
29. W. E. G. Muller, H. J. Rohde, R. Beyer, A. Maidhof, M. Lachmann, H. Taschenr, and R. K. Zahn, Cancer Res., 35, 2160 (1975).
30. R. Koshiura and G. A. LePage, Cancer Res., 28, 1014 (1968).
31. D. L. Chao and A. P. Kimball, Cancer Res., 32, 1721 (1972).
32. H. J. Schaeffer and C. F. Schwender, J. Med. Chem., 17, 6-8 (1974).
33. G. A. LePage, L. S. Worth, and A. P. Kimball, Cancer Res., 36, 1481 (1976).
34. F. M. Schabel, Jr., M. W. Trader, and W. R. Laster, Jr., Proc. Am. Assoc. Cancer Res., 17, 46 (1976).
35. G. A. LePage, Adv. Enzyme Regul., 8, 323 (1970).
36. D. B. Ellis and G. A. LePage, Mol. Pharmacol., 1, 231 (1965).
37. W. Plunkett and S. S. Cohen, Proc. Am. Assoc. Cancer Res., 16, 27 (1975).

38. H. C. Neu and L. A. Heppel, J. Biol. Chem., 239, 2927 (1964).
39. Y. Tseng and M. Goulian, J. Mol. Biol., 99, 317 (1975).
40. E. Farber, Fed. Proc., 32, 1534 (1973).

Chapter 8

THE DETERMINATION OF DI- AND POLYAMINES BY
HIGH-PRESSURE LIQUID AND GAS CHROMATOGRAPHY

Mahmoud M. Abdel-Monem

Department of Medicinal Chemistry
College of Pharmacy
University of Minnesota
Minneapolis, Minnesota

I. INTRODUCTION 249

II. HIGH-PRESSURE LIQUID CHROMATOGRAPHY 251

    A. Ion-Exchange Chromatography of the Polyamines 251
    B. Separation of Polyamines after Derivative Formation 256

III. GAS CHROMATOGRAPHY 261

    A. Separation of Polyamines 261
    B. Separation of Polyamines after Derivative Formation 262
    C. Gas Chromatography/Mass Spectrometry 264

IV. CONCLUSIONS 264

    ACKNOWLEDGMENTS 265

    REFERENCES 265

I. INTRODUCTION

    The polyamines are a group of low-molecular-weight aliphatic amines which are widely distributed in living cells. The term polyamines is usually used to describe not only spermidine [N-(3-aminopropyl)-1,4-butanediamine: $H_2NCH_2CH_2CH_2NHCH_2CH_2CH_2CH_2NH_2$] and spermine [N,N^1-bis(3-aminopropyl)-1,4-butanediamine: $H_2NCH_2CH_2CH_2NHCH_2CH_2CH_2CH_2NHCH_2CH_2CH_2NH_2$] but also the diamines putrescine (1,4-butanediamine: $H_2NCH_2CH_2CH_2CH_2NH_2$)

and cadaverine (1,5-pentanediamine: $H_2NCH_2CH_2CH_2CH_2CH_2NH_2$). Putrescine, spermidine, and spermine are present in all animal and plant tissues tested, and at least one of these is present in all microorganisms [1].

The history of the polyamines dates back to 1678 when Antony Van Leeuwenhoek observed the gradual formation of colorless crystals in a specimen of human semen [2]. These crystals were later observed by N. Vauquelin who, in 1791, erroneously concluded that they were calcium phosphate. The correct identification of these crystals as the phosphate salt of a simple organic nitrogen compound was attributed to P. Schreiner in 1878. The name spermine was given to this base in 1888 by Ladenburg and its chemical structure was unequivocally established by Rosenheim et al. in 1926.

The exact physiological role of the polyamines is not currently known. Studies of both normal and neoplastic rapid-growth systems indicate that the synthesis and accumulation of polyamines are elevated shortly after a stimulus-inducing proliferation [3]. This and the many effects in vitro of the polyamines on protein synthesis and on the synthesis and function of DNA and RNA suggest a causal relationship between the increase in polyamine levels and some aspects of cell growth [3].

Contemporary interest in the polyamines was probably triggered by the brief report of Russell describing the elevated levels of polyamines in the urine of cancer patients [4]. This preliminary report was followed by a more extensive study from the same laboratory demonstrating that patients with histologically diagnosed tumors and leukemias had average urinary polyamine levels that were several-fold greater than those detectable in normal volunteers [5]. Unfortunately, the analytical method used by these workers was not sufficiently specific for the polyamines and the urinary levels reported were considerably higher than those reported in other studies using more specific analytical methods. These early reports triggered interest in the possible use of urinary polyamine levels as biochemical markers of tumor presence and growth. A number of independent studies involving large numbers of patients have shown that the levels of at least two of these amines were elevated in the urine of a majority of the cancer patients [6-17]. However, one recent and comprehensive study indicates that the potential usefulness of the urinary polyamines as biologic markers is less promising than originally anticipated [18]. Another report described the usefulness of the polyamines as predictors of success and failure in cancer chemotherapy [19].

In the human urine the polyamines are present predominantly in the form of conjugates which produce the free polyamines after hydrolysis [5]. The monoacetyl derivatives of the polyamines are the only conjugates which have been identified in human urine [8, 20-24]. Recent studies have shown that both the N^1- [N-{3-[(4-aminobutyl)-amino]propyl}acetamide: $CH_3COHN(CH_2)_3NH(CH_2)_4NH_2$] and N^8-acetylspermidine [N-{4-[(3-aminopropyl)-amino]butyl}acetamide: $CH_3COHN(CH_2)_4NH(CH_2)_3NH_2$] are present in human urine [8, 24], contrary to previous reports of the presence of N^1-acetylspermidine only [22, 23]. The ratio of N^1-acetylspermidine to N^8-acetylspermidine in the 24-hr urine was considerably higher in the urine of cancer patients than it was in the urine of normal subjects [25, 26]. Recently, Bachrach [27] reviewed the literature on the use of polyamines as chemical markers of malignancy.

Various analytical approaches have been used for the determination of polyamines in biological samples. These approaches included paper electrophoresis [28], thin-layer [29-35], high-pressure liquid or gas-chromatographic separation of the amines or their derivatives, and radioimmunoassay [36]. This chapter will cover only the analytical procedures in which the separation of the polyamines was accomplished by high-pressure liquid or gas chromatography. It will attempt to provide a complete review of pertinent methods published prior to December 1976.

II. HIGH-PRESSURE LIQUID CHROMATOGRAPHY

A. Ion-Exchange Chromatography of the Polyamines

A number of methods for the determination of polyamines in biological fluids using ion-exchange resins has been reported in the literature. These methods are essentially modifications of the method described by Moore and Stein for the analysis of amino acids [37]. The majority of these methods utilize an automated amino acid analyzer for the performance of the analysis. The detection and quantitation of the amines in the column eluate are achieved colorimetrically after reaction with ninhydrin or fluorometrically after reaction with fluorescamine or o-phthalaldehyde.

Initially, weakly acidic resins containing carboxylic acid-exchange groups on an acrylic polymer lattice were used. Wall [38, 39] used Amberlite CG-50 which had been pulverized from large beads or ZeoKarb 226 beadform resins for the separation of a number of amines including putrescine and cadaverine. Morris et al. [40, 41] used columns packed with Bio-Rex 70 in conjunction with an amino acid analyzer for the separation and quantitative determination of the naturally occurring di- and polyamines and their acetyl derivatives. These authors used two elution buffers, a 0.33 M pyridinium acetate (pH 5.7) followed by a 0.38 M pyridinium acetate (pH 4.4). This method was capable of detecting as low as 5 nmol of the amines and the elution was complete in approximately 80 min.

The inadequate separation of the polyamines on the weakly acidic resins prompted some researchers to use the strongly acidic sulfonic acid resins. The sulfonated styrene-divinylbenzene copolymers of 8% cross-linkage were used in most of the reported studies. Hatano et al. [42] used a Hitachi Model KLA-3 amino acid analyzer equipped with a column (12 × 0.6 cm) packed with Aminex A-4 (spherical beads of sulfonated 8% cross-linked styrene-divinylbenzene copolymer, particle size 20 ± 4 μm, Bio-Rad Laboratories). The column temperature was maintained at 50°C. These authors used two stepwise elutions followed by a three-component gradient elution system for the separation of 16 amines, including putrescine and cadaverine, in a total of 7 hr. The method was capable of detecting as low as 0.1 μmol of each of the amines studied.

Aminex A-5 resin (similar to Aminex A-4 except for the small particle size of 13 ± 2 μm) was used by three groups of workers for the separation of the polyamines [43-45]. Bremer et al. [43] used a model BC200 amino acid analyzer (Bio-Cal Instruments, Munich, Germany) equipped with a 10 × 0.9-cm column packed with Aminex A-5 resin at a column temperature of 55°C. The separation of polyamines was achieved by stepwise elution with three citrate buffers (pH 5.28) containing increasing concentrations of sodium chloride. Three modifications of this method were subsequently published by Tabor et al. [46], Marton et al. [47], and Gehrke et al. [44]. The major modifications reported by Tabor [46] involved the change of the composition of the buffer and the column temperature. These changes permitted the separation of the polyamines at a lower flow rate of eluate which resulted in increased sensitivity and lower operating pressures.

Marton et al. [47] described a modification of the method of Bremer et al. which allowed the analysis for putrescine, spermidine, and spermine in physiological fluids at the rate of one sample per hour, with excellent reproducibility and adequate sensitivity. They used a Beckman Model 121 automated amino acid analyzer equipped with two short columns (0.9 cm I.D.) packed to a height of 5 cm with Beckman PA 35 spherical resin. The interference from free amino acids was eliminated by column elution with a citrate buffer (0.35 M sodium citrate and 0.35 M sodium chloride, pH 5.08) for 20 min followed by a 15-min elution with a second citrate buffer (0.35 M sodium citrate and 2.0 M sodium chloride, pH 4.68). The column effluent was discarded during this 35-min period. The polyamines eluted during the 35 to 95 min after sample injection using the second citrate buffer.

Another modification of the method of Bremer et al. was published by Gehrke et al. [44]. These authors studied the effects of pH and ionic strength of the buffer on the affinities of the polyamines for the resin. They observed that all the polyamines were strongly retained by the resin at pH < 11 and that elution with buffers containing increased salt concentration greatly decreased the retention time of the polyamines. These authors developed a rapid elution scheme in which other ninhydrin-positive components of urine, such as amino acids, which have a weaker affinity for the resin than the polyamines, were eluted using a high-pH (8.80), low-ionic-strength (0.35 N Na$^+$) buffer. The polyamines were eluted with a citrate buffer (pH 5.8) of increasing ionic strength followed by 0.4 N NaOH containing 0.25% EDTA. These authors used 3,3'-diaminodipropylamine as an internal standard for the quantitative estimation of the polyamines. The analysis was performed on a Bio-Cal Model BC-200 amino acid analyzer (Bio-Cal, Richmond, California) with a 7.5 × 0.9-cm column packed with Aminex A-5. The column was maintained at 65°C throughout the analysis. The total analysis time and column regeneration time were reported to be 120 min/sample. The applicability of this method in the routine analysis of urine samples was demonstrated by the authors. The separation of a mixture of the polyamines is illustrated in Fig. 1.

Fluorescamine reagent was used for the monitoring of the polyamines in column eluates by Veening et al. [45]. These authors separated the polyamines on a 15 × 0.45-cm nickel column packed with Aminex A-5 at 70°C employing a combined pH-salt gradient. The separation of the polyamines was complete in approximately 90 min. The sensitivity of the method was acceptable for the analysis of 1,3-diaminopropane, putrescine, and cadaverine in biological fluids. However, the sensitivity for spermidine and spermine was not as good as the authors had anticipated and the method was not suitable for the routine determination of these two amines in physiological fluids.

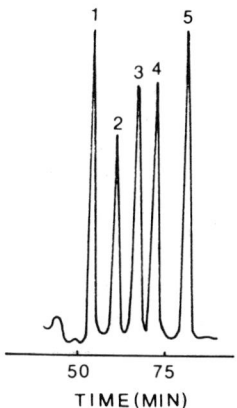

FIG. 1. Ion-exchange chromatography separation of a standard solution of 1, putrescine; 2, cadaverine; 3, the internal standard 3,3'-diaminodipropylamine; 4, spermidine; and 5, spermine. (Adapted from Ref. 44 with the permission of the publisher and author.)

A very significant advancement in the analysis of polyamines using ion-exchange resin chromatography is represented by the method developed by Marton et al. [48]. These authors described an automated micromethod for the determination of polyamines using the Durham D-500 amino acid analyzer. In this method a narrow pore (1.75 mm I.D.) stainless steel column was used. The packing material was a Durrum DC-4A resin (a sulfonated polystyrene resin with 8% cross-linkage and a bead diameter of 8.0 ± 0.5 µm). A citrate buffer was used for the elution of the polyamines at a flow rate of 18.5 ml/hr. The column pressure generated was approximately 2000 psi. The elution time for the polyamines from standard solutions or tissue samples was 35 min. The method was capable of detecting as low as 25 pmol of putrescine, 50 pmol of spermidine, and 100 pmol of spermine. The same authors reported a modification of this method which is suitable for the analysis of polyamines in the cerebrospinal fluid [49]. In this modification they used the Durrum DC-PA resin (sulfonated polystyrene resin with a bead diameter of 10 ± 1 µm). The resin was packed to a height of 8 cm in a stainless steel column (1.75 mm I.D.). This resin allowed the use of a multibuffer elution system with almost no change in column pressure, which resulted in a more stable baseline and permitted

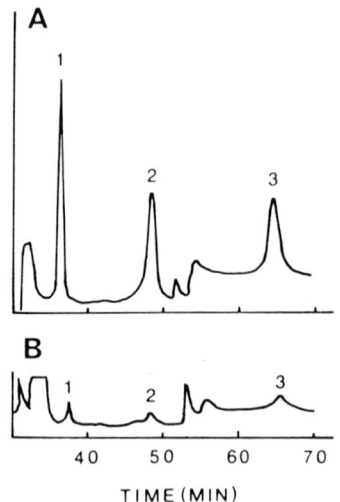

FIG. 2. Comparative chromatogram for the separation of a standard solution of 200 pmol each of 1, putrescine; 2, spermidine; and 3, spermine. (A) Monitoring fluorescence after reaction with o-phthalaldehyde. (B) Monitoring absorbance at 590 nm after reaction with ninhydrin. (Adapted from Ref. 50 with the permission of the publisher and author.)

detection at higher sensitivity. More recently, these authors used o-phthalaldehyde in place of ninhydrin for monitoring the polyamines in column eluates [50]. This modification resulted in a six- to tenfold increase in sensitivity. Using this modification, the authors were able to measure accurately as low as 3 to 6 pmol of putrescine or spermidine and 12 to 15 pmol of spermine. The elution of the polyamines from physiological fluids was complete in 70 min. Figure 2 demonstrates the increased sensitivity achieved by using the fluorescence method.

Recently, Gehrke et al. [51] reported a modification of their previously published method. In this modification they used an internal standard to improve the reliability of the method. The separation of the polyamines was achieved on a narrow pore (2.8 mm I.D.) column packed with Beckman AA-20 resin. The method is capable of detecting as low as 50 pmol of putrescine and cadaverine and 25 pmol of spermidine and spermine. The elution of the polyamines was complete in approximately 60 min (Fig. 3).

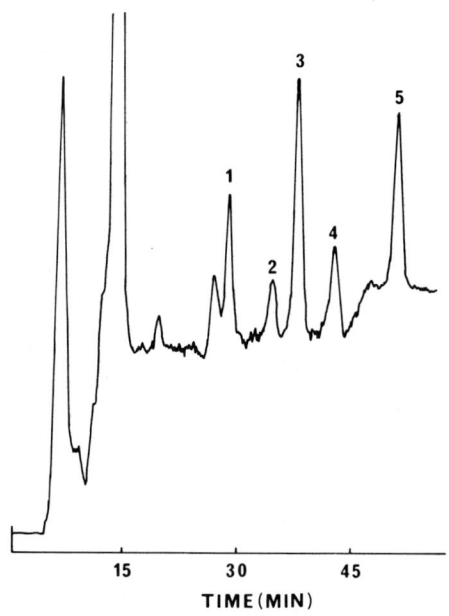

FIG. 3. Ion-exchange chromatography of a standard solution containing 50 pmol each of 1, putrescine; 2, cadaverine; 3, the internal standard 3,3'-diaminodipropylamine; 4, 25 pmol of spermidine; and 5, 16.7 pmol of spermine. (Adapted from Ref. 51 with the permission of the publisher and author.)

The separation of polyamines using ligand-exchange chromatography has also been reported [52]. This method will have very limited practical application in the determination of polyamines in biological fluids because of its low sensitivity.

B. Separation of Polyamines after Derivative Formation

The separation of derivatives of the polyamines by HPLC has been described by several investigators. Samejima [53] described the formation of the fluorescamine derivatives of aliphatic diamines and polyamines and their separation by HPLC. At pH 8.0 and room temperature, the reaction between fluorescamine and the primary amines was complete within a few minutes. The difluorescamine derivatives of general structure I were

I

formed under these reaction conditions and were stable for several hours. The fluorescamine derivatives of mixtures of the diamines or polyamines were separated using gradient elution in the reverse-phase partition mode. The stationary phase used was octadecylsilane chemically bonded porous layer beads 30 to 44 μm (Vydac reversed-phase, Varian Aerograph, Walnut Creek, California). The author studied the various factors affecting the elution profile of the fluorescamine derivatives and found that a mobile phase of methanol in 0.1 M borate buffer (pH 8.0) was the most acceptable. Modification of the composition of the gradient allowed the separation of mixtures of the different amines. A baseline separation of a mixture of the fluorescamine derivatives of putrescine, spermidine, and spermine was achieved within 20 min. The limit of detection was in the picomole range. The application of this method to the analysis of polyamines in biological samples has not been described.

Sugiura et al. [54] described the HPLC separation of the tosyl derivatives of putrescine, spermidine, and spermine. The tosylation of the polyamines proceeds in an aqueous acetone media (1:1) in the presence of $NaHCO_3$. The reaction mixture was heated at 70°C for 1 hr to remove an interfering substance. The tosylated polyamines were separated on a column (1 m × 2.1 mm I.D.) packed with aliphatic ether chemically bonded porous layer beads, 25 to 37 μm (Permaphase ETH, E. I. duPont de Nemours & Company, Instrument Products Division, Wilmington, Delaware). The mobile phase was water-acetonitrile (6:4) at a column temperature of 30°C. The column eluate was monitored by a uv detector but the wavelength at which the eluate was monitored was not indicated. The separation was complete in less than 20 min. The lower detection limits were not given; however, the lowest amounts measured were 3 μg putrescine, 4 μg spermidine, and 10 μg spermine. These figures indicate the much lower sensitivity of this method compared to other available methods. The application of this method to the analysis of polyamines in biological samples has not been reported.

The HPLC separation of the dansyl derivatives of the polyamines and their monoacetyl conjugates was studied in much detail by Abdel-Monem et al. [24, 25, 55, 56]. These authors synthesized and unequivocally characterized the dansyl derivatives of the diamines and polyamines as well as the dansyl derivatives of the monoacetyl diamines and polyamines. A number of chromatographic systems were investigated including the adsorption and partition modes using isocratic solvent elution or programmed gradient elution.

The separation of the dansyl derivatives of the diamines and polyamines can be achieved using isocratic solvent elution both in the adsorption and partition modes. However, using isocratic solvent elution it was necessary to purify the dansyl derivatives obtained from tissues and biological fluids by TLC prior to separation by HPLC. This was not necessary when programmed solvent elution was used in the partition mode.

A mixture of the dansyl derivatives of six di- and polyamines and ammonia was well separated within 18 min on a microparticle alumina column (Micropak Al-5, 25 × 0.2 cm I.D., Varian Aerograph, Walnut Creek, California) using chloroform-isopropanol (100:1) as the mobile phase. Although excellent separation was obtained with the alumina column, the adsorbent was very readily deactivated after a few samples and reproducible results using this column were not obtained. The use of a column (120 cm × 2.2 mm I.D.) packed with Corasil II (Waters Associates, Inc., Milford, Massachusetts) and elution with chloroform (containing 0.75% ethanol)-triethylamine (100:1.8) provided acceptable separation of the dansyl derivatives of spermine, spermidine, putrescine, and diaminopropane (Fig. 4). The elution solvent was not compatible with the use of a uv detector, but excellent sensitivity was obtained with a fluorescence detector.

The isocratic separation of the dansyl polyamines was also achieved on a bonded-phase column. Figure 5 represents the separation of the dansyl derivatives of putrescine, diaminopropane, spermidine, and spermine on a Micropak-CN column (Varian Aerograph, Walnut Creek, California) and

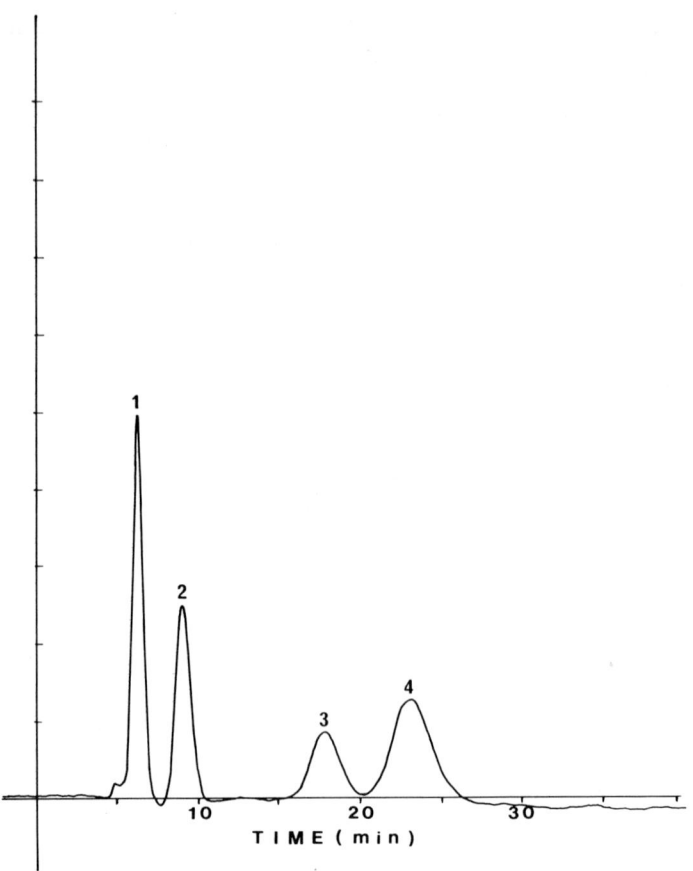

FIG. 4. High-pressure liquid chromatography separation of a standard solution containing 50 pmol of each of the dansyl derivatives 1, spermine; 2, spermidine; 3, putrescine; and 4, diaminopropane. Column: Corasil II (120 cm × 2.2 mm I.D.); detector: fluoromonitor; eluent: chloroform-triethylamine (50:1); flow rate: 24.0 ml/hr; and temperature: ambient.

elution with a solvent of cyclohexane-methylene chloride-isopropanol (88.5:7.5:4.0). This chromatographic system has proven to be very useful for the analysis of polyamines in tissues [57]. The crude dansyl derivatives obtained from tissue extracts were separated by TLC on silica gel plates in one dimension and the areas of the plates corresponding to the dansyl polyamines were scraped and eluted. The concentrated extracts were then separated by HPLC using the Micropak-CN column. 1,3-Diaminopropane was used as an internal standard and was added to the tissue extracts prior

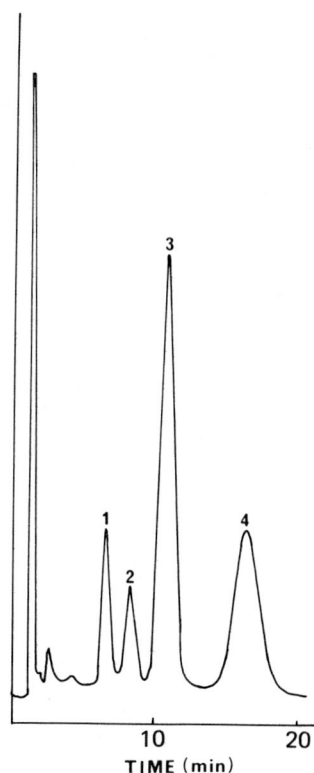

FIG. 5. High-pressure liquid chromatography separation of the dansyl derivatives of the polyamines obtained from the spleen of mice inoculated with L-1210 leukemic cells. Compounds tested were the dansyl derivatives of 1, diaminopropane (internal standard); 2, putrescine; 3, spermidine; and 4, spermine. Column: Micropak CN-10 (23 cm × 2.5 mm I.D.); detector: uv monitor at 254 nm; eluent: cyclohexane-methylene chloride-isopropanol (88.5:7.5:4.0); flow rate: 60 ml/hr (1600-1700 psi); and temperature: ambient.

to derivative formation. The reproducibility of the chromatographic system was excellent and hundreds of samples were analyzed, over a period of 6 months, using one column. The column showed no signs of deterioration and its efficiency remained virtually unchanged.

The use of programmed solvent elution with the bonded-phase column (Micropak CN-10) allowed the analysis of dansyl polyamines obtained from tissues directly and eliminated the TLC purification step. The separation

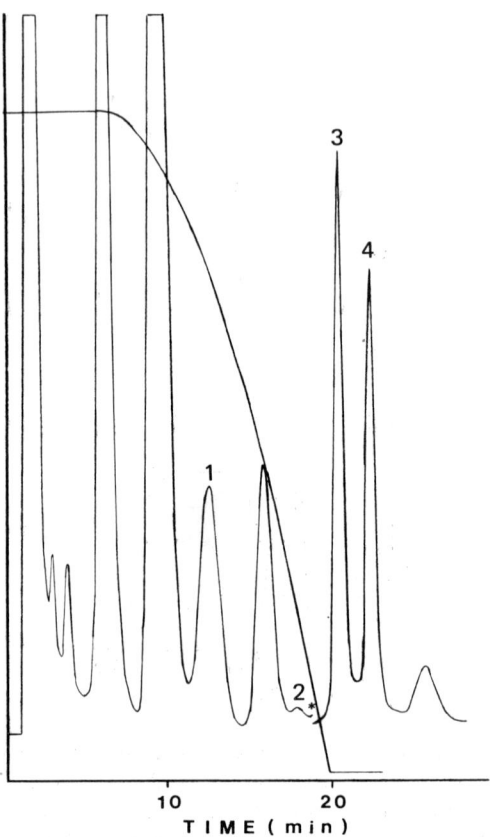

FIG. 6. High-pressure liquid chromatography separation of the dansyl derivatives obtained from the rat brain extract to which 1,2-diaminoethane (internal standard) had been added. Peaks were identified as the dansyl derivatives of 1, the internal standard 1,2-diaminoethane; 2, putrescine; 3, spermidine; and 4, spermine. Column: Micropak CN-10 (25 × 2.5 mm I.D.); detector: fluoromonitor; solvent: isocratic elution with cyclohexane-isopropanol (49:1) for 5 min and then programmed gradient elution using concave curve number 7 (Waters 660 solvent programmer), a second solvent of cyclohexane-methylene chloride-isopropanol (21:3:1), and a gradient time of 15 min; flow rate, 180 ml/hr (4500 psi, initially); temperature: ambient. The sensitivity of the fluoromonitor was reduced by a factor of 2 at the point marked with an asterisk.

of the dansyl derivatives of the polyamines obtained from rat brain is shown in Fig. 6. A sensitive method for the determination of polyamines in tissues was developed based on this chromatographic separation. The analysis was completed in approximately 25 min and the method was sufficiently sensitive to allow the determination of as low as 40 pmol of putrescine and 20 pmol of spermidine and spermine [56]. The application of this method to the determination of polyamines in L1210 leukemic cells of mice in culture was recently described [58].

The separation of the dansyl derivatives of the monoacetyl polyamines by HPLC has also been described by Abdel-Monem et al. [55, 24, 25]. The separation was performed in the adsorption mode using isocratic solvent elution, a column (50 cm × 2.2 mm I.D.) packed with microparticles of silica gel (Micropak Si-10, Varian Aerograph, Walnut Creek, California), and a mobile phase of chloroform-isopropanol (50:3). The dansyl derivatives of the monoacetyl polyamines were well separated from those of the polyamines. Although the dansyl derivatives of the N^1- and N^8-acetylspermidines were well separated from each other, the separation of the dansyl monoacetyl cadaverine and monoacetyl diaminopropane was not satisfactory [55]. The use of a Corasil II column (100 cm × 0.2 mm I.D.) and a solvent of chloroform-isopropanol (50:3) provided separation very similar to that obtained with the microparticle column [25]. The determination of the monoacetyl polyamines in human urine using HPLC was also reported by these authors [25]. The polyamines and the monoacetyl polyamines were extracted from the basified urine with isoamyl alcohol. The dansyl derivatives of the urinary polyamines were formed from the concentrated extracts. Two-dimensional TLC was used to purify these dansyl derivatives prior to their HPLC analysis.

III. GAS CHROMATOGRAPHY

A. Separation of Polyamines

The high polarity and basicity of the di- and polyamines results in peak tailing and partial irreversible adsorption of these compounds on gas-chromatographic column packing materials. The modification of stationary phases to overcome these difficulties in the separation of diamines, including putrescine and cadaverine, was reported by Smith and Radford [59] in 1961. These authors reported satisfactory separation of mixtures of the diamines using a stationary phase of 10% Carbowax 20M and 5% potassium hydroxide. The application of this chromatographic separation to the determination of the amines in biological samples was not described. In a brief communication, Arad et al. [60] described an extension of the method of Smith and Radford to the determination of amines in aqueous solutions. Aqueous solutions of aliphatic amines, including putrescine, were injected directly on a column packed with 20% Carbowax 20M and 5% potassium hydroxide on Chromosorb W. The aliphatic monoamines were eluted isothermally at 70°C and then the column temperature was raised to 180°C at a rate of 10°C/min. Putrescine eluted prior to the water peak. The appearance of ghost peaks after the injection of water on columns containing potassium hydroxide was later reported by Simonaitis and Guvernator [61].

The determination of putrescine in plant tissues using gas chromatography was described by Smith [62, 63]. This author used a column packing which is similar to that previously described by Johnson and Markham [64] and was made of 10% Carbowax 20M and 5% potassium hydroxide.

B. Separation of Polyamines after Derivative Formation

To overcome the disadvantages usually encountered in the direct chromatography of aliphatic amines, Vanden Huevel et al. [65] carried out a systematic study of the preparation and gas-chromatographic properties of amine derivatives, including derivatives of putrescine and cadaverine. The derivatives studied included the diacetyl, dipropionyl, dibutyryl, ditrifluoroacetyl, dipentafluoropropionyl, diheptafluorobutyryl, and the enamines with cyclopentanone, cyclohexanone, and cycloheptanone. These authors also studied a variety of stationary phases and made some useful observations with respect to the relationship between gas-chromatographic behavior and the structure of the amines and their derivatives.

The gas-chromatographic determination of the trifluoroacetyl derivatives of the polyamines obtained from biological samples has been described by several authors in the past decade. Brooks and Moore [66] described a method for the determination of putrescine, cadaverine, spermidine, and spermine, among other amines, in whole cultures of several species of Clostridium. The amines were extracted with chloroform from the basified residual broth obtained after extraction of acids and neutral components from whole cultures. The amines were converted to their trifluoroacetyl derivatives and the analysis was performed on a stationary phase of 3% SE30. The separation was achieved by temperature programming from 80 to 225°C during 25 min. Using these conditions, the trifluoroacetyl derivatives of tyramine and cadaverine, as well as those for spermidine and tryptamine, were not separated.

The use of gas chromatography for the determination of polyamines in human urine was reported by two groups in 1973. Gehrke et al. [67] used a small cation-exchange resin column for the cleanup of hydrolyzed urine samples. The column eluate was concentrated and treated with trifluoroacetic anhydride to form the trifluoroacetyl derivatives. Phenanthrene was used as an internal standard and was added to the sample before gas-chromatographic analysis. The separation of the derivatives was performed on a mixed-phase column of 2% OV-17 and 1% SP-2401 using temperature programming from 80 to 255°C with an increase rate of 8°C/min. A flame ionization detector was used in these studies. These authors reported a linear response for the three polyamines over the range of 10 to 600 ng injected. The application of this method to the analysis of urine samples from both normal volunteers and cancer patients was also reported. The authors also compared the data obtained by using this gas chromatography method and the ion-exchange chromatography method [44] for the analysis of polyamines in the urine of cancer patients. After the data were corrected for the difference in recovery between the two methods, the results agreed very well.

Denton et al. [68, 69] also used gas chromatography of the trifluoroacetyl derivatives for the analysis of polyamines in human urine. The separation was carried out on a 5% OV-17 stationary phase using an instrument equipped with a flame ionization detector. 1,7-Diaminoheptane was used as an internal standard. The application of the method in the routine analysis of urine samples was reported.

A rapid and convenient method for the determination of urinary polyamines by gas chromatography was described by Makita et al. [70]. These authors prepared the pentafluorobenzoyl derivatives of the amines by treating the hydrolyzed urine samples with pentafluorobenzoyl chloride and sodium bicarbonate. The separation of the pentafluorobenzoyl derivatives of the diamines was carried out on a 1.5% Carbowax 20M stationary phase at 250°C. The derivative of spermidine was not eluted from this column even at higher temperatures, but it was eluted within 10 min from a mixed-phase column containing 1.0% OV-1 and 0.25% SP-1000 at 260°C. A 1.5% OV-1 column at 290°C was used for the analysis of the spermine derivative. An electron capture detector was used in this analysis and gave a linear response for the derivatives of 1,3-diaminopropane, putrescine, cadaverine, and spermidine in the subnanogram range. The analysis of the spermine derivative using the electron capture detector in the subnanogram level was not successful, evidently because of excessive tailing. However, the use of the flame ionization detector permitted the analysis of this derivative in the microgram range. The detailed application of this method to the analysis of urine samples was not well documented by the authors.

The gas-chromatographic analysis of the polyamines as their N-permethylated derivatives was described recently by Giumanini et al. [71]. These authors showed that treatment of an aqueous solution of the polyamines with formaldehyde and sodium borohydride resulted in the N-permethylation of the polyamines in yields of 70 to 93%. The separation of the N-permethylated polyamines was successfully carried out on a variety of stationary phases including 5% versamid-5% potassium hydroxide and 3% OV-17. These authors also reported the relative abundance of fragments of a number of the N-permethylated polyamines using gas chromatography/mass spectrometry (GC/MS). Unfortunately, a systematic study of the application of this method to the analysis of polyamines in biological samples was not described.

C. Gas Chromatography/Mass Spectrometry

The first application of GC/MS in the qualitative and quantitative analysis of polyamines in biological fluids was described by Walle [72]. The gas-chromatographic analysis used by this author was essentially that previously described by him in collaboration with Denton et al. [68]. He used single-ion monitoring for the detection of 1,3-diaminopropane, putrescine, cadaverine, spermidine, spermine, and monoacetylspermidine in a crude extract of unhydrolyzed urine from a cancer patient. This represented the first report of the analysis of urine from cancer patients without prior acid hydrolysis.

Recently, Smith and Daves [73] described a sensitive method for the GC/MS determination of polyamines using deuterated analogs as internal standards. The enhancement of the sensitivity of this method was accomplished by the use of large excesses of the deuterated analogs to improve the chromatographic properties of the amines. The trifluoroacetyl derivatives of the polyamines were prepared prior to the gas-chromatographic separation. The method was suitable for the analysis of as little as 1 pmol of the amine injected and a linear response was obtained over the range 1 to 100 pmol. The application of this method to the determination of polyamines in serum samples has also been reported by these authors [74].

IV. CONCLUSIONS

Among the various analytical approaches which have been used for the determination of polyamines in biological samples, this chapter dealt primarily with those methods which involve high-pressure liquid chromatography or gas chromatography. The choice of an analytical method will depend on the specific application and availability of instruments. In general, the ion-exchange chromatography methods of Marton et al. [49, 50] and Gehrke et al. [51] are probably the most suitable for the analysis of polyamines in hydrolyzed urine samples, serum, or cerebrospinal fluid. The only method available for the determination of polyamine conjugates in unhydrolyzed urine samples is that of Abdel-Monem and Ohno [25]. This method requires lengthy sample preparation steps and personnel of high technical ability to perform the analysis. The determination of polyamines in tissues can probably be best performed using the method of Newton et al. [56]. The recently published GC/MS method using deuterated polyamines as internal standards [73] is potentially a very useful method pending the commercial availability of the deuterated polyamine standards.

ACKNOWLEDGMENTS

We wish to thank C. W. Gehrke and R. G. Smith for providing preprints of their manuscripts. We also thank the office staff of the College of Pharmacy, University of Minnesota, for preparation of the manuscript.

REFERENCES

1. T. A. Smith, Endeavour, 31, 22 (1972).
2. S. S. Cohen, Introduction to the Polyamines, Prentice-Hall, Englewood Cliffs, New Jersey, 1971.
3. U. Bachrach, Function of Naturally Occurring Polyamines, Academic Press, New York, 1973, pp. 47-95.
4. D. H. Russell, Nature (London), 233, 141 (1971).
5. D. H. Russell, C. C. Levy, S. C. Schimpff, and I. A. Hawk, Cancer Res., 31, 1555-1558 (1971).
6. A. Lipton, L. M. Sheehan, and G. F. Kessler, Cancer, 35, 464 (1975).
7. D. C. Tormey, T. P. Waalkes, D. Ahmann, C. W. Gehrke, R. W. Zumwalt, J. Snyder, and H. Hansen, Cancer, 35, 1095 (1975).
8. M. Tsuji, T. Nakajima, and I. Sano, Clin. Chim. Acta, 59, 161 (1975).
9. Y. Takeda, T. Tominaga, M. Kitamura, T. Taguchi, T. Takeda, and T. Miwatani, Gann, 66, 445 (1975).
10. W. R. Fair, N. Wehner, and U. Brorsson, J. Urol., 114, 88 (1975).
11. D. H. Russell and S. D. Russell, Clin. Chem., 21, 860 (1975).
12. F. Dreyfuss, R. Chayen, G. Dreyfuss, R. Dvir, and J. Ratan, Israel J. Med. Sci., 11, 785 (1975).
13. A. Lipton, L. M. Sheehan, and H. A. Harvey, Cancer, 36, 2351 (1975).
14. K. Fujita, T. Nagatsu, K. Maruta, M. Ito, H. Senba, and K. Miki, Cancer Res., 36, 1320 (1976).
15. L. J. Marton, O. Heby, V. A. Levin, W. P. Lubich, D. C. Crafts, and C. B. Wilson, Cancer Res., 36, 973 (1976).

16. O. Rennert, T. Miale, J. Shukla, D. Lawson, and J. Frias, Blood, 47, 695 (1976).
17. A. Lipton, L. Sheehan, R. Mortel, and H. A. Harvey, Cancer, 38, 1344 (1976).
18. T. P. Waalkes, C. W. Gehrke, D. C. Tormey, R. W. Zumwalt, J. N. Hueser, K. C. Kuo, D. B. Lakings, D. L. Ahmann, and C. G. Moertel, Cancer Chemother. Rep., P. 1, 59, 1103 (1975).
19. D. H. Russell, B. G. M. Durie, and S. E. Salmon, Lancet, ii, 797 (1975).
20. T. L. Perry, S. Hansen, and L. MacDougal, J. Neurochem., 14, 775 (1967).
21. T. L. Perry, S. Hansen, and L. MacDougal, Nature (London), 214, 484 (1967).
22. T. Nakajima, L. F. Zack, Jr., and F. Wolfgram, Biochim. Biophys. Acta, 184, 651 (1969).
23. T. Walle, Polyamines in Normal and Neoplastic Growth (D. H. Russell, ed.), Raven Press, New York, 1973, pp. 355-365.
24. M. M. Abdel-Monem and K. Ohno, J. Pharm. Sci., 66, 1089 (1977).
25. M. M. Abdel-Monem and K. Ohno, J. Pharm. Sci., 66, 1195 (1977).
26. M. M. Abdel-Monem, K. Ohno, I. E. Fortuny, and A. Theologides, Lancet, ii, 1210 (1975).
27. U. Bachrach, Ital. J. Biochem., 25, 76 (1976).
28. A. Raina and S. S. Cohen, Proc. Natl. Acad. Sci. U.S.A., 55, 1587 (1966).
29. J. E. Hammond and E. J. Herbst, Anal. Biochem., 22, 474 (1968).
30. N. Seiler and M. Weichmann, Z. Physiol. Chem., 348, 1285 (1967).
31. A. S. Dion and E. J. Herbst, Ann. N.Y. Acad. Sci., 171, 723 (1970).
32. K. Igarashi, I. Izumi, K. Hara, and S. Hirose, Chem. Pharm. Bull., 22, 451 (1974).
33. G. Dreyfuss, R. Dvir, A. Harell, and R. Chayen, Clin. Chim. Acta, 49, 65 (1973).
34. J. H. Fleisher and D. H. Russell, J. Chromatogr., 110, 335 (1975).
35. F. Abb and K. Samejima, Anal. Biochem., 67, 298 (1975).
36. D. Bartos, R. A. Campbell, F. Bartos, and D. P. Grettie, Cancer Res., 35, 2056 (1975).

37. S. Moore and W. H. Stein, J. Biol. Chem., 192, 663 (1951).
38. R. A. Wall, J. Chromatogr., 37, 549 (1968).
39. R. A. Wall, J. Chromatogr., 60, 195 (1971).
40. D. R. Morris, K. L. Koffron, and C. J. Okstein, Anal. Biochem., 30, 449 (1969).
41. D. R. Morris, Methods Enzymol., 178, 850 (1971).
42. H. Hatano, K. Sumizu, S. Rokushika, and F. Murakami, Anal. Biochem., 35, 377 (1970).
43. H. J. Bremer, E. Kohne, and W. Endres, Clin. Chim. Acta, 32, 407 (1971).
44. C. W. Gehrke, K. C. Kuo, R. W. Zumwalt, and T. P. Waalkes, J. Chromatogr., 89, 231 (1974).
45. H. Veening, W. W. Pitt, Jr., and G. Jones, Jr., J. Chromatogr., 90, 129 (1974).
46. H. Tabor, C. W. Tabor, and F. Irreverre, Anal. Biochem., 55, 457 (1973).
47. L. J. Marton, D. H. Russell, and C. C. Levy, Clin. Chem., 19, 923 (1973).
48. L. J. Marton, O. Heby, C. B. Wilson, and P. L. Y. Lee, FEBS Lett., 41, 99 (1974).
49. L. J. Marton, O. Heby, C. B. Wilson, and P. L. Y. Lee, FEBS Lett., 46, 305 (1974).
50. L. J. Marton and P. L. Y. Lee, Clin. Chem., 21, 1721 (1975).
51. C. W. Gehrke, K. C. Kuo, R. L. Ellis, and T. P. Waalkes, J. Chromatogr., 143, 3454 (1977).
52. J. D. Navratil and H. F. Walton, Anal. Chem., 47, 2443 (1975).
53. K. Samejima, J. Chromatogr., 96, 250 (1974).
54. T. Sugiura, T. Hayashi, S. Kawai, and T. Ohno, J. Chromatogr., 110, 385 (1975).
55. M. M. Abdel-Monem and K. Ohno, J. Chromatogr., 107, 416 (1975).
56. N. E. Newton, K. Ohno, and M. M. Abdel-Monem, J. Chromatogr., 124, 277 (1976).
57. C. E. Weeks and M. M. Abdel-Monem, J. Pharm. Sci., in press.
58. N. E. Newton and M. M. Abdel-Monem, J. Med. Chem., 20, 249 (1977).

59. E. D. Smith and R. D. Radford, Anal. Chem., 33, 1160 (1961).
60. Y. Arad, M. Levy, and D. Vofsi, J. Chromatogr., 13, 565 (1964).
61. R. A. Simonaitis and G. C. Guvernator, J. Gas Chromatogr., 5, 49 (1967).
62. T. A. Smith, Anal. Biochem., 33, 10 (1970).
63. T. A. Smith, Phytochemistry, 9, 1479 (1970).
64. M. W. Johnson and R. Markham, Virology, 17, 276 (1962).
65. W. J. A. Vanden Heuvel, W. L. Gardiner, and E. C. Horning, Anal. Chem., 36, 1550 (1964).
66. J. B. Brooks and E. C. Moore, Can. J. Microbiol., 15, 1433 (1969).
67. C. W. Gehrke, K. C. Kuo, R. W. Zumwalt, and T. P. Waalkes, Polyamines in Normal and Neoplastic Growth (D. H. Russell, ed.), Raven Press, New York, 1973, p. 343.
68. M. D. Denton, H. S. Glazer, T. Walle, D. C. Zellner, and F. G. Smith, Polyamines in Normal and Neoplastic Growth (D. H. Russell, ed.), Raven Press, New York, 1973, p. 373.
69. M. D. Denton, H. S. Glazer, D. C. Zellner, and F. G. Smith, Clin. Chem., 19, 904 (1973).
70. M. Makita, S. Yamamoto, and M. Kono, Clin. Chim. Acta, 61, 403 (1975).
71. A. G. Giumanini, G. Chiavari, and F. L. Scarponi, Anal. Chem., 48, 484 (1976).
72. T. Walle, Polyamines in Normal and Neoplastic Growth (D. H. Russell, ed.), Raven Press, New York, 1973, p. 355.
73. R. G. Smith and G. Doyle Daves, Jr., Biomed. Mass Spectrom., 4, 146 (1977).
74. F. Bartos, D. Bartos, D. P. Grettie, R. A. Campbell, L. J. Marton, R. G. Smith, and G. D. Daves, Jr., Biochem. Biophys. Res. Comm., 75, 915 (1977).

AUTHOR INDEX

Numbers in parentheses are reference numbers and indicate that an author's work is referred to although his name is not cited in the text. Underlined numbers give the page on which the complete reference is listed.

A

Abb, F., 251(35), 266
Abdel-Monem, M. M., 250(24), 251(24,25,26), 257, 258(57), 261(58), 264(56), 266, 267
Ahmann, D., 250(7,18), 265, 266
Ajisaka, K., 181(40), 205
Alderlieste, E., 68(38), 71(38), 73
Alford, C. A., Jr., 215(12), 246
Alvares, A. P., 4(22,23), 21(44), 34, 35
Amos, R., 202(98), 208
Anderson, R. W., 178(44), 181(44), 182(44), 205
Andersonn, K., 95, 97, 111
Andersson, O., 114(3), 147
Andersson, S.-I., 143(54), 149
Angelici, R. J., 200(70), 206
Arad, Y., 261, 268
Asano, K., 167(67), 174
Ashley, J. W., 153(13), 155(13), 171
Atkinson, M. R., 221(22), 247
Awata, N., 180(33), 205

B

Bach, R. D., 177(25), 204
Bachrach, U., 250(3), 251, 265, 266
Baczuk, R. J., 199(71), 200(71), 206
Baird, W. M., 27(54), 35
Baranov, V. A., 197(66), 198(66), 199(66), 206
Barefield, E. K., 188(121), 209
Barner, H. D., 223(26), 247
Bartels, H., 192(58), 206
Bartos, D., 251(36), 264(74), 266, 268
Bartos, F., 251(36), 264(74), 266, 268
Baumann, F., 212(2), 246
Beare, S. D., 180(42), 181(40), 184(50), 205, 206
Bengtsson, L., 138(49), 149
Benson, J. V., Jr., 130(30), 148
Berenblum, I., 32(59), 36
Berezkin, V. G., 152, 171
Bergot, B. J., 180(35,36,37), 205
Bernauer, K., 200(74), 207
Berry, I., 39(5), 70(5), 72
Berry, L., 202, 208
Beyer, R., 228(29), 247
Beyer, W. F., 176(13), 204
Bishop, C. T., 114(1), 147
Blair, H., 51(20), 72
Blaschke, G., 193(59,60,62), 194(59), 195(60), 206
Bloch, A., 215(20), 247
Blossey, E. C., 183(46,49), 205

Bodey, G. P., 215(13), 240(13), 246
Booth, J., 4(17), 17(42), 22(52), 34, 35
Borgen, A., 6(29), 25(29), 34
Bosanquet, C. H., 154(20), 172
Boshart, G., 186(130,131), 187(130, 131), 209
Boyd, G. E., 138(48), 148
Boyland, E., 4(14,17), 25, 33, 34
Boyle, P. H., 176(16), 177(16), 204
Bowman, B., 215(19), 247
Brand, L., 21(43), 35
Bremer, H. J., 252(43), 267
Breslow, R., 187(52), 206
Bresnick, E., 7(31), 34
Brewer, P. I., 202(98), 208
Brink, J. J., 223(23), 247
Brookes, P., 27(54,55), 35
Brooks, J. B., 262, 268
Brooks, R. V., 67(36), 73
Brorsson, U., 250(10), 265
Brown, C., 88(9), 109(9), 111
Brown, J., 51(20), 72, 94, 111
Brown, P. R., 39, 71, 202, 208, 212(1,5,7), 214(5), 246
Brown, W., 114(3), 122(22), 137(22), 147
Buchanan, R. A., 215(12), 246
Buchta, R. C., 202(92), 207
Burgoyne, L. A., 221(22), 247
Burnham, W. S., 10(32), 34
Busbee, D. L., 21(47), 50, 35
Bush, J. E., 39, 72
Buss, D. R., 176(5), 203
Butterfield, A., 54(21), 71(21), 72
Buu-Hoi, N. P., 10(32), 34
Byre, S. H., 27(58), 36

C

Cadogan, D. F., 167(72), 174
Caldwell, W., 201, 207
Campbell, R. A., 251(36), 264(74), 266, 268
Cantrell, E. T., 21(47,50), 35
Carlsson, B., 139(51), 149
Carstensen, H., 39, 72
Carubelli, R., 131(36), 148
Caserio, M. C., 176(14), 180(14), 204
Castagnoli, N., 6(29), 25(29), 34
Castells, R. C., 167, 174
Chan, K. K., 179(29), 180(29), 204
Chandler, C. D., 203, 209
Chang, S. H., 187(135), 209
Chao, D. L., 230(31), 247
Chao, K. C., 167(68), 174
Chappelear, P. S., 152(3), 171
Chasseaud, L. F., 4(25), 34
Chayen, R., 250(12), 251(33), 265, 266
Chen, C. J., 153(16), 162(16), 165(16), 172
Chen, S.-M. L., 180(33), 205
Chiavari, G., 263(71), 268
Chibata, I., 177(21), 200(83), 204, 207
Chitombo, K., 122(22), 137(22), 147
Clare, B., 89, 111
Clausen, B., 59, 73
Cohen, D., 137(46), 140, 148
Cohen, S. S., 214(8,9,11), 215(15, 17), 216(9), 217(9), 218(9,16), 219(9), 220(9), 221(9), 222(9), 223(15,21,24,25,26), 224(11), 225(8), 226(8), 227(8,16), 228(27,28), 229(8), 230(8,15), 231(11,24), 232(11), 233(9,11, 24,25), 234(24), 235(9,11,16, 24), 237(11,37), 238(8,11), 245(11), 246, 247, 250(2), 251(28), 265, 266
Conder, J. R., 153(12), 154, 155(12), 156(19), 159(52), 162(19), 165(52), 171, 172, 173
Conney, A. H., 3(7), 6(27,28), 21(43, 44,46), 27(56), 33, 35, 36

AUTHOR INDEX

Coon, M. J., 6(26), 34
Cooper, H. L., 21(48), 35
Cope, A. C., 177(25,26,27), 204
Cornish, D. W., 176(8), 204
Crafts, D. C., 250(15), 265
Cram, D. J., 188, 189(55,56,57), 190(56), 191(57,127), 192(57), 206, 209
Crocker, T. T., 6(29), 12(36), 25(29), 34, 35
Crosby, G. A., 183(47), 183(119), 205, 209
Croy, R. G., 3(10), 4(10,13,16), 5(57), 9(10), 10(13), 11(13), 14(16), 16(16), 17(39), 18(39), 19(39), 20(39), 23(53), 25(53), 26(53), 28(57), 29(57), 30(57), 33, 34, 35, 36
Curtin, D., 69(39), 73
Curtis, W. D., 188(124,125), 209

D

Daly, J. W., 6(28), 34
Dansette, P., 6(28), 15(38), 27(56), 34, 35, 36
Darvey, H., 6(29), 25(29), 34
Daudel, P., 4(21), 15(21), 22(21), 34
Davankov, V. A., 176(4), 197(4,65, 66,67,68,69), 198(65,66,67), 199(66,67), 203, 206
Dave, B. A., 147(63), 149
Daves, G. D., Jr., 264(74), 268
Deans, H. A., 152(3), 166(65), 167, 171, 173
De Clercq, H., 79(2), 80(2), 92(2, 14,31), 96(22), 97(22), 103 (25), 104(25), 110, 111
Dehm, H. C., 199(71), 200(71), 206
Delumeya, R., 105(26), 108, 111
Denton, M. D., 263, 264, 268
Derungs, A., 146, 149
DeStefano, J. J., 202, 208

Detaevernier, M., 109, 111
Detar, C. C., 200(79), 207
Deuel, B., 146(61), 149
Diamond, L., 4(24), 34
Dieckman, J. F., 50(16), 65(16), 66(16), 71(16), 72
Dietrichs, H. H., 143(56,57), 149
Dijkstra, A., 103(25), 104(25), 111
Dingman, J., 59, 66(29), 71(29), 73
Dion, A. S., 251(31), 266
Dixon, P. H., 202(96), 208
Doering, A., 215(15), 223(15), 230(15), 247
Doherty, R. F., 200(81), 207
Dolphin, R. J., 56, 71(22), 72
Done, J. N., 212(6), 246
Donnan, F. G., 132(42), 148
Donow, F., 193(59), 194(59), 206
Doyle Daves, G., Jr., 264, 265(73), 268
Drabowicz, J., 187(53), 206
Dreyfuss, F., 250(12), 265
Dreyfuss, G., 250(12), 251(33), 265, 266
Dryon, L., 92(31), 111
Dubois, R. J., 199(71), 200(71), 206
Dunham, L. L., 180(37), 205
DuPont, B., 214(10), 246
Dupuis, P. F., 103(25), 104(25), 111
Duquesesne, M., 4(21), 15(21), 22(21), 34
Durie, B. G. M., 250(19), 266
Durst, H. D., 188(122), 209
Dvir, R., 250(12), 251(33), 265, 266
Dyson, N., 157, 173

E

Eliel, E. L., 177(28), 179(28), 204
Ellis, D. B., 236(36), 237(36), 247
Ellis, R. L., 255(51), 264(51), 267

Endres, W., 252(43), 267
Engelhardt, H., 201(110), 208
Eon, C., 167(73), 174, 202(104), 208
Ericsson, T., 144(58), 145, 149
Ernst, R. E., 177(24), 204
Ernster, M., 21(46), 35
Eskes, A., 103(25), 104(25), 111
Evans, G. O., 183(45), 205

F

Fair, W. R., 250(10), 265
Fallick, G. J., 39(4), 72, 176(12), 202, 204, 208
Farber, E., 245(40), 248
Fazio, G., 110, 111
Fejes, P., 156(36), 172
Fieser, L., 39, 72
Fieser, M., 39, 72
Fike, W., 89(13), 111
Fitzpatrick, F. A., 59, 65(34), 66(29,34,35), 67(37), 71(29,34,37), 73
Fleisher, J. H., 251(34), 266
Flesher, J. W., 17(40,41), 35
Floridi, A., 131(38), 148
Flory, P. J., 158(48), 173
Fortuny, I. E., 251(26), 266
Freeman, M. L., 130(31), 148
Freireich, E. J., 215(13), 240(13), 246
Frias, J., 250(16), 266
Fujita, K., 250(14), 265
Furth, J. J., 215(16), 218(16), 227(16), 228(28), 235(16), 247
Fuson, R., 69(39), 73

G

Gaal, J., 200(72,73), 207
Gardiner, W. L., 262(65), 268

Gattuso, A. M., 110, 111
Gehrke, C. W., 250(7,18), 252(44), 253(44), 255(51), 262, 263(44), 264(51), 265, 266, 267, 268
Gelboin, H. V., 3(6,10), 4(10,12,13,16), 5(57), 9(10,12), 10(13,32), 11(13), 14(16), 16(16), 17(39), 18(39), 19(39), 20(39), 21(45,48,49,51), 23(53), 25(53), 26(53), 28(57), 29(57), 30(57), 33, 34, 35, 36
Gere, D., 212(2), 246
Gerhart, B. B., 187(132), 209
Gerritse, R. G., 153(18), 172
Giddings, J. C., 153, 171
Gil-Av, E., 176(2), 177, 178(2), 186(130,131), 187(130,131), 203, 209
Gindler, E. M., 129, 148
Gilpin, R. K., 201(90), 207
Ginzburg, B. Z., 137(46), 140, 148
Giumanini, A. G., 263(71), 268
Glazer, H. S., 263(68,69), 264(69), 268
Gleason, D. D., 176(13), 204
Goh, S. H., 15(37), 35
Gokel, G. W., 188(55,122), 189(55), 206, 209
Gommi, B. W., 21(46), 35
Gooding, K. M., 187(135), 209
Gotsevi, G., 188(57), 189(57), 191(57), 192(57), 206
Gottlieb, J. A., 215(13,14), 240(13), 246
Goulian, M., 245(39), 248
Gray, C. H., 202(96), 208
Greech, B., 58(27), 73
Green, G. W., 183(48), 205
Grettie, D. P., 251(36), 264(74), 266, 268
Grossman, D., 102(23), 111
Grover, P. L., 3(8), 4(20,21), 6(30), 12(35), 15(20,21), 22(20,21), 33, 34
Grunfeld, S., 105(26), 108, 111

AUTHOR INDEX

Guggenheim, E. A., 132(42), 148
Guilford, H., 200(78), 207
Guiochon, G., 154, 155(24,25), 156(37), 160(37), 162, 165(22), 167(73), 172, 173, 174, 202(104), 208
Gump, B. H., 165(57), 173
Guvernator, G. C., 262, 268

H

Habgood, H. W., 152, 171
Habig, W. H., 4(19), 34
Hadden, N., 212(2), 246
Haggerty, W. J., Jr., 176(11), 204
Halasz, I., 201(110), 208
Hall, M. S., 180(35,36,37), 205
Hammond, J. E., 251(29), 266
Hansen, H., 250(7), 265
Hansen, S., 250(20,21), 266
Hara, K., 251(32), 266
Harell, A., 251(33), 266
Harren, H., 68(38), 71(38), 73
Harrison, Y. E., 21(46), 35
Hartkopf, A., 102(23), 105(26), 108, 111
Harvey, H. A., 250(13,17), 265, 266
Harvey, R. G., 15(37), 27(54), 35
Hatano, H., 252, 267
Havlicek, J., 116, 118(17), 119, 120(10,17), 121(21), 122(17, 21), 123(10), 124(17), 125, 126(10,17), 128(23), 129(21), 130, 134(43), 135, 136(44), 137(10,43), 138(43), 139, 140(43), 141, 142, 143(44), 147, 148
Hawk, I. A., 250(5), 265
Hawkes, S. J., 96, 97, 102(23), 111
Hayashi, T., 256(54), 267
Haydel, J. J., 156(35), 172
Haynes, H. W., 164, 173

Heby, O., 250(15), 254(48,49), 264(49), 265, 267
Hecht, J. K., 177(26), 204
Heftman, E., 39, 72
Heidelberger, C., 3(8), 33
Heigher, I., 2(3), 33
Helfferich, F., 153(14,15), 154(15), 155(26), 172
Helgeson, R. C., 188(55), 189(55), 206
Henly, R. S., 202(113), 208
Henry, R. A., 27(58), 36, 50(16), 65(16), 66(16), 71(16), 72
Heppel, L. A., 239(38), 248
Herbst, E. J., 251(29,31), 266
Hermann, W., 51(19), 72
Hewer, A., 4(20), 6(30), 12(35), 15(20), 22(20,52), 34, 35
Hildebrand, G. P., 155(13), 171
Hirose, S., 251(32), 266
Ho, T. H., 164, 173
Hoekstra, M. S., 178(43), 181(43), 182(43), 205
Hoffman, D. H., 188(56), 189(56), 206
Holder, G., 27(56), 36
Holm, R. H., 177(24), 204
Hornig, E., 58(27), 73
Horning, E. C., 262(65), 268
Horvath, C., 212(4), 246
Hoste, J., 110(30), 111
Huber, J. F. K., 51(18), 68(38), 71(18,38), 72, 73, 153(18), 172
Huberman, E., 3(8), 33
Hubert-Habart, M., 223(21), 247
Hudson, D. R., 27(58), 36
Hueser, J. N., 250(18), 266
Huggins, M. L., 158(49), 173
Hulsman, J. A. R. J., 51(18), 71(18), 72
Hussey, C. L., 165(38), 173

I

Ichikawa, J., 2(2), 33
Igarashi, K., 251(32), 266

Ikekawa, N., 180(33), 205
Inczedy, J., 200(72,73), 207
Irreverre, F., 252(46), 267
Isaksson, T., 139(51), 149
Isenhour, T. L., 95(19), 102(23), 111
Ito, M., 250(14), 265
Izumi, I., 251(32), 266

J

Jacob, L., 162, 173
Jacobson, M., 6(27), 21(43), 34, 35
Jakoby, W. B., 4(19), 34
Janak, J., 165(61,63), 173
Janicki, C. A., 201(90), 207
Janini, G. M., 158(50), 168(76), 173, 174
Jeanneret, M.-F., 200(74), 207
Jennings, E. C., 63, 71(31), 73
Jennings, R. C., 180(35), 205
Jerina, D. M., 6(28), 15(38), 27(56), 34, 35, 36
Johansson, M., 120(20), 127(20), 147
Johnson, C. N., 147(64), 149
Johnson, H. W., Jr., 177(26), 204
Johnson, K. K., 179(29), 180(29), 204
Johnson, M. W., 262, 268
Johnson, W. S., 180(38), 205
Johnston, R. J., 82(4), 110
Jolley, R. L., 130(31), 131(35), 148
Jones, G., Jr., 252(45), 253(45), 267
Jones, G. H., 188(124,125), 209
Jones, I., 188(54), 206
Jones, J. K. N., 132(40), 148
Jonsson, P., 115(9), 116(15), 118(15), 119(18), 124(9), 147
Judy, K. J., 180(35,36,37), 205
Jupille, T. H., 202, 208
Justice, J. B., 95(19), 111

K

Kainosho, M., 181(40), 205
Kaloustian, M. K., 176(19), 204
Kaloustian, S. A., 176(19), 204
Kalyani, P. L., 17(42), 35
Kamiyama, Y., 140(53), 149
Kaplan, L., 188(56), 189(56), 190(56), 206
Kaplanova, B., 165(63), 173
Karasek, F. W., 202, 208
Karch, K., 201(110), 208
Karger, B. L., 39(5), 70(5), 72, 167(74), 174, 201, 202, 207, 208
Kauffman, G. B., 177(20), 204
Kaufman, L., 86(6), 110
Kawai, S., 256(54), 267
Keller, H., 177(26,27), 204
Keller, J., 215(15), 223(15), 230(15), 247
Kessler, G. F., 250(6), 265
Ketley, J. N., 4(19), 34
Keulemans, A. I. M., 160(53), 173
Keysall, G. R., 17(42), 22(52), 35
Khaliq, A., 215(14), 246
Khoury, F., 167(69), 174
Khum, J. X., 130(29), 131(29,35), 148
Kimball, A. P., 215(19), 230(31), 232(33), 247
King, H. S., 27(55), 35
King, J. W., 168(76), 174
Kingston, D. I., 187(132), 209
Kinoshita, N., 4(12), 9(12), 21(51), 33, 35
Kireeda, M., 180(34), 205
Kirkland, J. J., 39(3), 71, 202, 208, 212(3), 246
Kitamura, M., 250(9), 265
Klemm, L. H., 176(6), 203
Klyne, W., 39, 67(36), 72, 73
Knapp, D., 7(31), 34
Knox, J. H., 202, 208, 212(6), 246
Kobayashi, R., 152, 156(35), 166(66), 167(67), 171, 173, 173, 174

Kober, S., 51(17), 72
Koffron, K. L., 251(40), 267
Kohne, E., 252(43), 267
Kondo, L. E., 130(32), 148
Kono, M., 263(70), 268
Kooistra, D. A., 183(48), 205
Korpl, J. A., 201(90), 207
Koshiura, R., 230(30), 247
Kowalski, B. R., 110, 111
Kratchanov, C. G., 200(76,77), 207
Krige, G. J., 157(38,39,40,41), 172, 173
Kreutman, E., 58(26), 72
Kruskal, J. B., 86(5), 110
Kucera, P., 202, 208
Kuntcheva, M. J., 200(77), 207
Kuntze, K., 176(3), 203
Kuntzman, R., 4(22,23), 6(27,28), 21(43), 34, 35
Kuo, K. C., 250(18), 252(44), 253(44), 255(51), 262(67), 263(44), 264(51), 266, 267, 268
Kuroki, T., 3(8), 33

L

Labianca, D. A., 177(22), 204
Lachmann, M., 228(29), 247
Lafuma, F., 178(137), 209
Lakings, D. B., 250(18), 266
Laidler, D. A., 188(124,125), 209
Landesman, H., 89(13), 111
Landgraf, W. C., 63, 71(31), 73
Landram, G. K., 199(71), 200(71), 206
Lapi, L., 214(8), 225(8), 226(8), 227(8), 229(8), 230(8), 238(8), 246
Larmann, J. P., 203, 209
Larsson, K., 131(33), 132(39), 135(39), 140(52), 148, 149
Larsson, L.-I., 128(24), 148
Laster, W. R., Jr., 232(34)

Laub, R. J., 155(33,34), 172, 202, 208
Lauwereys, M., 79(3), 96(3), 97(3), 110
Lawson, D., 250(16), 266
Leary, J. J., 95, 111
Leary, S., 102(23), 111
Lee, P. L. Y., 254(48,49,50), 255(50), 264(49,50), 267
Leitch, R. E., 200(75), 207
Lenders, P., 79(3), 96(3), 97(3), 110
LePage, G. A., 215(14,19), 223(23), 230(30), 232(33), 236(35), 237(36), 246, 247
Lesec, J., 178(137,138), 209
Leutz, J. C., 4(24), 34
Levin, V. A., 250(15), 265
Levin, W., 4(22), 6(28), 21(43,44), 27(56), 34, 35, 36
Levy, C. C., 250(5), 252(47), 265, 267
Levy, M., 261(60), 268
Lewis, D. W., 179(31), 180(31), 205
Leznoff, C. C., 183(117), 208
Liao, H. L., 165(56), 173
Lichtenstein, J., 223(26), 247
Lim, C. K., 202(96), 208
Lipton, A., 250(6,13,17), 265, 266
Littlewood, A. B., 157, 173
Lindenbaum, S., 138(48), 148
Lochmuller, C. H., 176(1), 177(1), 178(1), 203
Locke, D. C., 153, 171, 201(93), 207
Lodge, B., 54(21), 71(21), 72
Loew, P., 180(38), 205
Loheac, J., 212(6), 246
Losse, G., 176(3), 203
Lowry, S. R., 95(19), 111
Lu, A. Y. H., 6(26,27,28), 27(56), 34, 36
Lubich, W. P., 250(15), 265

M

MacDonald, F., 212(2), <u>246</u>
MacDougal, L., 250(20,21), <u>266</u>
Maestas, P. D., 202, <u>208</u>
Maidhof, A., 228(29), <u>247</u>
Majors, R. E., 176(10), <u>204</u>
Makita, M., 263, <u>268</u>
Mallik, K. L., 153, <u>171</u>
Manduka, M. J., 223(25), 233(25), <u>247</u>
Marchese, S., 201(86), <u>207</u>
Mark, L. C., 21(43), <u>35</u>
Markham, R., 262, <u>268</u>
Marquardt, H., 3(8), <u>33</u>
Martin, M., 202, <u>208</u>
Martin, R. R., 21(50), <u>35</u>
Martinsson, E., 120(19), <u>147</u>
Martire, D. E., 155(29,30), 158(50), 165(56), 167(29,30), 168(29,30,76), <u>172</u>, <u>173</u>, <u>174</u>
Marton, L. J., 250(15), 250, 254(49,50), 255(50), 264(74), <u>265</u>, <u>267</u>, <u>268</u>
Maruta, K., 250(13), <u>265</u>
Mason, W., 58(26), <u>72</u>
Massart, D. L., 79(2,3), 80(2), 86(6), 88(7), 92(2,14,31), 96(3,22), 97(3,22), 103(25), 104(25), <u>110</u>, <u>111</u>
Mathur, D. S., 152, <u>171</u>
Matsuo, Y., 200(83), <u>207</u>
Mattisson, M., 137(45), 140, <u>148</u>
Mayo, B. C., 179(32), 180(32), <u>205</u>
McCloskey, D. H., 96, 97, <u>111</u>
McCredie, K. B., 215(13), 240(13), <u>246</u>
McCreary, M. D., 179(31), 180(31), <u>205</u>
McNair, H. M., 202, <u>208</u>
McReynolds, W. O., 95, 105, <u>111</u>
Meck, R. B., 200(70), <u>206</u>
Meijers, C. A., 51(18), 71(18), <u>72</u>
Meyers, R. D., 177(20), <u>204</u>
Miale, T., 250(16), <u>266</u>
Michaelis, A. F., 176(8), <u>204</u>
Mikes, F., 186(130,131), 187(130,131), <u>209</u>
Miki, K., 250(14), <u>265</u>
Mikolajczyk, M., 187(53), <u>206</u>
Miller, E., 67(36), <u>73</u>
Miller, H., 21(49), <u>35</u>
Mislow, K., 176(15,17,18), 178(18), 180(15), <u>204</u>
Miwatani, T., 250(9), <u>265</u>
Moertel, C. G., 250(18), <u>266</u>
Moffat, A. C., 88, 89, 109(9,10,11), <u>111</u>
Mollica, J. A., 63, 71(30), <u>73</u>
Moore, E. C., 215(17), <u>247</u>, 262, <u>268</u>
Moore, S., 251, <u>267</u>
Moore, W. R., 177(25), <u>204</u>
Mopper, K., 129, <u>148</u>
Morgan, G. O., 154(20), <u>172</u>
Mori, T., 200(83), <u>207</u>
Morisaki, M., 180(33), <u>205</u>
Morris, D. R., 251, <u>267</u>
Morrow, C. J., 201, <u>207</u>
Mortel, R., 250(17), <u>266</u>
Moyed, H. S., 215(18), <u>247</u>
Muller, W. E. G., 228(29), <u>247</u>
Munch-Petersen, B., 214(10), <u>246</u>
Munk, M., 212(2), <u>246</u>
Muntz, R. L., 180(41), <u>205</u>
Murakami, F., 252(42), <u>267</u>
Murrill, E. A., 176(11), <u>204</u>
Myers, G. E., 138(48), <u>148</u>

N

Nagatsu, T., 250(14), <u>265</u>
Nagel, C. W., 146, <u>149</u>
Nagy, L. G., 154(23), <u>172</u>
Nakahara, T., 167(67)
Nakajima, T., 250(8), 251(8), 250(22), 251(22), <u>265</u>, <u>266</u>
Nakanishi, K., 180(33,34), <u>205</u>

Navratil, J. D., 196(63), 197(63), 206, 256(52), 267
Nebert, D. W., 21(45), 35
Neckers, D. C., 183(46,48,49), 205
Neher, R., 39, 72
Nemoto, N., 4(18,19), 34
Neu, H. C., 239(38), 248
Newcomb, M., 188(55), 189(55), 206
Newton, N. E., 257, 261(56,58), 264, 267
Nichol, C. A., 215(20), 247
Noggle, G. R., 131(37), 148
Nonaka, A., 166, 173
Norden, B., 188(54), 206
Nurok, D., 176(2), 177, 178(2), 186(2), 203

O

O'Connor, M. J., 177(24), 204
Oesch, F., 3(9), 33
Ohms, J. I., 130(30), 148
Ohno, K., 250(24), 251(24,25,26), 257(24,25,55,56), 261(24,25, 55,56), 264(56), 266, 267
Ohno, T., 256(54), 267
Okawa, K., 140(53), 149
Okstein, C. J., 251(40), 267
Olson, M. C., 63, 71(32), 73
Olsson, L., 140(52), 149
Op de Beeck, J., 110(30), 111
Ortiz, P. J., 214(8), 223, 225(8), 226(8), 227(8), 229(8), 230(8), 233(25), 238(8), 246, 247
Orye, R. V., 158(47), 173
Overberger, C. G., 183(118), 209

P

Paart, E., 116(13), 120(13), 147
Paiaro, G., 177(23), 204

Pal, K., 4(20), 15(20), 22(20), 34
Papa, L. J., 202(92), 207
Parcher, J. F., 102(23), 111, 153(16), 154(21), 155(27,28), 162(16), 164, 165(16,58), 167(27,28), 169(28), 172, 173
Patel, I. B., 147(63), 149
Patterson, J. A., 130(30), 148
Paul, I. C., 180(41), 205
Pavan-Langston, D., 215(12), 246
Pawson, B. A., 177(27), 204
Pei, P., 202, 208
Perry, S. G., 202(98), 208
Perry, T. L., 250(20,21), 266
Persson, B.-A., 201(85,86), 207
Pescok, R. L., 165(57), 173
Peterson, D. L., 153(15), 154(15), 155(26), 172
Peterson, M. R., Jr., 179(30), 180(30), 205
Pirkle, W. H., 178(43,44), 180, 181(43,44), 182(42,43), 184(50,51), 185(51), 186(51), 205, 206
Pitt, W. W., Jr., 252(45), 253(45), 267
Pittet, A. O., 132(40), 148
Pittman, C. U., Jr., 183(45), 205
Plunkett, W., 214(8,9,11), 216(9), 217(9), 218(9), 220(9), 221(9), 222(9), 223(24), 224(11), 225(8), 226(8), 227(8), 228(9, 27), 229(8), 230(8), 231(11,24), 232(11), 233(9,11,24), 234(24), 235(9,11,24), 237(11,37), 238(8,11), 245(11), 246, 247
Pommier, C., 167(73), 174
Popova, M. I., 200(76,77), 207
Poppe, H., 68(38), 71(38), 73
Poppers, P. T., 21(46), 35
Pound, N., 54(21), 71(21), 72
Prausnitz, J. M., 158(47), 173
Pretorius, V., 157(38,39,40,41), 172, 173

Prijs, B., 192(58), 206
Pullman, A., 2, 33
Pullman, B., 2, 33
Purnell, J. H., 153(12), 154, 155(12,31,32,33,34), 156(19), 159(52), 162(19), 165(52), 167(71,72), 168, 169, 171, 172, 173, 174

Q

Quivoron, C., 178(137,138), 209

R

Raban, M., 176(17,18), 178(18), 204
Rabel, F. M., 202, 208
Racz, G., 154(23), 172
Radford, R. D., 261, 268
Raina, A., 251(28), 266
Ramachandram, S., 202(113), 208
Ramnas, O., 117(16), 147
Rapp, H. J., 21(49), 35
Rasmussen, R. E., 6(29), 12(36), 25(29), 34, 35
Ratan, J., 250(12), 265
Ratanasopa, V., 51(19), 72
Reed, D., 176(6), 203
Regnier, F. E., 187(135), 209
Reid, R. C., 157(44), 173
Reilley, C. N., 153(13), 155, 171
Rennert, O., 250(16), 266
Rezl, V., 165(60,61,62,63), 173
Rieman, W., 200(75), 207
Roberts, J. D., 176(14), 180(14), 204
Robinson, D. B., 167(69), 174
Robinson, R. L., 167(68), 174
Rogozhin, S. V., 176(4), 197(65, 66,67,68,69), 198(65,66,67), 199(66,67), 203, 206
Rohde, H. J., 228(29), 247

Rohrschneider, L., 94, 105, 108, 111
Rokushika, S., 252(42), 267
Roos, R., 67(33), 71(33), 73
Rothbart, H. L., 200(75), 207
Ruckert, H., 114(5), 116(11,12), 118, 126(11), 147
Russell, D. H., 250(5,11,19), 251(34), 252(47), 265, 266, 267
Russell, S. D., 250(11), 265

S

Sachkova, T. P., 197(67), 198(67), 199(67), 206
Saha, N. C., 152, 171
Sakai, Y., 140(53), 149
Sakakibara, K., 131, 148
Salmon, S. E., 250(19), 266
Samejima, K., 251(35), 256, 266, 267
Samuelson, O., 114(4,5,6,7), 115(9), 116(11,12,13,14,15), 117(16), 118(15,17), 119(18), 120(10,13,17,19,20), 121(21), 122(17,21,22), 123(10), 124(8, 9,17), 125, 126(10,11,17), 127(20), 128(23,24,25,26), 129(21), 130, 131(33), 132(39), 134(43), 135(39), 136(44), 137(10,22,43,45), 138(43,49, 50), 139(51), 140(43,52), 141, 142, 143(44,54,55), 144(58), 145, 146(59), 147, 148, 149
Sannes, K. N., 183(118), 209
Sannikova, G. S., 197(66), 198(66), 199(66), 206
Sano, I., 250(8), 251(8), 265
Sarhan, A., 200(80), 207
Sato, T., 200(83), 207
Saucy, G., 179(29), 180(29), 204
Saunders, R. M., 132(41), 148
Scarponi, F. L., 263(71), 268

Schaap, A. P., 183(46), 205
Schabel, F. M., Jr., 232(34), 247
Schaeffer, H. J., 230(32), 247
Schay, G., 152(1), 154, 156(36), 171, 172
Schilling, G. R., 4(22,23), 34
Schimpff, S. C., 250(5), 265
Schindler, A. E., 51(19), 72
Schleicher, R. G., 56, 72
Schmit, J. A., 50(16), 65(16), 66(16), 71(16), 72
Schoental, R., 32(59), 36
Schooley, D. A., 180(35,36,37), 205
Schubert, J., 167, 174
Schwanghart, A.-D., 193(60), 195(60), 206
Schwender, C. F., 230, 247
Scott, C. D., 131(35), 148, 201(84), 207
Scott, C. G., 179(29), 180(29), 204
Scott, R. P. W., 203, 204(116), 209
Sebastian, I., 201(110), 208
Seiler, N., 251(30), 266
Selkirk, J. K., 3(8,10), 4(10,13,16), 5(57), 9(10), 10(13), 11(13), 14(16), 16(16), 17(39), 18(39), 19, 20(39), 23(53), 25(53), 26(53), 28(57), 29(57), 30(57), 33, 35, 36
Semechkin, A. V., 197(65, 66, 67), 198(65,66,67), 199(66,67), 206
Senba, H., 250(14), 265
Seno, S., 191(128), 209
Sewell, P. A., 153(17), 172
Shaw, C. R., 21(47), 35
Shears, B., 4(12), 9(12), 33
Sheehan, L. M., 250(6,13,17), 265, 266
Sherwood, T. K., 157(44), 173
Shimizu, K., 143(55), 144, 146(59), 149
Shimizu, Y., 180(33), 205
Shoppee, C. W., 39, 72

Shriner, R., 69(39), 73
Shukla, J., 250(16), 266
Siddall, J. B., 180(36,37), 205
Siegel, B., 187(52), 206
Sievers, R. E., 179(64), 206
Siggia, S., 46(15), 47, 59, 65(34), 66(29,34), 69(40), 71(15,29,34), 72, 73
Sikkenga, D. L., 184(51), 185(51), 186(51), 206
Simatupang, M. H., 143(56,57), 149
Simonaitis, R. A., 262, 268
Sims, P., 3(8), 4(11,14,15,17,20,21), 6(11,30), 12(35), 15(20,21), 17(42), 22(20,21,52), 33, 34, 35
Sinner, M., 143(56,57), 149
Sjostrom, E., 114(4), 147
Slocum, S. A., 45(14), 71(14), 72
Smalldon, K. W., 88(8,9,10,11), 109(9,10,11), 111
Smith, E. D., 261, 268
Smith, F. G., 263(68,69), 264(68), 268
Smith, R. G., 264(74), 265(73), 268
Smith, T. A., 250(1), 262, 265, 268
Sneath, P. H. A., 78, 81, 84(1), 110
Snyder, J., 250(7), 265
Snyder, L. R., 176(7), 202, 203, 208
Snyder, R. V., 200(70), 206
Sogah, G. D. Y., 188(57), 189(57), 191(57,127), 192(57), 206, 209
Sokal, R. R., 78, 81, 84(1), 110
Soldano, B. A., 138, 148
Sousa, L. R., 188(56), 189(56), 190(56), 206
Souter, R. W., 176(1), 177(1), 178(1), 203
Sparangna, H., 58(26), 72
Stalkup, F. I., 166(65,66), 167, 173

Stead, A. H., 88(11), 109(11), <u>111</u>
Stein, W. H., 251, <u>267</u>
Stevensen, R., 212(2), <u>246</u>
Stewart, K. K., 200(81), <u>207</u>
Stock, R., 153(17), <u>172</u>
Stoddart, J. F., 188(124,125), <u>209</u>
Stoll, M. S., 202(96), <u>208</u>
Stoming, T., 7(31), <u>34</u>
Stout, M. G., 10(32), <u>34</u>
Strickhart, F. S., 12(34), <u>34</u>
Stromberg, H., 115, 124(8), 128(25,26), <u>147</u>, <u>148</u>
Strusz, R. F., 63, 71(30), <u>73</u>
Studebaker, J. F., 45(14), 71(14), <u>72</u>
Su, S. C., 201(86), <u>207</u>
Sugiura, T., 256, <u>267</u>
Sumizu, K., 252(42), <u>267</u>
Sunshine, I., 89, <u>111</u>
Swaisland, A., 4(20), 15(20), 22(20), <u>34</u>
Swenson, B., 115(7), <u>147</u>
Sydnor, K. L., 17(40,41), <u>35</u>

T

Tabor, C. W., 252, <u>267</u>
Tabor, H., 252, <u>267</u>
Taguchi, T., 250(9), <u>265</u>
Takahagi, H., 191(128), <u>209</u>
Takeda, T., 250(9), <u>265</u>
Takeda, Y., 250(9), <u>265</u>
Taschenr, H., 228(29), <u>247</u>
Thayer, A. L., 183(46), <u>205</u>
Thomas, H. L., 202, <u>209</u>
Thompson, M. H., 27(55), <u>35</u>
Timko, J. M., 188(55), 189(55), <u>206</u>
Tominaga, T., 250(9), <u>265</u>
Torii, M., 131, <u>148</u>
Tormey, D. C., 250(7,18), <u>265</u>
Tosa, T., 200(83), <u>207</u>
Toth, K., 179(29), 180(29), <u>204</u>
Touchstone, J., 56, 71(23), <u>72</u>

Trader, M. W., 232(34), <u>247</u>
Troetschler, R. G., 180(35), <u>205</u>
Tseng, Y., 245(39), <u>248</u>
Tsuge, S., 95(19), <u>111</u>
Tsuji, M., 250(8), 251(8), <u>265</u>
Tu, C. C., 114(2), <u>147</u>
Tyrsted, G., 214(10), <u>246</u>

U

Uhdeova, J., 165(60), <u>173</u>

V

Valentin, P., 154, 155(24,25), 162(22), 165(22), <u>172</u>
Valentine, D., Jr., 179(29), 180(29), <u>204</u>
Vanden Heuvel, W. J. A., 58(27), <u>73</u>, 262, <u>268</u>
Van Oudheusden, D., 96(22), 97(22), <u>111</u>
Vargas de Andrade, J. M., 168, 169, <u>172</u>
Vaughan, C. G., 187(134), <u>209</u>
Vaughn, R. H., 147(63), <u>149</u>
Veening, H., 176(9), <u>204</u>, 252(45), 253, <u>267</u>
Vermeulen, T., 176(5), <u>203</u>
Vestergaard, P., 59, <u>73</u>
Vigny, P., 4(21), 15(21), 22(21), <u>34</u>
Vivilecchia, R., 167(74), <u>174</u>, 176(8), <u>204</u>
Vofsi, D., 261(60), <u>268</u>
Vonderschmitt, D., 200(74), <u>207</u>

W

Waalkes, T. P., 250(7,18), 252(44), 253(44), 255(51),

AUTHOR INDEX

[Waalkes, T. P.]
 262(67), 263(44), 264(51), <u>265</u>, <u>266</u>, <u>267</u>, <u>268</u>
Wahl, G. H., 179(30), 180(30), <u>205</u>
Walborg, E. F., Jr., 130(32), <u>148</u>
Walker, H. G., 132(41), <u>148</u>
Wall, R. A., 132(40), <u>148</u>, 251, <u>267</u>
Waale, T., 250(23), 251(23), 263(68), 264(68), <u>266</u>, <u>268</u>
Wallenius, L.-O., 138(50), <u>149</u>
Walton, H. F., 196(63), 197(63), <u>206</u>, 256(52), <u>267</u>
Wang, I. Y., 6(29), 12(36), 25(29), <u>34</u>, <u>35</u>
Waqar, M. A., 221(22), <u>247</u>
Warr, G. A., 21(50), <u>35</u>
Waters, J. L., 176(12), 202, <u>204</u>, <u>208</u>
Weeks, C. E., 258(57), <u>267</u>
Weetal, H. H., 200(79), <u>207</u>
Wehner, N., 250(10), <u>265</u>
Weichmann, M., 251(30), <u>266</u>
Weinshanker, N. M., 183(47), <u>205</u>
Weiss, G., 180(34,34a), <u>205</u>
Welch, R. M., 21(46), <u>35</u>
Wernick, D. L., 179(31), 180(31), <u>205</u>
West, S., 6(27), <u>34</u>
Westlake, T. N., 155(27,28), 167(27,28), 169(28), <u>172</u>
Whang, J. J., 177(27), <u>204</u>
Wheals, B. B., 187(133,134), <u>209</u>
Whistler, R. L., 114(2), <u>147</u>
Whitehouse, M. J., 187(134), <u>209</u>
Whitesides, G. M., 179(31), 180(31), <u>205</u>
Whitlock, J. P., Jr., 17(39), 18(39), 19(39), 20(39), 21(48,49), <u>35</u>
Wiebel, F. J., 4(24), 10(32), 23(53), 25(53), 26(53), <u>34</u>, <u>35</u>
Wilke, C. R., 157, 158, <u>173</u>
Williams, K. T., 147(64), <u>149</u>
Williams, R. C., 203, <u>209</u>

Wilson, C. B., 250(15), 254(48,49), 264(49), <u>265</u>, <u>267</u>
Wilson, G. M., 158(46), <u>173</u>
Wilson, T. M., 146, <u>149</u>
Winkder, J., 21(45), <u>35</u>
Winkler, H. J. S., 177(25,26,27), <u>204</u>
Wold, S., 95, 97, 102(23), <u>111</u>
Wolfgram, F., 250(22), 251(22), <u>266</u>
Wong, J. Y., 183(47), <u>205</u>
Worth, L. S., 232(33), <u>247</u>
Wortmann, W., 56, 71(23), <u>72</u>
Wulff, G., 200(80), <u>207</u>

Y

Yagi, H., 6(28), <u>34</u>
Yamada, S., 177(21), <u>204</u>
Yamagiwa, K., 2(2), <u>33</u>
Yamamoto, M., 177(21), <u>204</u>
Yamamoto, S., 263(70), <u>268</u>
Yancey, J., 102(23), <u>111</u>
Yang, C. S., 12(34), <u>34</u>
Young, C. L., 153, <u>171</u>
Yudovich, A., 167(68), <u>174</u>

Z

Zabrocki, R. F., 200(80), <u>207</u>
Zack, L. F., Jr., 250(22), 251(22), <u>266</u>
Zahn, R. K., 228(29), <u>247</u>
Zamaroni, F., 212(2), <u>246</u>
Zec, J., 130(30), <u>148</u>
Zellner, D. C., 263(68,69), 264(68), <u>268</u>
Zill, L. P., 130, 131(37), <u>148</u>
Zimmerman, W., 57(25), <u>72</u>
Zumwalt, R. W., 250(7,18), 252(44), 253(44), 262(67), 263(44), <u>265</u>, <u>266</u>, <u>267</u>, <u>268</u>

SUBJECT INDEX

A

Acid yellow, 80
Addison's disease, 43, 57
Adenosine, structure of, 215
Adsorption isotherms, 165
Albumin, oral contraceptives and, 44
Alditols, partition chromatography of, 114–130
Aldonic acid group, terminal, anion-exchange chromatography of, 134–143
Aldosterone, 41
Amaranth, 80
Androgens, 71
 assays, 57–59
 review of, 42
Androsterone, 57
Anion-exchange chromatography
 of oligogalacturonic acids, 146–147
 of oligomers
  dicarboxylic, 144–146
  with one uronic acid moiety, 143–144
  with a terminal aldonic acid group, 134–143
Anomeric sugars, 117
Anterior pituitary hypofunction, 43
Anti-Langmuir isotherms, 153
9-β-D-Arabinofuranosyladenine, metabolism of, 215–235
Auramine, 80
Azo dyes, 3

B

Benzo(a)pyrene
 analysis of metabolism by high-pressure chromatography, 1–36
 enzyme inhibitors, effect of, 10–12
 epoxides as intermediates in, 12–16
 high-resolution (by recycling), 27–30
 in human tissue, 17–21
 in malignantly transformable cells, 21–27
 phenol identification, 30–32
 polycyclic hydrocarbons, 3, 6–10
 metabolism of, 4–6
Benzoates, 66
7,8-Benzoflavone (7,8-BF), 10–12
μ-Bondapak C18, 62, 69
Borate complexes, separations of oligosaccharides as, 130–131
Bordeaux B, 80
Bromsulphalein, oral contraceptives and, 44

C

Cadaverine, 253, 255, 264
Carbowax 1000, 102
Carmoisine, 80
Cellobionic acid series, 135

Chrisoine S, 80
Chrysoidine, 80
Cirrhosis of the liver, 57
Contraceptives, oral, biologically
　　indicative test results altered
　　by, 44
Corasil C18, 62
Corticosteroids, 71
　assays, 56
　disorders associated with, 43
　oral contraceptives and, 44
　pharmaceuticals, 63-64
　review of, 41-42
Cortisol, oral contraceptives
　　and, 44
Cortisone, separation of, 63-64
Crown ethers, 188-192
Cushing's syndrome, 43
Cyclohexane, 168

D

Dehydroepiandrosterone, 57
2'-Deoxyadenosine, 215
11-Deoxycortisone, 41
5'-Deoxynucleotide, 220
Derivatization, 65-67
3,3'-Diaminodipropylamine, 253,
　　255
1,3-Diaminopropane, 253, 258-259,
　　264
Diastereomers
　formation of, 178
　separation of, 177-183
Dicarboxylic oligomers, anion-
　　exchange chromatography of,
　　144-146
4,5-Dihydrodihydroxy-BP
　molecular weight, 10
　retention time, 10
7,8-Dihydrodihydroxy-BP
　molecular weight, 10
　retention time, 10

9,10-Dihydrodihydroxy-BP
　molecular weight, 10
　retention time, 10
Dihydrodiols, 4, 5
Dimethylbenz(a)anthracene, 4
2,4-Dinitrophenylhydrazone, 66
4,5-Diol, 7
　percentage of total metabolites, 24
7,8-Diol, 7
　percentage of total metabolites, 24
9,10-Diol, 7
　percentage of total metabolites, 24
DNA, 215-223, 227, 229, 235,
　　243-245

E

Eluents (active), direct resolutions
　　using, 184-186
Enantiomers, resolution of,
　　175-209
　direct resolutions using ion-pair
　　partition chromatography,
　　201-202
　direct resolutions using optically
　　active eluents, 184-186
　direct resolutions using optically
　　active substrate, 186-201
　potential optically active
　　resolving column, 183-184
　separation of, 176-177
　separation of diastereomers,
　　177-183
Enzyme inhibitors, 10-12
Enzyme-based supports, 200-201
Eosine, 80
Epoxides, 12-16
Erythro-9-(2-hydroxy-3-nonyl)
　　adenine (EHNA), 230-245
Erytrosine, 80
Estradiol, 55
Estriol, 55
Estrogens, 71

SUBJECT INDEX

[Estrogens]
  assays, 51-56
  flow chart for, 53
  pharmaceuticals, 64-65
  reversed-phase separation of, 55
  review of, 43
  separation of conjugates, 54
Estrone, 55
Ethanol, 118-125
Ethers, crown, 188-192
Etiocholanolone, 57

F

Fast Red E, 80
Flory-Huggins equation, 158
Fluorescamine, 256
Frontal chromatogram (frontalgram), factors influencing shape of, 153-154
Frontal chromatography, 151-174
  applications of, 165-170
  chromatogram (frontalgram), factors influencing shape of, 153-154
  investigations involving simplifying assumptions, 160-164
  modes of operation, 154-155
  theory of, 155-160

G

Gas chromatography, 6-7, 64
  polyamines, 261-264
Gas-liquid chromatography, 76
  classification of stationary phases in, 94-105
  classification of substances, 105-109
  pattern recognition procedure, 109-110
Gibbs-Donnan theory, 134-135

Globulin
  oral contraceptives and, 44
  thyroid-binding, 44
Glucocorticoids, 41
Gluconic acid, 142
Glucose, 118, 121, 123
  oral contraceptives and, 44
D-Glucose, 123, 125

H

Hallcomid M-18, 102
n-Heptane, 168
n-Hexane, 168
High-performance liquid chromatography of steroid hormones, 37-73
  in biological fluids, 51-60
  chemistry before, during, and after, 65-68
  development of separations, 61-62
  future of, 68-70
  initial developments, 46-51
  in pharmaceuticals, 61-65
  review of, 39-45
  scope of, 45-46
High-pressure liquid chromatography, 176
  analysis of benzo(a)pyrene metabolism by, 1-36
  enzyme inhibitors, effect of, 10-12
  epoxides as intermediates in, 12-16
  high-resolution (by recycling), 27-30
  in human tissue, 17-21
  in malignantly transformable cells, 21-27
  phenol identification, 30-32
  polycyclic hydrocarbons, 3, 6-10

[High-pressure liquid chromatography]
  polyamines, 251-261
  purine nucleoside analogs, research on, 211-248
    9-β-D-arabinofuranosyladenine, metabolism of, 215-235
    methods, 212-214
    9-β-D-xylofuranosyladenine, metabolism of, 236-245
Hormones, see Steroid hormones
Hydrocarbons
  polyaromatic, 2-3
  polycyclic, 3, 6-10
Hydrocortisone
  oral contraceptives and, 44
  separation of, 63-64
Hydrocortisone acetate, separation of, 64
3-Hydroxy-BP
  molecular weight, 10
  retention time, 10
9-Hydroxy-BP
  molecular weight, 10
  retention time, 10
17-Hydroxycorticosteroids, 43
Hydroxymethyl derivatives, 4
17-Hydroxypregnanolone, 42
Hypertension, 43
Hypogonadism, 57
Hypopituitarism, 57
Hyprose SP-80, 102

I

Inhibitors, enzyme, 10-12
Ion-exchange chromatography, 201
  polyamines, 251-256
Ion-exchange resins, 113-149
Ion-pair partition chromatography, 201-202
Isomaltose, separation of, 127, 128
Isomers, positional, 3

Isotherms
  adsorption, 165
  anti-Langmuir, 153
  Langmuir-type, 153
  linear, 153
  partition, 165

K

16-Ketoestradiol, 55
16-Ketoestrone, 55
Ketogenic steroids, oral contraceptives and, 44
17-Ketosteroids, 57-59
  diseases associated with excretion of, 57
  structure of, 58
K-region hypothesis, 2, 3

L

Langmuir-type isotherms, 153
Ligand-exchange resolutions, 196-199
Linear isotherms, 153
Liquid chromatography
  enantiomers, 175-209
    direct resolutions using ion-pair partition chromatography, 201-202
    direct resolutions using optically active eluents, 184-186
    direct resolutions using optically active substrate, 186-201
    potential optically active resolving column, 183-184
    separation of, 176-177
    separation of diastereomers, 177-183
  Varian Aerograph dual-column, 212, 213

SUBJECT INDEX

M

D-Maltose, 117
Mass spectrometry, 264
Mean character difference (MCD), 81
Metabolism
  of 9-β-D-arabinofuranosyladenine, 215-235
  of benzo(a)pyrene, 4-6
  of 9-β-D-xylofuranosyladenine, 236-245
  See also High-pressure liquid chromatography, analysis of benzo(a)pyrene metabolism by
Methycyclohexane, 168
Methylcholanthrene, 4
Metopirone, oral contraceptives and, 44
Metyrapone, oral contraceptives and, 44
Mineralocorticoids, 41
Monoacetylspermidine, 264
3'-Monophosphates, 230
5'-Monophosphates, 229
Myxedema, 57

N

Naphthol yellow S, 80
Nitrosamines, 3
Numerical taxonomy, 75-111
  applications, 87-110
    classification of stationary phases, 94-105
    classification of substances, 105-109
    pattern recognition procedure, 109-110
    selection of optimal sets, 87-94
  techniques, 77-87
    data, 78-79
    general considerations, 77-78
    graph theoretical procedures, 86-87
    linkage methods, 82-85
    measures of resemblance, 79-82

O

Oligogalacturonic acids, anion-exchange chromatography of, 146-147
Oligomeric acids, separation of, 139
Oligomeric sugars, 114-130
Oligomers
  anion-exchange chromatography of dicarboxylic, 144-146
  with one uronic acid moiety, 143-144
  with a terminal aldonic acid group, 134-143
  structure of, 119-124
Oligosaccharides, 113-149
  anion-exchange chromatography of oligogalacturonic acids, 146-147
  partition chromatography of alditols, 114-130
  partition chromatography of oligomeric sugars, 114-130
  permeation chromatography in aqueous solution, 131-134
  separation as borate complexes, 130-131
  xylan, 116
Operational taxonomic units (OTUs), 78-85, 105
  solvent systems, 80
Oral contraceptives, biologically indicative test results altered by, 44
Orange G, 80

Orange GGN, 80
Orange I, 80

P

Paired-ion chromatography, 201
Partition chromatography
  of alditols, 114-130
  of oligomeric sugars, 114-130
Partition isotherms, 165
Permaphase ODS, 62
Permeation chromatography of
    oligosaccharides in aqueous
    solution, 131-134
Phenanthrenedione, 67
Phenols, 4
  identification of, 30-32
Phenylcorasil, 62
Phosphoriboxyl pyrophosphate
    (PRPP), 236, 243, 245
Polyamines, 249-268
  gas chromatography of, 261-264
  high-pressure liquid chroma-
    tography of, 251-261
  ion-exchange chromatography of,
    251-256
  mass spectrometry, 264
  separation of, 261-262
    after derivative formation,
      256-264
Polyaromatic hydrocarbons, 2-3
Polycyclic hydrocarbons, 3, 6-10
Ponceau MX, 80
Ponceau 4R, 80
Ponceau 6R, 80
Ponceau SX, 80
Positional isomers, 3
Pregnancy, third trimester of,
  43
Pregnanetriol, 42
Procoagulants, oral contraceptives
  and, 44
Progesterone, 43-45
Progestins, 43-45

Purine nucleoside analogs,
    research on, 211-248
  9-β-D-arabinofuranosyladenine,
    metabolism of, 215-235
  methods, 212-214
  9-β-D-xylofuranosyladenine,
    metabolism of, 236-245
Putrescine, 253-255, 262, 264

Q

Quadrol, 102
3,6-Quinone, 8
6,12-Quinone, 8
1,6-Quinone-BP
  molecular weight, 10
  retention time, 10
3,6-Quinone-BP
  molecular weight, 10
  retention time, 10
6,12-Quinone-BP
  molecular weight, 10
  retention time, 10
Quinones, 4, 5, 7-8
  percentage of total metabolites, 24

R

Recycling, 27-30
Red 6B, 80
Red 10B, 80
Red FB, 80
Red 2G, 80
Resins, ion-exchange, 113-149
Rhodamine B, 80
RNA, 216-220, 227-228, 243-245
Rohrschneider index, 76

S

Scarlet GN, 80
Sephadex-type supports, 199-200

SUBJECT INDEX

Silica gels, 192
Siponate DS-10, 102
Spectrometry, mass, 264
Spermidine, 253, 254, 264
Spermine, 253, 254, 264
Steroid hormones
  basic structure and numbering system, 39
  high-performance liquid chromatography of, 37-73
  in biological fluids, 51-60
  chemistry before, during, and after, 65-68
  development of separations, 61-62
  future of, 68-70
  initial developments, 46-51
  in pharmaceuticals, 61-65
  review of, 39-45
  scope of, 45-46
Steroids, ketogenic, oral contraceptives and, 44
Stress, 43
Substrate (active), direct resolutions using, 186-201
Sugars
  anomeric, 117
  oligomeric, 114-130
Sulfobromophthalein, oral contraceptives and, 44
Sunset yellow, 80

T

Tartrazine, 80
Taxonomy, see Numerical taxonomy
Tergitol NPX, 102
Tetrahydrocortisol, 42
Tetrahydrocortisone, 42
Tetrasaccharide, 127
Thin-layer chromatography, 6, 47, 79, 109, 176
  selection of optimal sets in, 87-94

Thyroid-binding globulin, oral contraceptives and, 44
Thyroxine, oral contraceptives and, 44
1,2-epoxy-3,3-Trichloropropane (TCPO), 12
Triglycerides, oral contraceptives and, 44
Triiodothyronine, oral contraceptives and, 44
Trimer acid, 102
2,2,4-Trimethylpentane, 168

U

Uronic acid, 143-144

V

Vapor phase chromatography, 176
Varian Aerograph dual-column liquid chromatography, 212, 213
Virilism, 43

W

Wilson equation, 158

X

Xylan oligosaccharides, 116
Xylobionic acid, 132, 135, 142
Xylobiose, 127
9-β-D-Xylofuranosyladenine, metabolism of, 236-245
Xyloheptaonic acid, 142
Xylohexaonic acid, 142
Xylonic acid, 142
Xylooctaonic acid, 142
Xyloonaonic acid, 142

Xylopentaonic acid, 142
Xylopentaose, 136
Xylose, separation of, 133
D-Xylose, 120, 123, 125

Xylotetraonic acid, 142
Xylotetraose, 127
Xylotrionic acid, 142
Xylulose, 127